T0331109

The Geometrid Moths of Europe
Volume 1

Axel Hausmann

The Geometrid Moths of Europe

Volume 1

Axel Hausmann

Introduction

Archiearinae, Orthostixinae, Desmobathrinae, Alsophilinae, Geometrinae

Apollo Books, Stenstrup

2001

Editor: Axel Hausmann, Zoologische Staatsssammlung, München.

Text formatting and production management: Winther Grafik, Odense.

Printed by: Litotryk Svendborg A/S

This publication should be cited as:
Hausmann, A., 2001. Introduction. Archiearinae, Orthostixinae, Desmobathrinae, Alsophilinae, Geometrinae. *In* A. Hausmann (ed.): The Geometrid Moths of Europe 1: 1 – 282

ISBN 87-88757-35-8 (Vol. 1)
ISBN 87-88757-54-4 (Vols 1-6)

Published with financial support from:

AAGE V. JENSENS FONDE

Author's address:

Zoologische Staatssammlung München
Münchhausenstr. 21
D-81247 Munich
Germany

E-mail: Axel.Hausmann@zsm.mwn.de

Front cover: *Eucrostes indigenata* (Vill.). South-western Turkey, May 1999.
(Photo *Michael Leipnitz*)

Contents

PART I: Introduction to the series

PART II: Systematic account

Preface

To date more than 900 European geometrid species are known. This is about 4% of the world's fauna. They are often small and inconspicuous in wing pattern and colours, hence not aesthetically attractive. Geometrid moths are also said to be 'difficult', and there are, indeed, many similar species (e.g. Sterrhinae or genus *Eupithecia*) which cannot be properly identified without turning to refined techniques of anatomical preparation.

On the other hand, they are quite important to man: In Central Europe 87 species or roughly 15% of the geometrid fauna are known as pest species in forestry and agriculture (KUDLER 1978). Anyway, no comprehensive systematic work covering all European geometrid species has been compiled for nearly a century.

In 1995 the project 'The Geometrid Moths of Europe' (GME) was initiated. Editor and authors agreed on the following four principle **aims**:

(1) *Facilitate identification*: The book series is intended for both amateurs and professionals as well as for those engaged in applied entomology. The books will allow correct identification to species and subspecies level of all European Geometridae. For all taxa, even at the higher systematic levels of genera and tribes, differential diagnoses will be provided. Ample illustrations of differential features will facilitate identification.

(2) *Summarize current information*: Up-to-date information on the systematics, morphology, zoogeography and ecology of the European Geometridae to be presented in a concise and easily accessible manner.

(3) *Critical assessment* of the taxonomic status and nomenclature of all European taxa with the aim of achieving nomenclatural stability.

(4) *Initiate cooperation for continuous updating*: The project should not end with the publication of the books but rather establish a network for collaboration between lepidopterists interested in European Geometridae to yield a continuous updating of pertinent information. The availability of modern electronic communication facilities as the internet is envisaged for that purpose in the near future.

Geographically, the GME project defines **Europe** as including Iceland and the Mediterranean Islands but excluding Cyprus, northern Africa, the Asiatic part of Turkey, the Canaries, Madeira and the Azores. In the east, the Ural Mountains are included. The southern border follows the Ural river and the Caspian Sea, then turns westwards through the Don Kuma-Manitsh depression to the Black Sea and excludes the Ciscaucasian slopes.

Some extralimital species are, however, included in the systematic checklist at the end of this volume as they may have been overlooked in Europe or may sooner or later become part of its fauna.

The **GME book series** is planned to be published in 6 volumes:

Vol. 1: Introduction. - Archiearinae, Orthostixinae, Desmobathrinae, Alsophilinae, Geometrinae
Vol. 2: Sterrhinae
Vol. 3: Larentiinae I: all tribes except for Perizomini and Eupitheciini
Vol. 4: Larentiinae II: Perizomini and Eupitheciini
Vol. 5: Ennominae I: Abraxini - Bistonini
Vol. 6: Ennominae II: Boarmiini - Compsopterini

Although part of a uniform project the 6 volumes may not be strictly homogeneous in style and contents as the personality of authors is, to a certain degree, felt more relevant for the quality of the project than absolute formal rigidity.

The introductory chapters do not claim to cover all aspects of lepidopterology. For further information the reader is referred to the modern handbooks of lepidopterology by SCOBLE (1995) and KRISTENSEN (1999). The terminology follows SCOBLE (1992).

In the *systematic part*, descriptions of new taxa and other taxonomic changes are not a main objective. Consequently, for the scope of Volume 1 these items were dealt with separately (HAUSMANN 1994a; 1995a; 1997b; 2000). Also, the morphology of the early stages is not treated here in depth. The nomenclature has been thoroughly revised: for the 42 species of this first volume nearly 400 names had to be checked.

Munich, 01.01.2001

Axel Hausmann
(Editor-in-Chief)

Acknowledgements

The GME book series is founded upon the knowledge of numerous background contributors, often non-professionals, in a number of countries. Their assistance is greatly appreciated even if only those involved in the contents of this Volume 1 can be named here. A number of colleagues in Lepidopterology granted valuable help in the preparation, or were directly involved in the editorial process, of Volume 1. I owe them sincere thanks for constructive advice and useful comments. Especially I enjoyed the kind cooperation of:

Dr. Elena M. ANTONOVA, Moskow (RUS): Provision of innumerable faunistic data: Nearly all the eastern European dots in the maps have been inserted by her, I. Kostjuk and Dr. V. Mironov. Provision of literature.

Ulf BUCHSBAUM, München (D): Kind help in computer questions.

Kim BUCKMASTER, London (GB): Provision of some faunistic data.

BUNDESMINISTERIUM FÜR BILDUNG UND FORSCHUNG, Bonn (D): Funding of the project 'Inventur der Geometriden-Arten Europas [Inge]' in the BIOLOG-program as scientific basis for this publication.

Dr. Ernst-G. BURMEISTER, München (D): Precious comments and advice on some parts of the manuscript (introduction, conservation).

Dr. Jaroslaw BUSZKO, Torun (PL): Provision of many faunistic data (revision of the maps for Poland) and host-plant records.

Martin CORLEY, Oxfordshire (GB): Provision of many faunistic data. Loan of material. Linguistic editor, constructive revision of the manuscript.

Dr. Jorge DANTART, Valencia (E): Provision of faunistic data and literature.

Thomas DRECHSEL, Neubrandenburg (D): Provision of faunistic data and host-plant records.

Günter EBERT, Karlsruhe (D): Kind support of the author when visiting the Karlsruhe museum. Loan of material. Provision of faunistic data and host-plant records.

Dr. Ulf EITSCHBERGER, Marktleuthen (D): Permanent loan of about 200.000 geometrid moths from his collection.

Andrès EXPÓSITO HERMOSA, Madrid (E): Provision of faunistic data (revision of the maps for Spain).

FREUNDE DER ZOOLOGISCHEN STAATSSAMMLUNG MÜNCHEN, e.V. (association): Funding of students, guest scientists and visits to museums.

Dr. Jörg GELBRECHT, Berlin (D): Revision of the whole manuscript with valuable comments and advice. Provision of many host-plant records. Loan of material. Discussion of taxonomic problems.

Barry GOATER, Hampshire (GB): Provision of many faunistic data (France, Balearic Islands).

Stanislav GOMBOC, Ljubljana (SLO): Provision of numerous faunistic data (Slowenija). Donation of material.

Heinz HABELER, Graz (A): Provision of many faunistic data (Croatia, Greece). Loan of material.

Dr. Gerhard HASZPRUNAR, München (D): Support and encouragement of research on the GME series as the Director of the ZSM.

Dr. Christoph HÄUSER, Stuttgart (D): Kind support of the author when visiting the Stuttgart museum. Loan of material.

Claude HERBULOT, Paris (F): Partial revision of the manuscript with valuable comments and advice. Kind support during visits to his collection. Provision of literature and faunistic data. Loan of material (also types). Discussion of taxonomic problems.

Dr. Peter HUEMER, Innsbruck (A): Partial revision of the manuscript (introduction) with valuable comments and advice. Kind support of the author when visiting the Innsbruck museum. Provision of faunistic data.

Kurt KOSSNER, München (D): Funding of a support student for $^1/_2$ year.

Igor KOSTJUK, Kiyev (UKR): Responsible for elaboration of the eastern European parts of the distribution maps from literature and collection material: Nearly all the eastern European dots in the maps have been inserted by him, Dr. Antonova and Dr. Mironov. Provision of literature. Loan of material. Discussion of taxonomic problems.

Werner KRAUS, Kaiserslautern (D): Provision of literature and faunistic data. Loan of material.

Dr. Martin KRÜGER, Pretoria (SA): Discussion of some taxonomic problems.

Ruth KÜHBANDNER, München (D): Considerable and kind help with drawings and other graphic work, also computer based.

Anatolii KULAK, Minsk (BY): Provision of valuable faunistic data (Byelorussia).

Karl KUCHLER, München (D): Loan and donation of material.

Michael LEIPNITZ, Stuttgart (D): Provision of photographs, faunistic data and many host-plant records. Donation of material.

Dr. Hans LÖBEL, Sondershausen (D): Provision of faunistic data and host-plant records. Donation of material.

Dr. Martin LÖDL, Wien (A): Kind support of the author when visiting the Vienna museum. Loan of material (also types).

David LONG, Edinburg (GB): Assistance with plant nomenclature.

Dr. Hans MALICKY, Lunz (A): Donation of extensive material from Crete and other Mediterranean regions.

Risto MARTIKAINEN, Tampere (FIN): Donation of important literature.

Dr. Robert MAZEL, Perpignan (F): Provision of literature and faunistic data (revision of maps for southwestern France).

Dr. Wolfram MEY, Berlin (D): Kind support of the author when visiting the Berlin museum. Loan of material (also types).

Michael MILLER, München (D): Preparation of about 2.000 genitalia slides.

Dr. Vladimir MIRONOV, St. Petersburg (RUS): Revision of the whole manuscript with valuable comments and advice. Provision of innumerable faunistic data: Nearly all the eastern European dots in the maps have been inserted by him, E. Antonova and I. Kostjuk. Loan of material (also types). Discussion of taxonomic problems.

Dr. Bernd MÜLLER, Berlin (D): Revision of the whole manuscript with valuable comments and advice. Provision of many host-plant records. Loan of material.

Marianne MÜLLER, München (D): Considerable and kind phototechnical help (mainly concerning the monochrome photographs).

Dr. Clas M. NAUMANN, Bonn (D): Provision of literature.

Mag. S. ORTNER, Bad Ischl (A): Loan and donation of material.

Dr. Paolo PARENZAN, Palermo (I): Provision of faunistic data.

Mark PARSONS, London (GB): Provision of faunistic data. Examination of some BMNH types. Loan of material.

Norbert PÖLL, Bad Ischl (A): Provision of many faunistic data (mainly from Austria and the Czech Republic). Loan of material.

Dr. Valter RAINERI, Genova (I): Provision of literature and faunistic data.

Jim REID, Hertfordshire (GB): Provision of many faunistic data and photographs (revision of the maps for Great Britain).

Dr. Laszló RONKAY, Budapest (H): Provision of many valuable faunistic data (revision of the maps for Hungary).

Dr. Klaus SATTLER, London (GB): Revision of the whole manuscript with valuable comments and advice.

Dr. Stefano SCALERCIO, Donnici inf. (I): Provision of many faunistic data (Italy). Loan of material.

Paul SCHAIDER, Igoumenitsa (GR): Provision of faunistic data. Donation of much material (Spain, Greece).

Dr. Klaus SCHÖNITZER, München (D): Valuable comments and advice on some parts of the manuscript. Help with the SEM-photos.

Johannes SCHUBERTH, München (D): Loan of computer hardware.

Dr. Malcolm SCOBLE, London (GB): Revision of the whole manuscript with valuable comments and advice. Loan of material (also types).

Dr. Andreas SEGERER, München (D): Valuable comments and advice on the whole manuscript. Kind help in computer questions (digital photography, network).

Dr. Yuri SINEV, St. Petersburg (RUS): Loan of material (also types).

Bernard SKINNER, South Croydon (GB): Provision of many faunistic data (revision of the maps for Great Britain and Ireland). Provision of host-plant records.

Peder SKOU, Stenstrup (DK): Idea and first moves to initiate the GME-series, publisher risks. Revision of the whole manuscript with valuable comments and advice. Provision of photographs, literature, faunistic data and host-plant records.

Manfred SOMMERER, München (D): Valuable comments and advice. Revision of the whole manuscript, (with rearrangement of chapters and additions). Loan of material.

Dr. Dieter STÜNING, Bonn (D): Revision of the manuscript with valuable comments and advice. Information on type-material in ZFMK. Loan of material (also types).

Dr. Gerhard TARMANN, Innsbruck (A): Kind support of the author when visiting the Innsbruck museum.

Dr. Jaan VIIDALEPP, Tartu (EST): Revision of the whole manuscript with valuable comments and advice. Provision of literature. Loan of material. Discussion of taxonomic problems.

Hartmut WEGNER, Adendorf (D): Provision of many faunistic data and host-plant records. Donation of material.

Dr. Norbert ZAHM, Schmelz-Hüttersdorf (D): Provision of valuable faunistic data (Germany, C. Italy).

Dr. Alberto ZILLI, Roma (I): Provision of valuable faunistic data (Italy).

Finally - but not least - I thank my wife Silvia, whose encouragement has supported me during the long periods of preparation of this book.

PART I: Introduction to the series

1. Short history and state of exploration of European Geometridae

The Inventory

The current number of more than 900 geometrid species in Europe is the result of about 250 years' work by various lepidopterists. The following chronological list shows some important publications (with regard to the family Geometridae) since LINNAEUS and the number of valid species names due to their authors. When works were issued in parts, references are made only for the part relating to Geometridae.

Among the most productive contributors to geometrid taxonomy are to be mentioned HÜBNER with 102 species, DENIS & SCHIFFERMÜLLER with 82 species, LINNAEUS with 64 species, HERRICH-SCHÄFFER with 58 species and STAUDINGER with 56 species. These five authors alone described altogether 362 species (i.e. ca. 40 % of the European fauna). In the 20th century only 173 valid species names (i.e. ca. 19 % of the European fauna; see Text-fig. 7) were added.

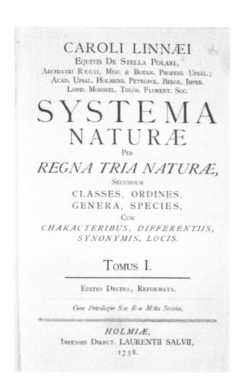
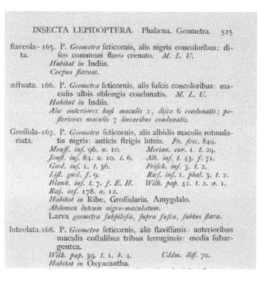

Text-figs 1-2: Title page and part of text page from 'Systema Naturae' (LINNAEUS, 1758).

1758: LINNAEUS ('Carl von LINNÉ'): The tenth edition of the 'Systema Naturae' is the start of the modern zoological (binominal) nomenclature (Lepidoptera in volume 1). Important subsequent works: Systema Naturae, ed. 12 (1767) and 'Fauna Suecica', (1761). Numerous original descriptions of European Geometridae (64 valid species names).

1759: CLERCK: 'Icones insectorum rariorum'. Type-catalogue: MIKKOLA (1985) (11 valid species names).

1763: SCOPOLI: 'Entomologia carniolica', dealing with the fauna of Carinthia, today Slovenia, Geometridae on pp. 214-232 (18 valid species names).

1767: HUFNAGEL: 'Vierte Tabelle (...) der Nachtvögel hiesiger Gegend', dealing with the Berlin distr. (31 valid species names).

1775: DENIS & SCHIFFERMÜLLER: 'Ankündung eines systematischen Werkes von den Schmetterlingen der Wienergegend'. This is a faunistic list (without illustrations of Geometridae and descriptions for most of the species) of the Lepidoptera of the Vienna district. It was the announcement of a planned comprehensive work which never came to fruition (SATTLER 1969). Geometridae are discussed on pp. 95-119, 276-285 and 314-316 (82 valid species names). Most of the names are usually considered as being available and valid although their nomenclatural availability is sometimes disputed (see chapter 5).

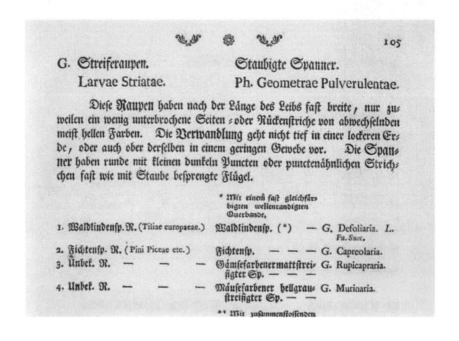

Text-fig. 3: Part of text page from 'DENIS & SCHIFFERMÜLLER', in the edition of 1776.

Text-fig. 4: Frontispiece from 'DENIS & SCHIFFERMÜLLER', in the edition of 1776.

1775-1798: FABRICIUS: Student of LINNAEUS. Six extensive publications (17 valid species names).

1781: GOEZE: 'Entomologische Beyträge ...', which is a catalogue of described European Lepidoptera species, Geometridae in vol. 3 (3): 274-439. It contains extensive bibliographic citations (3 valid species names).

1784ff.: THUNBERG: Various publications on Swedish Lepidoptera, e.g. 'Insecta suecica' (1784). Type-catalogue by KARSHOLT & NIELSEN (1985) (11 valid species names).

1785: (GEOFFROY in) FOURCROY: 'Entomologia parisiensis', dealing with the lepidopterous fauna of the Paris district. Many synonyms, no valid species names (according to LERAUT 1997 one valid species name).

1787ff.: ESPER: 'Esper's Schmetterlinge'. Type-catalogue: HACKER (1999) (7 valid species names).

1787ff.: HÜBNER: Various publications. Most important the 'Sammlung europäischer Schmetterlinge, vol. 5' (main parts [1799] and [1813]) which is the first well illustrated book on Geometridae of Europe (plates without text). Original references for numerous European Geometridae (102 valid species names). The 'Verzeichniss bekannter Schmettlinge' (1816-[1826]) introduces many generic names.

Text-fig. 5: Plate 16 of the 'Sammlung europäischer Schmetterlinge, vol. 5' (HÜBNER,[1799ff.])

1789: (DE) VILLERS: 'Caroli Linnaei Entomologica', which is a catalogue of European Insects, Geometridae in vol. 4, pp. 495-514 (8 valid species names).

1794: BORKHAUSEN: 'Systematische Beschreibung der Europäischen Schmetterlinge', Geometridae in vol. 5, dealing with the European Geometridae, with extensive descriptions and literature citations, but without illustrations. (7 valid species names).

1802: SCHRANK: 'Fauna Boica', the fauna of Bavaria, southern Germany. Geometridae in vol. 2 (2), 1-58 (2 valid species names). SCHRANK is the first to introduce a generic concept in the modern sense.

1809: HAWORTH: 'Lepidoptera Britannica': systematic list of the Lepidoptera of Great Britain with descriptions, without illustrations (16 valid species names).

1825ff.: CURTIS: 'British Entomology' (2 valid species names; many new generic names).

1825ff.: TREITSCHKE: 'Die Schmetterlinge von Europa', Geometridae in vol. 6 (1-2). This is the first comprehensive book on the Geometridae (Lepidoptera) of Europe with extensive text parts, but without illustrations (15 valid species names). Description of many genera.

1828ff.: FREYER: His main publication is the 'Neuere Beiträge zur Schmetterlingskunde' (1831-1858), which contains contributions without systematical order (11 valid species names).

1829: STEPHENS: 'Illustrations of British Entomology', a monograph on British Geometridae in vol. III: 142-328 with good colour plates (5 valid species names; a lot of generic descriptions and diagnoses).

1829ff.: DUPONCHEL (in GODART): 'Histoire naturelle des Lépidoptères ou Papillons de France', a comprehensive, well illustrated work on European Lepidoptera, Geometridae in volume VII/2 and VIII/1 (13 valid species names; description of various genera).

1831ff. GEYER: finished HÜBNER's 'Sammlung europäischer Schmetterlinge', the last plates being attributed to him (6 valid species names).

1833: RAMBUR: 'Catalogue de l'île de Corsica', i.e. the first fauna of Corsica, and some other papers. Type-catalogue: HERBULOT (1987) (16 valid species names).

1838ff.: HERRICH-SCHÄFFER: Most important the 'Systematische Bearbeitung der Schmetterlinge Europas', dealing with European Lepidoptera (in six volumes), Geometridae in volume 3 and 6. Numerous original descriptions of European Geometridae (58 valid species names). First extensive studies on the venation and other morphological features, establishing a system of lepidopteran families.

Text-fig. 6: Sample plate of the 'Systematische Bearbeitung der Schmetterlinge Europas'
(HERRICH-SCHÄFFER, 1838ff.)

1837ff.: EVERSMANN: Various publications, e.g. the 'Fauna Lepidopterologica Volgo-Uralensis (1944) (12 valid species names).

1840: BOISDUVAL: 'Genera et index methodicus europaeorum Lepidopterorum', a systematic checklist of European Lepidoptera (16 valid species names; some new genera).

1846ff.: ZELLER: Various papers, mainly concerning the fauna of southern Europe and Turkey. Some names in the first published paper in 1846 refer to LIENIG as author (12 valid species names; some new genera).

1852ff.: MANN: Various publications, mainly concerning the fauna of southern Europe and Turkey (5 valid species names).

1853ff.: LEDERER: Various publications, e.g. 'Versuch ... in möglichst natürliche Reihenfolge zu stellen' (1853), usually cited as 'Lepid. System', which is an attempt to re-arrange the European Lepidoptera in a 'natural order' (8 valid species names).

1853ff.: (DE) LA HARPE: Various publications, e.g. 'Faune Suisse, Lépidoptères' (1853), Geometridae in Pt. 4 (6 valid species names).

1857ff.: GUENÉE: 'Histoire Naturelle des Insectes, Species général des Lépidoptères', dealing with the Lepidoptera of the world with descriptions; Geometridae in the two volumes 'Uranides et Phalénites, I-II' (13 valid species names).

1857ff.: STAUDINGER: First catalogues on the Palaearctic fauna of Lepidoptera (editions 1861, 1871, 1901). Many other publications with numerous original descriptions of European Geometridae (56 valid species names).

1860ff.: WALKER: Many new descriptions of tropical Geometridae, one concerning a species with distribution in Europe (one valid species name).

1862ff.: MILLIÈRE: 'Iconographie et description de Chenilles et Lépidoptères inédits' and 'Catalogue des Lépidoptères des Alpes Maritimes', which mainly concern the fauna of southern France (16 valid species names).

1863ff.: MABILLE: Various publications, mainly on the fauna of southern Europe (13 valid species names).

1867ff.: CHRISTOPH: Various publications on 'Russian' Lepidoptera, e.g. 'Lepidoptera aus dem Achal-Tekke-Gebiete' in Mem. Lep. (ROMANOFF) (6 valid species names).

1876ff.: ALPHÉRAKY: Various publications, mainly on 'Russian' and Asiatic Lepidoptera in Mem.Lep. (ROMANOFF), e.g. 'Lépidoptères du district de Kouldjà' (1881) (4 valid species names).

1876ff.: OBERTHÜR: Various papers on Palaearctic and tropical Lepidoptera, mostly published in his well illustrated book series 'Lépidoptérologie Comparée' in 22 volumes and 'Etudes d'Entomologie', 21 volumes (9 valid species names).

19

1878ff.: BUTLER: Extensive taxonomic research on tropical Geometridae, a few descriptions of species with distribution in Europe (2 valid names).

1888ff.: PÜNGELER: Various papers on Palaearctic Lepidoptera (9 valid species names).

1895ff.: REBEL: 'Berge's Schmetterlingsbuch' (1910) and many other papers (1895-1939), which mainly concern the fauna of the Balkans (13 valid species names).

1902: DIETZE: Various papers on the genus *Eupithecia*, e.g. 'Biologie der Eupithecien' (1910-1913) (7 valid species names).

1906ff.: BANG-HAAS: Various papers, mainly on western Palaearctic Lepidoptera, e.g. edition of the 'Novitates Macrolepidopterologicae' (5 valid species names).

1909: PETERSEN (Wilhelm): First comprehensive description and illustration of the genital morphology of (more than 100) European species of the genus *Eupithecia*. (2 valid species names).

1909ff.: WAGNER (Fritz): Various papers, mainly on the Geometridae of south-eastern Europe and Turkey (4 valid species names).

1912ff.: PROUT (in SEITZ, 1912-1916; suppl. 1934-1938): 'The macrolepidoptera of the world', the Palaearctic Geometridae in volume 4 and suppl. 4. This is the first and to date only comprehensive monograph of Palaearctic Geometridae. Valuable morphological characterizations throughout, few data on biology, colour plates sometimes not adequate for unambiguous species identification. (8 valid species names; some generic names).

1913ff.: TURATI: Numerous papers on the Geometridae of northern Africa, but also Sicily and Sardinia (4 valid species names).

1914ff.: SCHAWERDA: Many papers, mainly on the Geometridae of southern Europe (10 valid species names).

1917ff.: WEHRLI: Numerous papers and revisions, most important the update of the Ennominae-part of 'SEITZ Macrolepidoptera' (supplement 4; 1939-1954). Many additional descriptions (11 valid species names; numerous generic names).

1927ff.: REISSER: Various papers, mainly on the Sterrhinae (7 valid species names).

1929ff.: HEYDEMANN: Extensive revisions of some genera and species complexes (2 valid species names, many subspecies and infrasubspecific forms).

1938ff.: HARTIG: Various publications, mainly concerning the fauna of Italy and Sardinia (one valid species name).

1939ff.: URBAHN: Various papers with differential diagnosis of difficult species complexes, fauna of 'Pommern' (1939; north-western Poland) (one valid species name).

1940f.: STERNECK: Revision of Palaearctic Sterrhinae, based on morphological characters of the males. The resulting sequence and grouping of species is maintained until today (one valid species name).

1945ff.: WOLFSBERGER: Important faunistic papers on the fauna of Bavaria and the Alps (one valid species name).

1945ff.: PINKER: Many papers on the Geometridae of the Canaries, but also on the fauna of the Balkan countries (2 valid species names).

1946ff. HERBULOT: More than 250 publications, mainly on tropical Geometridae. 'Mise à jour', i.e. the update of the Geometridae fauna of France (1961-1963). Revisionary work on some southern European species (3 valid species names).

1947ff.: AGENJO: Studies on Spanish Geometridae, e.g. the Checklist (1947) and the 'Faunula almeriense' (1952) (3 valid species names, 4 new genera).

1958ff.: SCHÜTZE: Many papers on the genus *Eupithecia* (2 valid species names).

1969ff.: VOJNITS: Many papers on the genus *Eupithecia*. Fauna of Hungary (Ennominae: 1980) (4 valid species names).

1976ff.: VIIDALEPP: Geometridae fauna of the former Soviet Union (1976-1979). The latest checklist (1996) offers a wealth of information. Many other taxonomic and faunistic papers (3 valid species names).

1979ff.: EXPOSITO HERMOSA: Various taxonomic papers on Spanish Geometridae. Provisional checklists of Spain. (2 valid species names).

1980ff.: MENTZER (VON): Taxonomic work on some southern European species complexes (4 valid species names).

1980ff.: LERAUT: Synonymous checklists of the Lepidoptera of France, Belgium and Corsica, second edition in 1997. These checklists are widely accepted as a source for correct species names. Some other publications on Geometridae (one valid species name).

1985ff. RAINERI: Various faunistic and taxonomic studies on southern European Geometridae. Checklist of the Geometridae of Italia (one valid species name).

1990f.: MIRONOV: Modern catalogue with new taxonomic arrangement of the Eupitheciini of the former Soviet Union (2 valid species names).

1990ff.: HAUSMANN: Various faunistic and taxonomic studies on southern European Geometridae (9 valid species names).

1994: PARENZAN: Fauna of the Geometridae of southern Italy, many new records for the region (one valid species name).

Most of the currently valid species names were described before 1880 (more than 3/4; see Text-fig. 7). After a gap in taxonomic publications during World War II the extent of studies in European Geometridae has become more encouraging in recent years with an average of one newly described species a year. Taken as a whole, the species inventory of the European Geometridae seems fairly complete. At least 95 % of the existing geometrid fauna of Europe is probably known. There may be unrecognized pairs of very similar sister species, and in southern Spain, the Urals and in south-eastern Europe some new species can still be expected.

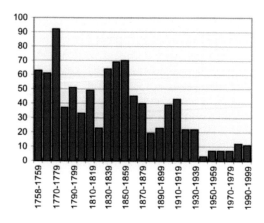

Text-fig. 7: Exploration of European Geometridae: Number of original descriptions (valid species names) per decade from 1758 until today.

The gaps

Recently SCOBLE et al. (1995:140) characterized the current situation of systematic research on western Palaearctic Geometridae: 'Revisionary work since then [i.e. middle of the 20th century] has been patchy. Few works cover taxa across the region and there are many subregional (often country) treatments. ... The need for co-ordination and revision is acute in the western Palaearctic subregion'. The following deficiencies and gaps are pointed out:

There is no top quality identification literature covering the complete fauna of Europe: Formerly widely used reference books on the Lepidoptera of Europe such as HOFMANN (1884) and SPULER (1908-1910) lack many species and are no longer compatible with modern nomenclatural and taxonomic standards. Due to its outstanding and accurate colour illustrations of the European Geometridae the work of CULOT (1917-1919), available now in an excellent reprint edition, is still a very useful tool for the identification of Geometridae although it shares with other entomological literature of that time the lack of differential diagnoses and an outdated nomenclature.

There are good more recent, but regionally restricted, identification books on the Geometridae of northern Europe (SKOU 1986), Sweden (NORDSTRÖM et al. 1941), Denmark (HOFFMEYER 1966), Central Europe (FORSTER & WOHLFAHRT 1981; FAJCÍK & SLAMKA 1996), Germany (KOCH 1984), Poland (BLESZYNSKI 1960-1966; BUSZKO 2000) and Great Britain (BROOKS 1991; SKINNER 1984; SOUTH 1961). They all, however, treat the morphology only partially and, due to their restricted geographical focus, cover between them no more than 560 species (about 60 %) of the European species.

Until recently the texts of PROUT in SEITZ (1912-1916) including the supplements (PROUT and WEHRLI in SEITZ suppl. 1934-1954) were the only diagnostic information source on many European Geometridae. However numerous taxonomic and nomenclatural changes have occurred since and must be considered. The illustrations in 'SEITZ' are often misleading, being too small or inaccurate. There are no illustrations of anatomical details (e.g. wing venation), and there is practically no discussion of the morphology of the genitalia which are so important in the classification and diagnostics of many groups.

A comprehensive work on the geometrid morphology, especially on the diagnostically important genitalia is lacking. Morphological detail has to be extracted from numerous different papers published in the 20th century, whilst the genitalia of many species, mainly those in southern Europe, have never been figured. Consequently the correct identification of many species is possible only for a few specialists.

For just a few countries the genitalia of all, or most, Geometridae are figured: PIERCE (1914: Great Britain), BLESZINSKI (1960-1966: Poland), and FAJČÍK & SLAMKA (1996: central Europe). Important and well illustrated publications exist for several difficult groups, such as the Sterrhinae by STERNECK (1940-1941: males only) or the genus *Eupithecia* by PETERSEN (1909), JUUL (1948), WEIGT (1976-1991), SKINNER et al. (1981), SKOU (1986), and others.

Data on biology and habitats are not easily accessible: Much information about host-plants and habitats, phenology, competition strategies, etc. is published piecemeal for the European Geometridae. For practical purposes in conservation, pest control, systematic or phylogenetic studies the summarizing of such scattered information is essential. Moreover, such an overview should offer the opportunity to eliminate observations which are not authentic and reliable but were merely drawn from other authors.

No overall co-ordinated zoogeographic survey exists for Europe, although biodiversity surveys are expected from all signatories to the convention on Biodiversity (UNCED 1992). The faunistic data on Geometridae have to be compiled from a large number of regional checklists and inventories for individual countries. The first comprehensive list of European Geometridae by SCHMIDLIN (1964) is largely outdated. The recent 'Checklist of European Lepidoptera' (KARSHOLT & RAZOWSKI (eds) 1996: Geometridae by MÜLLER) combines at least the species inventories of the European countries. Recording merely the presence or absence within (artificial) political units of rather unequal size is too crude to reflect accurately the distribution of a given species. For example contrast Malta with 'eastern Europe' which covers the entire European part of the former USSR (without the Baltic states).

Amongst the more important national or regional faunistic inventories with distributional and/or biological data the following are worth mentioning:

Northern Europe: SKOU 1986; OPHEIM 1972 (N); GUSTAFSSON 1987 (S); SOTAVALTA 1995 (FIN); KARSHOLT et al. 1985 (DK); KARSHOLT & STADEL NIELSEN 1998 (DK); FIBIGER & SVENDSEN 1981 (DK); SULCS & VIIDALEPP 1972 (Baltic states); SAVENKOV et al. 1996 (LV); LUIG & KESKÜLA 1995 (EST); VIIDALEPP 1995 (EST); IVINSKIS 1993 (LIT); DERZHAVETSCH et al. 1986 (RUS: St. Petersburg district); KOZLOV & JALAVA 1994 (RUS: Kola Peninsula).

Central Europe: FORSTER & WOHLFAHRT 1981; BLESZINSKI 1960-1966 (PL); MALKIEWICZ & SOSINSKI 1999 (PL); BUSZKO 2000 (PL); GAEDIKE & HEINICKE 1999 (D); GELBRECHT 1999 (D); WOLF 1988 (D); GELBRECHT & MÜLLER 1987 (eastern D); WEGNER 1999 (north-western D); STAMM 1981 (western D); KRAUS 1993 (D: Pfalz); ERLACHER &

FRIEDRICH 1994 (D: Thüringen); EBERT 2000-2001 (south-western D); ANE 1988 (D: northern Bavaria); HUEMER & TARMANN 1993 (A); MACK 1985 (A: north-eastern Alps); EMBACHER 1990 (A: Salzburg); STERZL 1967 (A: Niederösterreich); HUEMER 1994 (A: Vorarlberg); LASTUVKA 1993; 1998 (CZ, SK); HRUBY 1964 (SK); REIPRICH & OKAL 1989 (SK); REZBANYAI-RESER 1993 (CH: Ticino); RAPPAZ 1979 (CH: Vallesia); WOLFSBERGER 1966/1974; 1971 (I: Lake Garda, Monte Baldo); HELLMANN 1987; 1999 (I: Brenta; Valle d'Aosta); VOJNITS 1980 (H); KOVACS 1965 (H).

Western Europe: BAYNES 1970; 1973 (IRL); SKINNER 1984 (GB); SUTTON & BEAUMONT 1989 (GB: Yorkshire); LHOMME 1923f. (F); HERBULOT 1961-1963 (F); LERAUT 1980 & 1998 (F, B, Corsica); RUNGS 1988 (F: Corsica); CHAPELON 1992 (F: Saone-et-Loire); DUFAY 1961 (F: eastern Pyrenees); MAZEL & PESLIER 1997 (F: eastern Pyrenees); LEMPKE 1949 (NL); JANSSEN 1985 (NL); DE PRINS 1998 (B)

Southern Europe: VIVES MORENO 1994 (E); GOMEZ DE AIZPURUA 1988 (northern E); ABOS-CASTEL 1995 (E: Huesca); REDONDO & GASTÓN 1999 (E: Aragon); DANTART & ROCHE 1992 (E: Andorra); DANTART, PEREZ DE-GREGORIO & VALLHONRAT 1993 (E: Balears); SILVA CRUZ & GONCALVES 1977 (P); 1984 (northern P); PASSOS DE CAR-VALHO 1984 (northern P); MONTEIRO & PASSOS DE CARVALHO 1984 (P: Algarve); MARIANI 1943 (I); RAINERI & ZANGHERI 1995 (I); HARTIG & AMSEL 1951 (I: Sardinia); FIUMI & CAMPORESI 1988 (I: Romagna); PROLA & RACHELI 1979 (central I); PARENZAN 1994 (southern I); VALLETTA 1973 (Malta); ZECEVIC & RADOVANOVIC 1974 (eastern YU); URBAHN 1966 (ALB); CAPUSE & KOVACS 1987 (RO); POPESCU-GORJ 1987 (RO); GANEV 1983 (BG); NESTOROVA 1998 (BG); PINKER 1968 (MK); SEVEN 1991 (European TR).

Eastern Europe: VIIDALEPP 1976-1979, 1996 (RUS, former USSR); ANTONOVA 1972-1998 (RUS: various regions); BUDASHKIN & KOSTJUK 1987 (UKR: southeastern Crimea); KOSTJUK (in EFETOV & BUDASHKIN) 1990 (UKR: Crimea); KOSTJUK 1985, 1986 (central UKR: Dnieper region); KOSTJUK et al. 1998 (northern UKR: Chernigov-region); MERZHEEVSKAYA et al. 1976 (BY); BUSZKO 2000 (PL).

As synonymous checklists LERAUT (1980; 1997), WOLF (1988) and VIVES MORENO (1994) include many nomenclatural informations. The above enumeration does not claim to be complete. Nevertheless it gives an impression of the time-consuming task of gathering the necessary data from a multitude of publications. So the need for co-ordination is still acute (SCOBLE et al. 1995:140).

2. The importance of Geometridae to man

Pests and pest control

Insect pests are often responsible for considerable damage to agriculture and forestry. The uncontrolled increase in abundance of a population (**'gradation'**), is mainly caused by a disturbed natural balance due to e.g. monocultural plantations and absence of predators, even the latter often being anthropogenously caused. Such gradations usually break down after one or two years - without any use of pesticides - through an increase of harmful micro-organisms, predators and parasitoids. It may be that general climatic changes ('global warming') and the anthropogenous alteration of flora and fauna are the reason why various species have recently shifted out of the area of their original distribution. Under the altered conditions those species may become pests.

A prerequisite of effective and fast control of pests is that the insect should be clearly identified. The identification of pest species, however, is often problematic, as most pest guides and keys limit themselves to just a few species and ignore the existence of further similar species. Moreover other potential pests may also occur on the same plant. Better biological knowledge will enable us to reduce the environmental risks by replacing chemical pesticides with 'soft' biological control measures such as parasitoids or pheromone traps.

Sound taxonomic research and exact observation in the field are essential for that purpose. To give just one example: The polyphagous geometrid *Phaiogramma stibolepida* (BUTLER, 1879) is reported as a pest on cotton in Africa (ZHANG 1994). Recent taxonomic research revealed the species to be closely related to, or perhaps conspecific with, the Mediterranean *Phaiogramma faustinata* (MILLIÈRE, 1868) which means that it could also become a pest in the Mediterranean area (see p. 206).

In the GME series an attempt is made to give accurate and detailed information on the phenology of all stages of a species and to indicate its specific parasitoids as far as known.

KUDLER (1978) listed 87 'geometrid pest species in forestry', i.e. 16 % of the about 550 central European species. ZHANG (1994) recorded 351 geometrid species worldwide as economically important. Common European pest species in the Geometridae are for instance:

In **forestry** and orchards the caterpillars of *Alsophila aescularia* ([DENIS & SCHIFFER-MÜLLER], 1775), *Operophtera brumata* (LINNAEUS, 1758), *Erannis defoliaria* (CLERCK, 1759), and *Bupalus piniaria* (LINNAEUS, 1758) are feared as serious pests. The scientific name 'defoliaria' refers to the fact that these species often totally defoliate infested trees.

In **agriculture** very few geometrid species are of significant economic importance: *Idaea seriata* (SCHRANK, 1802) and other *Idaea* species are found as minor pests on tobacco. Greater losses are caused by several Noctuidae species in agriculture.

In **gardens** *Abraxas grossulariata* (LINNAEUS, 1758) was formerly known (and denounced by the popular moth books) as a common pest on gooseberry and currant shrubs but since the middle of the 20th century the abundance of that species has decreased to such an extent that this pretty moth has become extinct in many parts of Europe.

In **stored foods** damage is mainly caused by Pyralidae. No European geometrid moth is known as a serious pest. The larvae of some *Idaea* species, e.g. *Idaea inquinata* (SCOPOLI, 1763), are reported feeding on dry tea, dried herbs, and even in herbariums (WEIDNER 1993).

Endangered species and conservation

Geometridae are useful environmental **indicators**. Natural or man-induced changes in the environment cannot be assessed by monitoring each and every organism living in a specific habitat. Indicator groups which are particularly sensitive to such changes, have to be selected. The Geometridae are considered to be 'of particular value as indicators given their relatively restricted mobility' (SCOBLE 1995:179; MINET & SCOBLE 1999:313). The monitoring of environmental processes requires more often than not the capture and identification of the insect components of an ecosystem. Lamentably few junior entomologists are capable of recognizing the species, and the full potential of the species-rich Geometridae as indicators of the character of local habitats is not taken into account.

Public awareness of the threats to the environment in Europe has resulted in various concepts of nature and species protection. Exceeding the scope of the international Conventions of Berne and Washington, which listed only a limited number of Lepidoptera, some countries have placed many more species on 'red data lists' and classified them as 'endangered', 'potentially endangered' or even as threatened with extinction. Thus in Germany, a number of Geometridae is now legally protected, whereas the law in countries such as Spain generally protects any butterfly or moth as well as any other living animal.

How can the threat to Geometridae be assessed realistically? The threat to nocturnal insects is generally hard to verify. Some species have been notoriously rare for more than a century. Do they need protection now? 'Rareness' may to a considerable degree depend on the chosen sites, methods, and frequency of observation (HAUSMANN 1993b). Amongst the moths (especially Sphingidae and Noctuidae) a number of long-range travellers are known, and there is the migrant *Orthonama obstipata* (FABRICIUS, 1794) in the Geometridae, but many Geometrids virtually never leave their immediate habitat (e.g. *Theria primaria* (HAWORTH, 1809)). Light-sources attract many moths, but there are species (or one of their sexes) that are repelled by strong lights and remain hidden in nearby vegetation rather than approach a light-trap. It is also obvious that species restricted to barely accessible habitats (peat-bogs, rocky summits) are usually observed less frequently than others.

As there are only scant faunistic data for many European regions, and as appropriate long-term observations are mostly wanting, the reliability and genuine value of 'red lists' and species protection laws may be challenged to a major extent. But it is certain that insects - much more than vertebrates - depend on microclimatic factors, host-plants, and interaction with other organisms, and thus are threatened directly by risks to their habitats. **There is no protection of a species without the protection of its habitat**. If protection of habitats is on the agenda, monitoring by sampling insects, especially those with specific indicator quality such as Geometridae, should be encouraged rather than restricted by regulation.

There can be no doubt that in the 20th century the populations of many European Lepidoptera have been seriously disturbed by man's activities, except perhaps in the extreme northern and certain parts of southern Europe. Wetland drainage, monocultural afforestation and deforestation of autochtonous woods in the Mediterranean area and elsewhere, are just some examples. The noticeable decrease in species affects most species living in open habitats, i.e. moors and heathlands in western, central and eastern Europe. Such habitats are mainly lost because of the modern use of land by agriculture and forestry (see BLAB & KUDRNA 1982). As a consequence species such as *Lithostege farinata* (HUFNAGEL, 1767), *Lithostege griseata* ([DENIS & SCHIFFERMÜLLER], 1775), *Cataclysme riguata* (HÜBNER, [1813]), *Eupithecia pauxillaria* (BOISDUVAL, 1840) in Germany or *Thetidia smaragdaria* (FABRICIUS, 1787) and *Thalera fimbrialis* (SCOPOLI, 1763) in Great Britain are suffering severe habitat losses.

Collecting and collections

Public concern

Whereas game hunting enjoys a certain esteem, the gentleman trying to net a butterfly on a Sunday morning usually encounters his share of mockery and comment. The passion for the beauty and fascinating miracles that can be detected in the insect world was nevertheless shared by a wide range of personalities who were famous and successful in 'normal' life. To name just a few: King Leopold of Belgium, Princess Therese of Bavaria, Lord Rothschild, Lord Walsingham, Prince Caradja, Prince Romanoff, the famous writers Ernst Jünger and Vladimir Nabokov, and Ludwig Osthelder, then president of the district government of Upper Bavaria, who is renowned for his complete (i.e. Macro- and Microlepidoptera) treatment of the Lepidoptera fauna of southern Bavaria.

For many decades, having an insect collection was acceptable in public perception. In our time, insect collectors, be they amateurs or professional entomologists, are increasingly confronted with social harrassment in the same way as smokers or those working for an atomic power plant. Moreover, the complicated species protection laws add to the prejudice that collectors are wrongdoers who are to blame for the impoverishment of the environment.

Emotions are one thing, facts are another.
Some simple calculations can help to understand the dimensions. It should be borne in mind that insects constitute the largest quantity of animal biomass on earth. Insects are the nutritional basis for many other animals, especially birds, bats and other insectivorous mammals. One single **bat** eats 1,8-3,6 kg insects per year (SCOBLE 1995:181) which includes more than 10,000 moths (a middle-sized live moth weighs about 0.1-0.2 g). Hence a small number of bats remove more moths than all European collectors together.

Actually, in Europe many moths die by collisions with **cars** (LANDRY 1980; DONATH 1987). In an Austrian highway-sector of just <u>one kilometer</u>, some 3,000 insects were reported killed by <u>one car</u> (GEPP 1973). But in Europe the network of roads consists of about five million kilometers, with some hundred million cars, making billions of kilometers each year. The number of moths killed in traffic in one <u>minute</u> exceeds by far the number taken by the few hundred European collectors in one <u>century</u>!

Industrial and public **illuminations** attract and kill moths or disturb their reproduction. The strong lamps illuminating a monumental statue in **southern** Italy attract about 5 million Macrolepidoptera per year (HAUSMANN 1992). Deliberate **burning** of wildland in southern Europe probably destroys billions of moths and their early stages year after year.

Text-fig. 8: Insects killed on the front of a car.

Text-fig. 9: Moths killed at the illumination of a monument in southern Europe.

In the light of these figures, collecting is negligible as an ecological factor in the survival of moth populations. This is generally true even for rare species, a qualification that is certainly not respected by bats, cars, and public mercury lamps. If a collector can extinguish a moth species in a certain locality, the population has already shrunk to a degree that the next natural calamity (late frost, wet and cold summer or a drought) would have extinguished it anyway. Of course, this is not an excuse for irresponsible collecting, and, notwithstanding specific legislation, every lepidopterist is urged to observe the general rule, as incorporated in the laws of many European countries, that unreasonable killing of animals is unlawful.

But there is no reason to ban collecting or collectors in general. On the contrary, for the sake of effective nature conservation, more entomologists are needed to clarify numerous unresolved questions.

In some countries legislation on species protection also affects private moth collections. In Germany for instance, it is unlawful to possess a private collection containing explicitly protected species without authorisation by the appropriate administrative bodies.

Techniques

Most geometrid moths are nocturnal, and many species can be obtained by the use of light traps (MUIRHEAD-THOMPSON 1991). In the bright summer nights of northern Scandinavia and northern Russia however light sources are very ineffective in attracting moths (SKOU 1986). Automatically run killing traps should only exceptionally be operated for that purpose as they attract many other insects, too, and mostly do not yield specimens of a quality suitable for identification.

The techniques of serious **collecting** of the adults and of rearing moths from their early stages are explained in general lepidopterological literature, and are largely applicable also to Geometridae (BOURGOGNE 1963; COMMON 1990; FRIEDRICH 1984, English version 1986 with notes by J. REID concerning European Geometridae; KOCH 1984; SOUTH 1961; WEIGT 1980).

Modern biochemical methods (sex pheromone traps, as already available for some Lymantriidae and Sesiidae) have not yet been focussed on Geometridae. They could play an important role for the monitoring of some diurnal species (e.g. *Epirranthis diversata* ([DENIS & SCHIFFERMÜLLER], 1775).

The **preparation** of geometrid moths for a collection does not differ much from that of other (Macro-)Lepidoptera. A special problem is the green pigment of the Emerald moths (subfamily Geometrinae) which in most European representatives fades under the influence of humidity (acids according to COOK et al. 1994) so quickly that setting of an already dried and then relaxed specimen usually results in severe discoloration. No really satisfactory technique has as yet been found. Aesthetically perfect specimens therefore require immediate setting after killing.

As some morphological features are essential for the identification of species, care should be taken to preserve legs, antennae and the abdomen. The common way of preparing the genitalia consists of spreading the valves (ventral view) after maceration and removal of the aedeagus. The genitalia may be stored in glycerine tubes or mounted on glass slides, the latter being preferred by most lepidopterologists. Everting of the vesica is very difficult in the smaller Geometridae (e.g. Sterrhinae), and at present is not standard practice for genitalia preparations of Geometridae (see BOUR-GOGNE 1963; ROBINSON 1976; WEIGT 1979).

Public collections

Public moth collections are often stored in 'Natural History Museums' and are important parts of our cultural heritage as well as databases of evolution. These collections are usually not accessible to the general public because the scientific and historical assets, particularly irreplaceable type-specimens, must not be exposed to light, shocks or other unnecessary risks. For scientific work visits may be arranged with the relevant authorities (more and more museums host web-sites on the internet). Many museums have initiated projects for electronic cataloguing of their holdings. Some moth catalogues are already available on the internet (e.g. ZSM: www.zsm.mwn.de).

As far as the Geometridae of Europe are concerned, the most important collections are preserved in London and Berlin, where the types of about 20 % of the European geometrid moth species are deposited: i.e. the Natural History Museum of London (**BMNH**) with the types of HAWORTH, PROUT, OBERTHÜR (partim), ZELLER, HERRICH-SCHÄFFER (partim) and others, the collection of the Linnean Society (**LSL**) with the types of LINNAEUS and the Museum für Naturkunde der Humboldt Universität zu Berlin (**MNHU**), with the original material of STAUDINGER, PÜNGELER, LEDERER (partim), HERRICH-SCHÄFFER (partim). The Zoologisches Forschungsinstitut und Museum Alexander Koenig (**ZFMK**) in Bonn holds the large geometrid collection of WEHRLI (including also some types of BOISDUVAL, GUENÉE, and OBERTHÜR) which served as a basis for WEHRLI's revisions in the supplement to SEITZ. Large collections of Geometridae from eastern Europe (types of ALPHÉRAKY, CHRISTOPH (partim), ERSCHOFF, EVERSMANN are found in the Zoological Institute and Academy of Science at St. Petersburg (**ZISP**). One of the largest collections of Geometridae (about one million specimens) with much material from southern Europe, the famous collection HERBULOT and important historical types (ESPER, RAMBUR, MABILLE) is housed at the Zoologische Staatssammlung München (**ZSM**) in Munich. The Naturhistorisches Museum Wien (**NHMW**) in Vienna keeps large collections with many types (e.g. MANN, PINKER, REBEL, SCHAWERDA, WAGNER, ZERNY). The type material of HÜBNER, and DENIS & SCHIFFERMÜLLER which was originally there, was, however, consumed by fire in 1848. Also the collections HUFNAGEL and SCOPOLI were destroyed, thus the type material of 233 species, i.e. one quarter of the European fauna, is lost.

Other noteworthy museum collections relevant for European Geometridae are (in alphabetical order): Amsterdam (NL), Basel (CH), Bucarest (RO), Budapest (H; **HNHM**; types of TREITSCHKE), Copenhagen (DK; **ZMUC**; types of FABRICIUS), Dresden (D), Eberswalde (D); Edinburgh (GB), Firenze (I), Frankfurt (D), Genève (CH; **MHNG**; types of GUENÉE, partim), Genova (I), Hamburg (D), Helsinki (FIN), Innsbruck (A), Karlsruhe (D; **SMNK**; types of REISSER), Kiew (UKR), Leiden (NL), Lund (S), Luzern (CH), Luxembourg (L), Madrid (E), Moscow (RU), Oslo (N), Oxford (GB), Paris (F; **MNHN**, Muséum National d'Histoire Naturelle, types of DUPONCHEL), Praha (CZ), Rome (I), Sofia (BG), Stockholm (S; **SMNS**; types of CLERCK), Stuttgart (D), Torino (I),

Uppsala (S; Dept. Zool. Univ.; types of THUNBERG, LINNAEUS partim), Warszawa (PL), Wiesbaden (D; **LMW**, types of ESPER, partim).

For further information see the detailed lists in KUDRNA & WIEMERS (1990) and HORN et al. (1990).

3. The geometrid moth in its environment

Behaviour and reproduction

Activity patterns

Most European Geometridae species are nocturnal, but they often fly preferably at dusk. Others are said to be active mainly after midnight, such as *Abraxas grossulariata* (LINNAEUS, 1758), *Crocallis elinguaria* (LINNAEUS, 1758) and *Eulithis pyraliata* ([DENIS & SCHIFFERMÜLLER], 1775) (PROUT 1912a:IV). Some species are diurnal (genera *Archiearis, Heliothea, Ematurga, Bichroma, Eurranthis, Athroolopha, Psodos, Glacies*), others both day- and night-active (*Rheumaptera undulata* (LINNAEUS, 1758), *Odezia atrata* (LINNAEUS, 1758), *Chiasmia clathrata* (LINNAEUS, 1758), genera *Epirranthis, Chlorissa, Rhodostrophia, Rhodometra, Casilda*). Certain species, which are nocturnal in central Europe become diurnal in northern Europe, as e.g. *Entephria caesiata* ([DENIS & SCHIFFERMÜLLER], 1775).

Resting position

Each species is characterized by a specific resting position. There are three different basic types:

Most European Geometridae rest in **planiform** position (Text-figs 10-11), i.e. the wings are appressed flatly to the substrate. Some species rest with wings in a delta-like position (Text-fig. 10) with the hindwings totally covered (e.g. genera *Pseudoterpna* and *Erannis*, many Larentiinae), others spread their wings at right angles to the body (Text-fig. 11). Continuous transitions occur between these extremes. In the delta-like resting species the hindwings are bright, nearly without wing pattern, lacking the transverse fascia of the forewing. The species of the genus *Alsophila* rest with crossed forewings, *Apeira syringaria* (LINNAEUS, 1758) slightly folds them. *Apochima flabellaria* (HEEGER, 1838) rolls its forewings into tubes and raises them in the air over the body.

In **veliform** resting moths (Text-fig. 12) the wings are vertically folded over the abdomen as is common in butterflies, and the upperside of the wings is invisible. Examples: *Euchoeca, Narraga, Bichroma* and *Bupalus. Chiasmia clathrata* (LINNAEUS, 1758), *Ennomos, Deuteronomos* and others usually rest with slightly opened wings. In *Selenia* and *Siona* the wings are either slightly opened or closed.

In the **tectiform** position (Text-fig. 13) the wings are positioned like a roof over the abdomen as is common in Noctuidae. Examples: *Heliothea discoidaria* BOISDUVAL, 1840, *Rhodometra sacraria* (LINNAEUS, 1767), genera *Compsoptera* and *Chesias*. The latter slightly roll their forewings. In the genera *Archiearis, Gypsochroa, Epirranthis* and *Trichopteryx* the forewings are rolled around twigs.

Some species, when resting, raise their abdomen upwards (genera *Eulithis* and *Pelurga*). The reason for this behaviour is unknown.

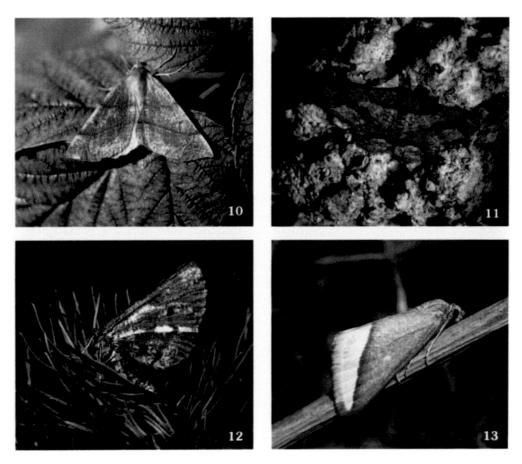

Text-figs 10-13: Resting positions. Text-fig. 10: *Colotois pennaria* (LINNAEUS, 1761): planiform, 'delta-type'. Text-fig. 11: *Menophra japygiaria* (COSTA, 1849): planiform, 'spread type'. Text-fig. 12: *Bupalus piniaria* (LINNAEUS, 1758): veliform. Text-fig. 13: *Casilda antophilaria* (HÜBNER, [1813]): tectiform (photo 13 LEIPNITZ).

Feeding (adult)

Most geometrids suck nectar from flowers and drink water. Some European species (e.g. genera *Ennomos, Deuteronomos, Eumera, Theria,* many Bistonini, some *Idaea* species etc.) however have secondarily lost the proboscis and do not feed or drink anything. Though it is well known that nectar sucking by Lepidoptera is very important for **pollination** of many plant species, much research is still needed to get a better impression of the role of Geometridae for plant pollination. *Aplocera* species (Larentiinae) not rarely bear pollinia of Orchidaceae on their head, eyes or proboscis (see Text-fig. 14). The extremely long proboscis of the southern European *Glossotrophia* species may be understood as the product of co-evolution with the flower-tubes of Caryophyllaceae, mainly *Silene*, on which their larvae also feed. Some tropical species, e.g. genus *Scopula*, are known to imbibe tears, blood and sweat (BÄNZIGER & FLETCHER 1985; SCOBLE 1995), but that phenomenon has not been reported from European species.

Text-fig. 14: *Aplocera annexata* (FREYER, 1830), eastern Turkey, head with attached pollinium.

Mating behaviour and reproduction

Geometridae usually copulate immediately in the first days after emerging from the pupa. Obligatory egg maturation before copulation or egg deposition is unknown in European species. In diurnal species visual attraction may, exceptionally, play a role in mating, but the chemical scent attraction of males by the female's sex **pheromones** is always the elementary mechanism which facilitates the sexes finding each other. Female scent glands are situated between the posterior abdominal segments (A7-A8 or A8-A9). The chemistry of the female pheromones is species-specific.

Even the males are characterized by a large diversity of pencils (on legs and abdomen) and membranous eversible '**coremata**'. They relate to additional scent mechanisms, but are, unfortunately, little studied for the species of the study area. Coremata typically occur on abdominal segments 7-8 of Xanthorhoini (Larentiinae), on the valvae (male genitalia) of many Geometrinae species, and on the saccus (male genitalia) of the genus *Gymnoscelis* (Larentiinae).

Text-fig. 15: Copulation in *Eupithecia intricata* (ZETTERSTEDT, 1839)

Text-fig. 16: Copulation in *Operophtera brumata* (LINNAEUS, 1758) (photo LEIPNITZ)

Egg positioning

Soon after fertilization, the females begin to deposite their eggs. They are laid either singly or in batches, usually well attached to a particular part of the host-plant. The eggs of *Hemistola chrysoprasaria* (ESPER, 1795) are stacked in 'towers' of about 5-20 eggs, resembling plant tendrils (PEKING 1953; SCOBLE 1995). Eggs are often covered with hair-like scales from the anal tuft of the female (Text-fig. 151). Many short-winged or wingless females, such as the winter moths *Alsophila* and *Chondrosoma* lay their eggs in broad ribbons around twigs (Text-fig. 17). Females of some species (e.g. some *Idaea* species) scatter the eggs into the vegetation without attaching them. On average egg numbers range from 100 to 400 per female. The caterpillars usually hatch about one week after the egg positioning, in some Mediterranean Sterrhinae species even after 2-3 days. In the genera *Philereme* and *Lygris* however the eggs hibernate.

Text-fig. 17: Egg batch of *Chondrosoma fiduciaria* ANKER, 1854 (photo LEIPNITZ).

Text-fig. 18: Egg positioning in *Dyscia penulataria* (HÜBNER, [1819]) (photo RÖDEL).

Life cycle and phenology

Moth populations are characterized by typical patterns of seasonal appearance ('**phenology**'). Species with one generation per year are called '**univoltine**', those with two generations '**bivoltine**', those with more generations '**plurivoltine**' (=polyvoltine). The larvae of univoltine populations can, exceptionally (in warm summers), grow at different rates, the quick ones developing as a 'partial second brood'. In correlation to temperature and length of the vegetation period bi- or polyvoltine phenology is common south of the Alps, univoltinism in the mountains and north of the Alps. In some '**protandrous**' species the females emerge later than the males, e.g. *Archiearis parthenias* (LINNAEUS, 1758) and genus *Glossotrophia*. Overlying pupae of various species can be facultatively kept back as a 'reserve' (e.g. *Archiearis* species, *Eupithecia venosata* (FABRICIUS, 1787)) or obligatorily wait for certain favorable conditions for emergence (e.g. *Bichroma famula* (ESPER, 1787)).

The adult geometrid moth usually lives 1-3 weeks, sometimes a few days only, sometimes up to two months. Species with hibernating adults, e.g. the females of *Chloroclysta siterata* (HUFNAGEL, 1767), *Chloroclysta miata* (LINNAEUS, 1758) or the species of the genus *Triphosa* can extend their life span up to about 9 months. The last often aggregate in caves for hibernation. In most species, however, the overwintering stage is the larva or the pupa, exceptionally the egg. An aestival diapause ('oversummering') is reported for *Camptogramma grisescens* (STAUDINGER, 1892) from south-eastern Europe and Turkey: Sometimes thousands of specimens aggregate in caves.

The 'winter moths', i.e. species of the genera *Alsophila, Operophtera, Agriopis, Erannis, Theria* and others, fly exclusively in the cold season, when their enemies, such as bats are sleeping or, as migrant birds, are away in the south. Obviously, the females of these species do not need their wings for escape behaviour. The material of the wings and even that of the tympanal organ is economized allowing more eggs to be produced (DIERL & REICHHOLF 1977; PELLMYR 1980).

Defence strategies

Dangers

Predators such as birds, bats and other insectivorous mammals are the most important natural enemies of adult geometrid moths. Day after day many moths die after collisions with traffic or get distracted by billions of light sources. The immature stages are, given their comparatively long 'life' (stage) and their low mobility, more threatened than the adults, e.g. by predators, parasitoids, diseases and man's activities. Caterpillars are predated mainly by birds, ants etc. Predation by small mammals (e.g. Soricidae) has been reported on the cocoons of *Operophtera brumata* (LINNAEUS, 1758) in Great Britain (BUCKNER 1969). The main parasitoids belong to the families Ichneumonidae and Braconidae (Hymenoptera). Man's activities such as habitat alteration, habitat destruction, air pollution, insecticides etc. also considerably threaten the early stages of moths. In response to natural enemies many different survival strategies have evolved.

Chemical defence

Various Lepidoptera species are unpalatable to predators, even in the adult stage, due to internally accumulated chemical substances. These are either assimilated from the larval food-plant, or sucked up by the adult from plants or other substrates, or synthetized by the insect itself (e.g. Zygaenidae). Such species usually wear warning colours ('**aposematic coloration**'), are mostly diurnal, do not move quickly, nor try to escape when disturbed.

The chemistry of European Geometridae is poorly studied. *Abraxas grossulariata* (LINNAEUS, 1758) and *Calospilos sylvata* (SCOPOLI, 1763) are known to be highly toxic (ROTHSCHILD 1985). *A. grossulariata* contains high concentrations of cyanoglucoside sarmentosin, an "endogenous biosynthesis" is suggested (NISHIDA et al. 1994). There is almost no escape behaviour in the genera *Abraxas* and *Calospilos* and their adults are aposematically coloured as are their caterpillars and pupae. The Mediterranean *Eucrostes indigenata* (VILLERS, 1789) is beautifully coloured, but not in an aposematic manner. The red green caterpillars feed on *Euphorbia*, which may provide a certain protection against predators, as with some other unpalatable *Euphorbia*-feeding moths.

Deception

Many species use 'deception' for their survival strategies. The best known phenomenon in Lepidoptera and other insects is '**mimicry**', a complex dynamical phenomenon of co-evolution. Two types of mimicry are distinguished: '**Batesian mimicry**', i.e. unpalatable species copied by palatable species, and '**Müllerian mimicry**', i.e. several unpalatable species choosing the same 'outfit' to share the efficiency of their coloration. In European Geometridae no example of mimicry has yet been reported.

A false head is perhaps exhibited by *Ourapteryx sambucaria* (LINNAEUS, 1758): There are not only small eye spots on the hindwings, but also false 'antennae', i.e. the tailed veins M3 of the hindwings. Furthermore the crosslines of all the wings lead attention to this false head. When predators pick out small parts of the hindwing, the moth can survive; specimens with such injuries are often commoner in nature than intact ones.

Concealment

When geometrid moths are disturbed during daytime, they usually escape for a certain distance and then try to hide again. Hiding is easy when the coloration of the moth is **cryptic**, i.e. similar to the substrate. Most geometrids are cryptically coloured (Text-figs 19-21). The predominant colours are brown, mixed with grey, green and ochreous. Often the wavy transverse lines are present also on the hindwing forming a continuous pattern. In 'substrate races', local populations become similar to the typical colour of the soil or rocks in their distribution area.

During the 20th century the percentage of melanistic (=darkened or black) specimens increased in many Lepidoptera species, mainly geometrids (URBAHN 1971; DOUWES et al. 1976). Dark specimens on darkened backgrounds are said to have certain advantages against predators. The most famous example for '**melanism**' is *Biston betularia* (LINNAEUS, 1758). The phenomenon of increasing number of black specimens ('f. *carbonaria*') has often been associated with the darkened tree trunks in industrial zones and the decreasing density of (white) lichens. Selection surely plays an important role, but there are still many unresolved questions of detail (MIKKOLA 1979; SCOBLE 1995).

Text-figs 19-21: Cryptic coloration. Text-fig. 19: *Glossotrophia mentzeri* HAUSMANN, 1993. Text-fig. 20: *Chloroclysta truncata* (HUFNAGEL, 1767). Text-fig. 21: *Alcis repandata* (LINNAEUS, 1758) (photo 19 LEIPNITZ; 21 WEGNER).

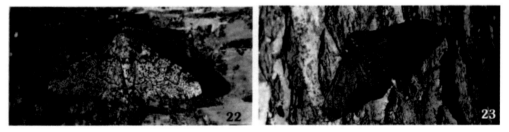

Text-figs 22-23: *Biston betularia* (LINNAEUS, 1758). Text-fig. 22: Nominate form. Text-fig. 23: Melanistic form (photo 22 FOGH NIELSEN).

Larva

Nearly all geometrid moth larvae are cryptically coloured. Many of them are brown with warts and resemble twigs. Others are simply green like the young fresh twigs or the leaves on which they feed. Many Coniferae-feeders such as *Heterothera firmata* (HÜBNER, [1822]), *Bupalus piniaria* (LINNAEUS, 1758) and *Macaria signaria* (HÜBNER, [1809]) have camouflaged larvae with longitudinal stripes and resemble the needles of the host-plant. The caterpillars of *Tephronia* species and *Cleorodes lichenaria* (HUFNAGEL, 1767) are excellently hidden by their white dorsal markings and warts, which are extremely similar to the lichens on which they feed. *Geometra papilionaria* (LINNAEUS, 1758), *Hemistola chrysoprasaria* (ESPER, 1795) and others are brown in autumn and in winter, when they have to hide on the ground or on wood and change their colour to green in the springtime, when they sit on the leaves or on young fresh green twigs.

Text-figs. 24-29: Cryptic and camouflaged larvae. Text-fig. 24: *Hemithea aestivaria* (HÜBNER, 1789). Text-fig. 25: *Thalera fimbrialis* (SCOPOLI, 1763). Text-fig. 26: *Scopula imitaria* (HÜBNER, [1799]). Text-fig. 27: *Perizoma sagittata* (FABRICIUS, 1787). Text-fig. 28: *Eupithecia phoeniceata* (RAMBUR, 1934). Text-fig. 29: *Nychiodes waltheri* (WAGNER, 1919) (photo 24 STÜNING; 25, 27 WEGNER; 28 REID; 29 LEIPNITZ).

The caterpillars of many geometrid species rest in a rigidly immobile position and stick out twig-like from the surface. When they are disturbed most of them remain rigid allowing them to fall to the ground. Others similarily rest attached with the last two pairs of prolegs only, but they 'loop' the head and the first segments under the abdominal segments. Special protection is gained by hiding between interwoven leaves (many species), by covering the body with small pieces of the host-plant (Comibaenini; see Text-fig. 192) or by choosing concealed places, e.g. in flower-cups (many *Eupithecia* species) or inside fruit capsules (*Perizoma*).

Mining in the leaves of *Primula* and other plants is reported for the first instars of *Perizoma incultaria* (HERRICH-SCHÄFFER, 1848) (HERING 1937:408). No case of active or passive chemical defence has been reported, even if there are certainly some unpalatable species (see above 'chemical defence'). Many caterpillars however are capable of secreting a dark green liquid out of the mouth. It is unknown if that is linked with a chemical defence effect.

Pupa

The last larval instar looks for a convenient place for **pupation**. The survival of the pupa as an immobile stage usually depends on whether it is well hidden or not. Many species pupate on the ground between leaves etc., some go 'underground' (e.g. *Operophtera brumata* (LINNAEUS, 1758) and *Lycia hirtaria* (CLERCK, 1759)), others are hidden between loosely interwoven leaves and twigs of the foodplant. Some species build hard, stable cocoons (e.g. *Archiearis notha* (HÜBNER, [1803])). The genus *Cyclophora* is characterized by a 'girdle pupa' (Text-fig. 114), which is known also from Papilionidae, Pieridae, Hedylidae and Elachistidae. The caterpillars of *Ourapteryx sambucaria* (LINNAEUS, 1758) pupate in a hanging web.

Niche and habitats

Each species occupies a characteristic ecological niche in the ecosystem. This niche depends not only on the abiotic and the biotic factors, which act upon the species (specimens) from outside, but also on the ecological plasticity and the preadaptative characteristics of the species. **Biotic factors** are e.g. the spectrum of available host-plants, enemy pressure and abundance of parasitoids. The niche also depends on **abiotic influences**, such as mean annual rainfall, mean annual temperature and altitude (which influence the availability of certain suitable host-plants), furthermore humidity, moonlight and wind may favour or obstruct the flight of the adult insects.

Host plant spectrum

The niche of a given species is usually mainly determined by the host-plant spectrum available to their caterpillars. The width of the niche is often strongly correlated with the number of different plant taxa that can be attacked by the larvae. The terminology of the feeding types follows that established by HERING (1950) and HUEMER (1988).

Monophagous species are specialized on one single host-plant species only ('monophagy 1'), on some ('monophagy 2') or on all the species of the same plant genus ('monophagy 3'). Examples of strict monophagy ('1'): *Baptria tibiale* (ESPER, 1791), *Eupithecia immundata* (LIENIG, 1846) and *Eupithecia actaeata* (WALDERDORFF, 1869) all of them feeding exclusively on *Actaea spicata*, *Asthena anseraria* (HERRICH-SCHÄFFER, 1855) specialized on *Cornus sanguinea*, and *Eupithecia gelidata* MÖSCHLER, 1860 on *Ledum palustre*. Examples of monophagy 3: *Aplasta ononaria* (FUESSLY, 1783) feeding on *Ononis* species, or *Comibaena bajularia* ([DENIS & SCHIFFERMÜLLER], 1775) on various species of the genus *Quercus*.

Oligophagous species can feed on a few plant species, which either belong to the same plant family (oligophagy 1) or on single representatives of different families, which however are more or less closely related to each other (oligophagy 2-3). 'Disjunctive oligophagous' species live on a few plant species which are not closely related to each other. The host-plants of oligophagous species often bear the same phytochemical substances. Examples: *Pseudoterpna pruinata* (HUFNAGEL, 1767), living on various genera of the Papilionaceae (oligophagous 1), *Bupalus piniaria* (LINNAEUS, 1758) feeding on Pinaceae and Cupressaceae (oligophagous 2) or *Eulithis populata* (LINNAEUS, 1758), on Ericaceae and Salicaceae (disjunctive oligophagous).

Polyphagous species have a very wide host-plant spectrum, the host-plants belonging to the same plant class (polyphagy 1) or not (polyphagy 2). Examples: *Gymnoscelis rufifasciata* (HAWORTH, 1809), *Peribatodes rhomboidaria* ([DENIS & SCHIFFERMÜLLER], 1775), *Alcis repandata* (LINNAEUS, 1758) and many others.

The caterpillars of European Geometridae usually feed on flowering plants (Angiospermae), but some species are specialized also on Gymnospermae: Coniferae-feeders are restricted to the subfamilies Larentiinae and Ennominae, such as the species of the genera *Thera*, *Heterothera*, some species of the genera *Eupithecia* and *Menophra*, *Macaria signaria* (HÜBNER, [1809]), *Peribatodes secundaria* (ESPER, 1797), *Pungeleria capreolaria* ([DENIS & SCHIFFERMÜLLER], 1775), *Adalbertia castiliaria* (STAUDINGER, 1899), the *Hylaea* species and *Bupalus piniaria* (LINNAEUS, 1758). Certain Indo-Pacific Orthostixinae and the species of the tribe Lithiini (worldwide) are 'pteridophagous', i.e. they feed on ferns (Pteridophyta) (WEINTRAUB et al. 1995). *Alcis jubata* (THUNBERG, 1788), *Cleorodes lichenaria* (HUFNAGEL, 1767) and the *Tephronia* species are obligatorily lichenivorous. Carnivory is characteristic for most of the *Eupithecia* species in Hawaii (MONTGOMERY 1982), while from the study area accidental insectivory is reported for larvae of *Eupithecia centaureata* ([DENIS & SCHIFFERMÜLLER], 1775) (HAWKINS 1942). Cannibalism may rarely occur in times of overcrowding. The species of the genus *Idaea* are 'detritus-feeders', i.e. they prefer moulding or dry leaves of low plants. This ability to live on dry plant material enables many Sterrhinae to colonize hot habitats, such as oases, semideserts and deserts.

Species belonging to the same genera or tribes are often specialized on host-plants that are closely related to each other (e.g. Lithiini see above). This phenomenon is more likely due to similarity of plant chemistry rather than to 'co-evolution', i.e. the 'history of congruent evolution between plants and insects at the species level' (SCOBLE 1995:174; POWELL et al. 1999:416f.).

Obligatory **host-plant changes** are characteristic for some bivoltine species, e.g. *Eupithecia tripunctaria* (HERRICH-SCHÄFFER, 1852), feeding on the flowers of *Sambucus nigra* in the spring generation and on Umbelliferae in the summer. Similarly, in central Europe, *Eupithecia innotata* (HUFNAGEL, 1767) often changes from deciduous trees and shrubs to *Artemisia* species in the second generation. The early instars of *Archiearis* species feed in the catkins of *Salix* or *Populus* and later change to the leaves. Similarly a lot of other species first feed on the flowers and then on the leaves of the host-plant.

Habitat types

The preferred habitat type is mainly determined by the presence of suitable larval host-plants. Most geometrid larvae feed on trees, thus they are typical species of woods, forest fringes, hedges etc.. For this reason in southern Europe geometrid moths tend to prefer the mountainous zone, where there are extensive woods. The higher one goes the higher the proportion of Geometridae species to other ,Macrolepidoptera' becomes (up to the timberline). Examples of species strongly specializing in habitats inside woods (**silvicolous**): *Eupithecia conterminata* (LIENIG, 1846), *Acasis appensata* (EVERSMANN, 1842) and *Nothocasis sertata* (HÜBNER, [1817]).

A lot of species however live in very characteristic habitats outside woods: **Xerothermophilous** species specialize in warm and dry habitats, e.g. open heathlands (examples: species of the genera *Lithostege*, *Narraga*), or the dry and hot coastlands of southern Europe. Examples: *Pingasa lahayei* (OBERTHÜR,1887), *Eucrostes indigenata* (VILLERS, 1789), *Phaiogramma faustinata* (MILLIÈRE, 1868), *Idaea alicantaria* REISSER, 1963.

Text-figs 30-31: Extreme habitats of European Geometridae. Text-fig. 30: Rocky slopes in the high Alps (about 2,500-2,800 m), Bolzano-district, northern Italy, typical habitat for montane species like *Psodos, Glacies, Sciadia tenebraria* etc. Text-fig. 31: Surroundings of Almeria, southern Spain, one of the hottest and driest localities in Europe, habitat of some extremely xerothermophilous species (photo 30 TARMANN; 31 RÖDEL).

Casilda antophilaria (HÜBNER, [1813]), *Scopula emutaria* (HÜBNER, [1809]), and some others are '**halophilous**', i.e. specializing in salt marshes.

'**Hygrophilous**' species prefer wet localities. Many authors distinguish between hygrophilous and tyrphophilous species, the latter associated with moors, mainly peat-bogs (BLAB & KUDRNA 1982), such as *Arichanna melanaria* (LINNAEUS, 1758), *Eupithecia gelidata* MÖSCHLER, 1860 and *Carsia sororiata* (HÜBNER, [1813]).

Montane species inhabit hills and mountains, up to an extreme specialization on high alpine rocky slopes up to 4,000 m (species of the genera *Psodos* and *Glacies, Pygmaena fusca* (THUNBERG, 1792), *Sciadia tenebraria* (ESPER, 1806)).

Ubiquitous species do not show strong habitat specializations, they can, more or less, occur everywhere.

Geographical shifts in ecological niche or in strategy are not rare in Europe: Species that are xerothermophilous and locally distributed in central Europe, may be ubiquitous in southern Europe, as e.g. *Idaea degeneraria* (HÜBNER, [1799]), *Idaea rufaria* (HÜBNER, [1799]), or *Cyclophora porata* (LINNAEUS, 1767). *Idaea seriata* (SCHRANK, 1802), is found nearly exclusively near human buildings in central Europe, but is common and ubiquitously distributed in southern Europe. *Microloxia herbaria* (HÜBNER, [1813]), which is restricted to the coastal plains of the Mediterranean, inhabits oases and irrigated lands in central Asia, often up to 2,300 m in the mountains (xeromontane).

Certain subfamilies and tribes of European Geometridae tend to show characteristic habitat preferences. The Sterrhinae, many Eupitheciini, *Psodos* and *Glacies* species inhabit open habitats, while the Geometrinae and the Ennominae are mainly silvicolous. Also many Larentiinae inhabit woods and forest fringes, even if they do not feed on the trees, but are herbivorous and live - cold adapted - in the shade of the forests. For this reason southern European landscapes are characterized by strong gradations between the lowland (with predominantly open habitats) and the wood-covered mountains. In lowland light catches (0-400 m) in southern Italy, the Sterrhinae are dominant with an average of 47% of the Geometridae species flying on a mid-summer night. In the mountains of the same district (prov. Cosenza; over 1,300 m) their proportion decreases sharply to an average of 9%, whilst the dominant subfamily are the Larentiinae with 61%.

The attribution of species to plant associations is not adopted in this book series, because many European plant associations are degenerated. Man-made plantations can facilitate colonization and the niche of many species does not include the whole complex of a plant association. Other species inhabit, under favourable conditions, many different plant associations (WEGNER pers. comm.).

Imaginal habitat and larval habitat are not always the same. As usually little is known about such differentiations, in this book series the term 'habitat' refers to the imaginal habitat only.

Mobility and bionomic strategies

Flight and phototaxis

The flight of most geometrid moths is fluttering, slow and zigzagging. Some robust-bodied species have a whirring and linear flight. Females of many species have large and heavy abdomens, their flight is slow and clumsy.

Nocturnal Lepidoptera are usually attracted by light, some species more than others. The reasons why moths approach light sources are still controversial. The females of many species are less attracted to light sources than the males ('low ♀ ratio at light'), or they are simply less active. In *Idaea seriata* (SCHRANK, 1802) there is a geographical gradient in phototaxis, readily being attracted to light in southern Europe, less often however in central Europe.

Mobility

Animal movements can be undirected, the specimens gradually spreading from the natal area ('dispersal') or directed ('migration'). The **dispersal** varies around a specific mean value in the population of a given species. As Geometridae are generally bad flyers, short-range dispersal (**stenotopy**) is typical for many species. Mark-recapture experiments in southern Germany revealed on average much higher stenotopy in Geometridae than in Noctuidae (HAUSMANN 1990). Many silvicolous species hardly leave the woods towards open land for more than 50 meters. Some stenotopic species (sedentars), however, can exceptionally fly over remarkable distances, such as *Arichanna melanaria* (LINNAEUS, 1758) over more than 20 km (HACKER 1981; HAUSMANN 1990:184). Perhaps such wide-range dispersal is occasionally induced by the destruction of the habitat. Dispersal is often higher in males than in females, e.g. in *Idaea aversata* (LINNAEUS, 1758) and *Scotopteryx chenopodiata* (LINNAEUS, 1758) (HAUSMANN 1990:200). Sometimes dispersal is greater at the end of the flight period or in the second generation of bivoltine species, e.g. in *Cyclophora punctaria* (LINNAEUS, 1758), *Epirrhoe tristata* (LINNAEUS, 1758) and *Chiasmia clathrata* (LINNAEUS, 1758) (HAUSMANN 1990:201).

Wide-range dispersal is characteristic for certain species, such as *Rhodostrophia calabra* (PETAGNA, 1786), *Gymnoscelis rufifasciata* (HAWORTH, 1809) or *Peribatodes rhomboidaria* ([DENIS & SCHIFFERMÜLLER], 1775). Species with very low dispersal are e.g. *Gypsochroa renitidata* (HÜBNER, [1817]), *Lignyoptera thaumastaria* (REBEL, 1901) or *Eupithecia conterminata* (LIENIG, 1846).

Man's influences become more and more important, since travel activities 'exploded' in Europe during the 20th century. Illuminated trains and ships attract moths and transport them far from their native localities. Some interesting observations have recently been published by HABELER (1996) and TABBERT (2000).

Mobility of immature stages

Egg and pupa are immobile stages. The mobility of the caterpillars, however, is often underestimated. Even if active walking usually occurs only over short distances, transport of the first larval instars by wind is rather important, e.g. for the distribution strategy of species with wingless females (PELLMYR 1980; HAUSMANN 1990:203). The small caterpillars often rope down on silken threads and can be carried far away by the wind or by birds.

Migration

Migration and dispersal can not always be sharply separated from each other. Migration can be defined as the directed movement from one place to another, which is usually seasonally determined and predictable (BAKER 1978). EITSCHBERGER et al. (1991) present a wider definition and include also 'emigrants' and 'dismigrants', i.e. species, whose movements are neither predictable nor periodic. They enumerate three 'emigrant species' in Geometridae, *Cyclophora puppillaria* (HÜBNER, [1799]), *Rhodometra sacraria* (LINNAEUS, 1767) and *Orthonama obstipata* (FABRICIUS, 1794). These species are known to undertake flights over one thousand and more kilometers from southern Europe to central Europe, in the north sometimes up to southern Scandinavia and Ireland. However this is a non-periodic, non-predictable phenomenon and in none of these species has the southwardly directed return of the descendants been reported. *Cyclophora puppillaria* is reported from the eastern Alps at 3,200 m (MAZZUCCO 1966).

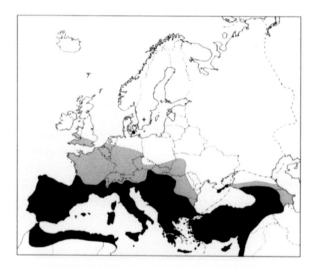

Text-fig. 32: Distribution area (propagating at site; black) and 'emigratory flights' (grey) of *Cyclophora puppillaria* (HÜBNER, [1799]).

Text-fig. 33: *Cyclophora puppillaria*, ♀, resting on a rock in southern Italy.

Bionomic strategy

For the characterization of the bionomic strategy of a species the concept of the 'r-K-continuum' has been developed (see SPITZER & LEPS 1988; SPITZER et al. 1984; REJMANEK & SPITZER 1982). 'r-strategists' pursue a 'dynamic strategy', which is adapted to rapid and frequent changes of the environmental conditions. They are able to utilize resources in a large radius, typically moving far from the native area. Usually they produce many eggs, and often have a wide spectrum of host-plants (polyphagy). Examples: many southern European ubiquists such as *Gymnoscelis rufifasciata* (HAWORTH, 1809) or *Cyclophora puppillaria* (HÜBNER, [1799]).

'K-strategists' are adapted to constant environmental conditions. They are able to utilize resources in a very restricted radius only, typically not moving far from the native area and thus being bad colonizers of new habitats. They usually produce a comparatively small number of eggs, and often have a restricted spectrum of host-plants (mono- or oligophagy). Such species are more vulnerable to environmental changes, and their extinction from habitat is hardly reparable. Examples: *Gypsochroa renitidata* (HÜBNER, [1817]) and *Baptria tibiale* (ESPER, 1791).

Between these extremes there is a continuum of transitions without sharp borders. In this book series the bionomic strategy is mentioned only in some cases of clear 'K-' or 'r-strategy'.

Zoogeography

In the following chapter some examples of dynamic processes and distribution types will be presented. A valuable introduction to the biogeography of Lepidoptera has recently been published in the 'Handbook of Zoology, vol. 35' (HOLLOWAY & NIELSEN 1999).

Dynamics

The distribution area of a species is no static or final state, it is not more than a momentary 'photograph', the actual consequence of a lot of dynamic processes (extinctions and colonizations) in the past.

During the 20th century **areal expansions** have been observed in *Hemistola chrysoprasaria* (ESPER, 1795) and *Horisme corticata* (TREITSCHKE, 1835), both correlated with the expansion of their host-plant, *Clematis*. *Eupithecia sinuosaria* (EVERSMANN, 1848), expanded from eastern Europe to central Europe (REZBANYAI-RESER 1989). *Eulithis mellinata* (FABRICIUS, 1787) colonized many parts of southern Germany and in recent years *Gymnoscelis rufifasciata* (HAWORTH, 1809) has extended its range northwards, perhaps correlated with increasing mean annual temperatures.

'**Neozoa**' are species that have been, more or less, artificially introduced into completely new regions, such as *Hemithea aestivaria* (HÜBNER, 1789) in North America or *Stegania trimaculata* (VILLERS, 1789), which has been recorded in the centre of Berlin since 1987 and now extends to the suburbs (GELBRECHT pers. comm.). In some cases there is no sharp border to the areal expansion.

Areal regressions have been observed in many species, e.g. *Archiearis puella* (ESPER, 1795), *Epirranthis diversata* ([DENIS & SCHIFFERMÜLLER], 1775), *Scopula caricaria* (REUTTI, 1853), *Eupithecia pauxillaria* BOISDUVAL, 1840 and *Thetidia smaragdaria* (FABRICIUS, 1787) from many central European regions, the last also from England. These regressions are mainly due to habitat destruction by man.

Distribution types

A lot of attempts have been made to achieve a unique system and terminology of **'distribution types'**. We will largely follow that of LATTIN (1967) including some elements of the systems of VIGNA TAGLIANTI et al. (1992) and LA GRECA (1963) in this book series. The most important distribution types are represented in the species of the present volume, the respective maps may serve for illustration:

Table 1: Some important distribution types of European Geometridae with examples.
Notes: (1) Also called 'Circum-Mediterranean' or 'Holo-Mediterranean'. (2) 'Atlanto-Mediterranean' according to some authors. The term 'West-Mediterranean' is used also for distributions including Italy (=Circum-Tirrhenic).

Distribution type	Example	map on p.
Holarctic	*Hemithea aestivaria* (HÜBNER, 1789)	192
Palaearctic	*Hemistola chrysoprasaria* (ESPER, 1795)	160
West-Palaearctic	*Alsophila aescularia* ([DENIS & SCHIFFERMÜLLER], 1775)	104
European	*Archiearis puella* (ESPER, 1787)	87
Euro-Caucasian	*Alsophila aceraria* ([DENIS & SCHIFFERMÜLLER], 1775)	107
Eurasiatic	*Thalera fimbrialis* (SCOPOLI, 1763)	179
Euro-Siberian	*Chlorissa viridata* (LINNAEUS, 1758)	195
European-Westasiatic	*Pseudoterpna pruinata* (HUFNAGEL, 1767)	123
Mediterranean (1)	*Xenochlorodes olympiaria* (HERRICH-SCHÄFFER, 1852)	166
Sub-Mediterranean	*Chlorissa cloraria* (HÜBNER, [1813])	200
West-Mediterranean (2)	*Thetidia plusiaria* BOISDUVAL, 1840	148
East-Mediterranean	*Proteuchloris neriaria* (HERRICH-SCHÄFFER, 1852)	146
Adriato-Mediterranean	*Hemistola siciliana* PROUT, 1935	164
Mediterranean-Turanian	*Phaiogramma etruscaria* (ZELLER, 1849)	203
Sarmatian	*Thetidia correspondens* (ALPHERAKY, 1883)	150

Long-term isolation, often on islands or mountains, or areal regression can lead to fairly restricted distribution areas of certain taxa, such as *Liodesina homochromata* (MABILLE, 1869) which is **'endemic'** to Corsica or *Glacies wehrlii* (VORBRODT, 1918), an endemic of a few mountains in Switzerland.

Some species are not continuously distributed, but range in two or more areas, that are widely separated from each other (**disjunct distribution area**), e.g. *Gypsochroa renitidata* (HÜBNER, [1817]), occurring in a limited area in France and from Bulgaria to the Urals. The 'boreo-montane' or 'arctic-alpine' species disjunctly inhabit Scandinavia and the Alpine regions and sometimes occur also on some higher mountains between (KRAMPL 1992). Examples: *Colostygia turbata* (HÜBNER, [1799]), *Eupithecia undata* (FREYER, 1840), *Pygmaena fusca* (THUNBERG, 1792) and *Glacies coracina* (ESPER, 1805).

Mapping

The distribution **maps** are presented in this book series as a combination of the extrapolated (hypothetical) distribution area and dots referring to localities from where specimens have been actually examined by the author(s) and some additional colleagues.

The distribution areas are mainly based on the published national and regional faunistic inventories (see chapter 1). The eastern European distributions have been compiled for Vol. 1 with the enormous help of V. MIRONOV, E. ANTONOVA and I. KOSTJUK. Most dots of

Poland have been added by J. BUSZKO. Much collection material (ZSM with coll. HERBULOT; NHMW; ZFMK; MNHU) has been evaluated. In addition the maps have been discussed with some specialist colleagues from various countries: Baltic States (J. VIIDALEPP); United Kingdom (J. REID and B. SKINNER); Germany (H. WEGNER and J. GELBRECHT); Switzerland (L. REZBANYAI-RESER); Austria and Czech Republic (N. PÖLL); France (C. HERBULOT and R. MAZEL); Spain (A. EXPOSITO); Hungary (L. RONKAY).

4. The geometrid moth: a successful product of evolution

Morphology of the adult

Size

The largest European geometrid moth is *Hypomecis roboraria* ([DENIS & SCHIFFERMÜLLER], 1775) with a wingspan exceptionally of 60 mm. The smallest ones are *Idaea troglodytaria* (HERRICH-SCHÄFFER, 1852) and *Idaea deitanaria* REISSER & WEISERT, 1977 with a wingspan of sometimes only 8 mm.

Text-fig. 34: *Hypomecis roboraria* ([DENIS & SCHIFFERMÜLLER], 1775) and *Idaea deitanaria* REISSER & WEISERT, 1977, the largest and smallest European geometrid moths. Natural size (photo JØRGENSEN).

In the populations of some species size shows large **variation**, in others it is fairly constant. In this book series the wingspan is expressed as the interval between the lower and upper limit from measurements of usually more than 100 specimens per species. Nevertheless extreme variations can occur, mainly beyond the lower limit, due to starvation.

Many species show **seasonal variation** ('dimorphism') in size, mainly in southern Europe, where the summer and the autumnal generation(s) are considerably smaller than the first one (Text-figs 39-40).

Wingspan can also be subject to **sexual dimorphism**: Extreme examples are wingless (apterous) or short-winged (brachypterous) females, e.g. the species of the genera *Alsophila, Operophtera, Malacodea, Ithysia, Lignyoptera, Chondrosoma, Microbiston, Apocheima, Lycia* (partim), *Agriopis, Erannis, Theria, Elophos* (partim) and *Itame loricaria* (EVERSMANN, 1837) (see SATTLER 1991). Considerably smaller, but well developed wings are characteristic e.g. for the females of *Epirranthis diversata* ([DENIS & SCHIFFERMÜLLER], 1775), *Pygmaena fusca* (THUNBERG, 1792), *Chemerina caliginearia* (RAMBUR, 1833), the species of the genera *Myinodes* and *Elophos* (partim) (Text-figs 38-39). In the species of the genera *Pingasa, Hemistola, Xenochlorodes, Campaea, Angerona* and others, the males are smaller than the females.

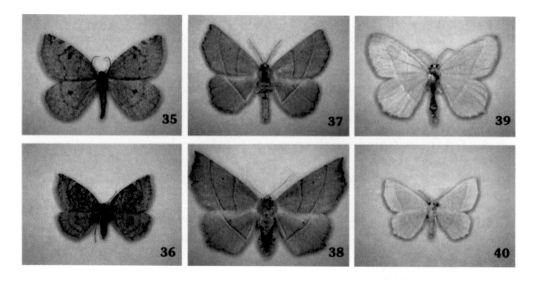

Text-figs 35-40: Dimorphism. Text-figs 35-36: Sexual dimorphism in *Epirranthis diversata* ([DENIS & SCHIFFERMÜLLER], 1775), top ♂, bottom ♀. Text-figs 37-38: Sexual dimorphism in *Campaea honoraria* ([DENIS & SCHIFFERMÜLLER], 1775), top ♂, bottom ♀. Text-figs 39-40: Seasonal dimorphism in *Campaea margaritata* (LINNAEUS, 1767), top ♂ of first generation, bottom ♂ of second generation (photo JØRGENSEN).

Wings

Wing shape

Wings are broad and delicate in most geometrid moths. The forewing termen is usually smooth, rarely jagged, e.g. in the *Ennomos* species, or concave near the apex in the genera *Artiora, Macaria* and *Eilicrinia*. Some genera, such as *Timandra, Toulgoetia, Ennomos* and *Eilicrinia*, have a pointed forewing apex. The hindwing termen is smooth in most species, sometimes angled or tailed at the M3-vein (many Geometrinae, genera *Ennomos, Ourapteryx* and others), or arched between the veins (e.g. genus *Triphosa*, some species of the *Gnophos* group). The hindwings are extremely small in the Larentiinae species *Celonoptera mirificaria* LEDERER, 1862.

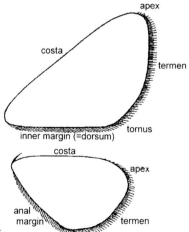

Text-fig. 41:
Terminology of wing margins.

Some **special structures** are worth mentioning, such as the androconium (hair tuft) on the underside of the forewing base in *Eulithis* and *Anticollix*, transparent areas or spots on the wing (some extralimital species), lobes on the anal margin of the hindwing, e.g. in *Lobophora halterata* (HUFNAGEL, 1767), and the 'fovea' in the males of many Ennominae species (rarely also in females), which is a blister-like small area at the base of the forewing upperside.

In nearly all the species fore- and hindwings are linked by the 'frenulo-retinacular wing-coupling system', i.e. by the '**frenulum**' which arises from the base of the hindwing and inter-locks with the '**retinaculum**' at the underside of the forewing. In the males the frenulum is developed - if present - as a single stout bristle, in the females as a brush of weak bristles.

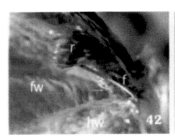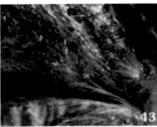

Text-figs 42-44: Underside of wings, frenulo-retinacular wing coupling system. Text-fig. 42: *Hydriomena impluviata* ([DENIS & SCHIFFERMÜLLER], 1775), ♂. Text-fig. 43: *Deileptenia ribea-ta* (CLERCK, 1759), ♀. Text-fig. 44: *Aplasta ononaria* (FUESSLY, 1783), ♂, frenulum absent. r=reti-naculum, f=frenulum, fw=forewing, hw=hindwing.

Venation

In Geometridae, the differential diagnosis of genera, tribes and subfamilies is often based on venation characters. The forewing veins are subdivided into subcostalis (Sc), radials (R1-R5), medial veins (M1-M3) cubitan veins (CuA1-CuA2) and anal vein(s) (A). In the old literature (e.g. PROUT 1912a; SPULER 1910) there are different terms for Sc ('costalis' or 'I'), R1-R5 ('subcostalis 1-5' or 'II'), M1-M2 ('radials 1-2' or 'III/1-2'), M3 ('mediana 3' or 'III/3'), CuA ('mediana 1-2' or 'IV') and A2 ('submedia' or $\propto$). The venation of the hindwing is similar, but Sc and R1 are always totally fused (in the genus *Orthostixis* basally separate), R2-5 are fused into one single vein 'Rs'. In both wings the '**cell**' is delimited distally by the 'discocellular vein'. Frequently, when R1 turns back into R2 between cell and costa, there is a second smaller cell in the forewing, called the '**areole**', 'secondary cell' or 'accessory cell'. Areoles can be subdivided, some Ennominae species have three areoles. '**Connate**' veins arise from the cell at the same point (see M3 and CuA1 of forewing in Text-fig. 45), '**separate**' veins arise separately (see M3 and CuA1 of hindwing in Text-fig. 45), and '**stalked**' veins are fused into a common stalk distally from the cell before diverging (see Rs and M1 of hindwing in Text-fig. 45). Sometimes veins that are basally separate, are fused distally ('**anastomosis**').

The venation of the family Geometridae is characterized by the frequent presence of areoles in the forewing. A1 is usually lacking in the forewing, A2 and A3 are fused, but diverge near the base of the wing. In the hindwing of most species only one anal vein is expressed. Sometimes a short second anal vein (A3) is present, and in some old, 'primitive' groups (Archiearinae, Orthostixinae) there is a vein-like folding in the wing between A2 and CuA2 ('A1'), which how-ever is never tubular. M2 of the hindwing is developed as a tubular vein only distally from the cell (differential feature from Drepanidae) or is totally lacking in the Ennominae (Text-figs 46-53).

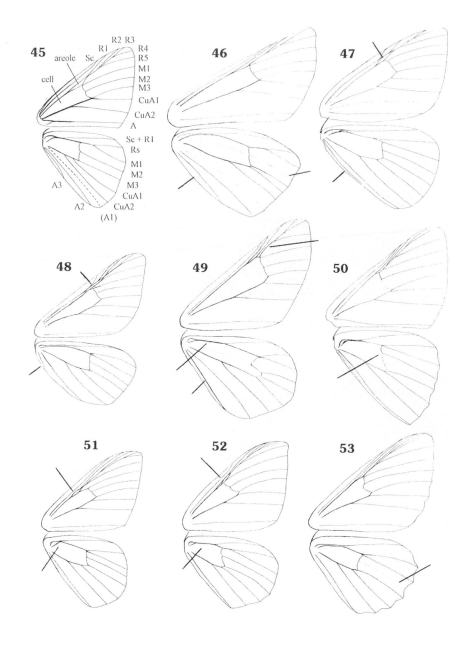

Text-fig. 45:   Terminology of venation.

Text-figs 46-53:   Diagnostic characters in geometrid venation.

Text-fig. 46:   *Archiearis parthenias* (LINNAEUS, 1761) (Archiearinae)

Text-fig. 47:   *Orthostixis cribraria* (HÜBNER, [1799]) (Orthostixinae)

Text-fig. 48:   *Gypsochroa renitidata* (HÜBNER, [1817]) (Desmobathrinae)

Text-fig. 49:   *Alsophila aescularia* ([DENIS & SCHIFFERMÜLLER], 1775) (Alsophilinae)

Text-fig. 50:   *Pseudoterpna coronillaria* (HÜBNER, [1813]) (Geometrinae)

Text-fig. 51:   *Scopula marginepunctata* (GOEZE, 1781) (Sterrhinae)

Text-fig. 52:   *Xanthorhoe fluctuata* (LINNAEUS, 1758) (Larentiinae)

Text-fig. 53:   *Peribatodes correptaria* (ZELLER, 1847) (Ennominae)

Wing pattern and colours

The coloration and the pattern of the wings are determined by the structure, pigmentation and arrangement of the scales. Interference colours are produced by 'opalescent scales', e.g. in the pretty eye spots of *Problepsis ocellata* (FRIVALDSZKY, 1845). Most colours in Lepidoptera, however, are a product of 'bile pigments' in the scales. The subfamily Geometrinae is characterized by its easily fading green pigment ('geoverdin'), which is not a bile pigment (COOK et al. 1994).

In this book series, the terminology of the wing pattern largely follows that of MCGUFFIN (1967-1990) (see Text-fig. 54). When there is one transverse line only between the basal and the medial area, it is called 'antemedial line', and accordingly 'postmedial line', when there is just one single line between the medial and the terminal area.

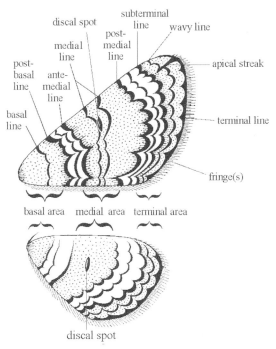

Text-fig. 54: Wing pattern.

Due to the characteristic resting position of Geometridae the upperside of the wing is usually more intensely coloured than the underside. Contrarily, in veliform resting species, such as the *Selenia* species or many Semiothisini, the underside often bears bright colours or characteristic patterns. Colour and wing pattern of most species are cryptic in strict correlation with their defence strategy (see chapter 3).

As no human being is exactly like another, even Lepidoptera vary in pattern and colour of the wings. Some species show small '**variation**', others vary a lot. In contrast to this variation around the typical form there are also totally different 'individual aberrations'. They can occur without transitions to the typical form (see chapter 5; Text-figs 55-60).

Text-figs 55-60: *Abraxas grossulariata* (LINNAEUS, 1758). Text-figs 55-56: 'Typical' form (55) and individual aberration (56) taken at light, Germany. Text-figs 57-60: Aberrations from rearings, England.

Text-figs 61-62: *Campaea margaritata* (LINNAEUS, 1767). Text-fig. 61: 'Typical' form. Text-fig. 62: Individual aberration (both taken at light, Germany).

Text-figs 63-66: Polychroism in *Hylaea fasciaria* (LINNAEUS, 1758). Text-figs 63-64: Typical red and green form (both from Austria). Text-fig. 65: Aberrant red form (Austria). Text-fig. 66: Dark form occurring at high altitudes in Valesia, Switzerland.

Some species are characterized by the occurrence of two or more forms, which are strongly different from each other, (almost) without being linked by transitional forms. Such patterns can become established in populations in certain proportions ('**polychroism**'). This phenomenon is characteristic for various species of the genus *Idaea* (Text-figs 67-68). *Hylaea fasciaria* (LINNAEUS, 1758) occurs in both green and red forms, in many populations in defined percentages or exclusively in one form only (Text-figs 63-66). In other *Hylaea* species only green forms are known. The green pigment of some Geometrinae species can occasionally be modified to red. A special case of polychroism is the '**melanism**', which has already been discussed in chapter 3 ('defence strategies'; Text-figs 22-23). The genetic backgrounds for polychroisms and melanisms are diverse and have been investigated for a very few species only.

Text-figs 67-68: Polychroism in *Idaea aversata* (LINNAEUS, 1758), both specimens from the same population in southern Denmark. Text-fig. 67: Nominotypical form. Text-fig. 68: 'f. *remutata*' (photo JØRGENSEN).

Sometimes the coloration is strongly different between the sexes ('**sexual dichroism**', often called 'sexual dimorphism'), e.g. in *Idaea pallidata* ([DENIS & SCHIFFERMÜLLER], 1775), *Nycterosea obstipata* (FABRICIUS, 1794), *Bupalus piniaria* (FABRICIUS, 1794) and *Crocota peletieraria* (DUPONCHEL, 1830) (Text-figs 69-70). '**Seasonal dichroism**' is characteristic for *Xanthorhoe biriviata* (BORKHAUSEN, 1794) (Text-figs 71-72) and some further species of the genera *Aplasta, Cyclophora, Lythria, Eilicrinia* and *Selenia*.

Text-figs 69-70: Sexual dichroism of *Crocota peletieraria* (DUPONCHEL, 1830). Text-fig. 69: ♂. Text-fig. 70: ♀ (Pyrenees) (photo JØRGENSEN).

Text-figs 71-72: Seasonal dichroism of *Xanthorhoe biriviata* (BORKHAUSEN, 1794). Text-fig. 71: first generation. Text-fig. 72: second generation (Funen, Denmark) (photo JØRGENSEN).

Head

Text-figs 73-74: Head parts of two Geometridae species. v = vertex; f = frons; ch = chaetosema; ce = compound eye; lp = labial palpus; p = proboscis. Text-fig. 73: *Scotopteryx bipunctaria* ([DENIS & SCHIFFERMÜLLER], 1775), ♂. Text-fig. 74: *Colotois pennaria* (LINNAEUS, 1761), ♂.

Geometridae have one pair of **compound eyes**, but no additional external ocelli, as occur in many other Lepidoptera families. The surface of each compound eye is divided into hexagonal facets (several thousand in Geometridae). Each facet is the outer part of a single eye, the '**ommatidium**'. The eyes of most Geometridae species are 'naked', i.e. without cilia between the facets. Such 'interfacetal hairs' are characteristic for various other lepidopterous families.

The paired 'palpi' (correctly '**labial palpi**') are often bushy scaled and comparatively short, rarely as long and 'naked' as in Noctuidae.

The **proboscis** (=tongue) is usually well developed, but in some species it is vestigial (*see Colotois pennaria*, Text-fig. 74) or absent (secondary loss). Its length can exceptionally approach the length of the forewing (genus *Glossotrophia*). In contrast to some other families a piercing proboscis is unknown in Geometridae.

The **frons** is flat, slightly convex, or, exceptionally, strongly convex. It is either smooth-scaled (Text-fig. 73), or rough-scaled (Text-fig. 74), the latter mainly in the species emerging in the late autumn (e.g. *Colotois pennaria* (LINNAEUS, 1761)), in the early spring (e.g. genus *Archiearis*), or inhabiting high mountains (e.g. genera *Psodos* and *Glacies*).

Paired **chaetosema(ta)** (='Jordan's organ') are present but sometimes rather small in Geometridae (absent in Drepanidae). The chaetosema is a cuticular patch dorsocaudad of the eye with tufts of hair-like scales.

The antenna ('**flagellum**') is subdivided into many 'segments' (better: 'flagellomeres' or 'antennomeres'). Their total number varies between the species and sometimes also within the species. In the species with uni- or bipectinate antennae each flagellomere bears one or two '**branches**' (latin: 'rami', sing. 'ramus'). Some species are known to have quadripectinate antennae (e.g. genera *Xanthorhoe* and *Rhodostrophia*). The pectination may be complete to the tip or partial (Text-figs 75-76). On the flagellum, and on the branches, there are sensory areas with sensilla (sing.: 'sensillum'), which enable olfactory perception. As the males usually search for females by olfactory means, their antennae are often better developed (i.e. with longer braches, or with larger or denser cilia) than the antennae of the females. Ciliate antennae can be subdivided into various types, e.g. the ciliate-setose type, when setae are short and numerous (Text-fig. 79), the ciliate type, when they are longer and less numerous and the fasciculate type, when cilia arise in bunches (Text-fig. 80).

Text-figs 75-76: Bipectinate antennae. Text-fig. 75: Bipectinate to tip, *Apochima flabellaria* (HEEGER, 1838), ♂. Text-fig. 76: Partially bipectinate to 3/4 of length, *Biston betularia* (LINNAEUS, 1758), ♂.

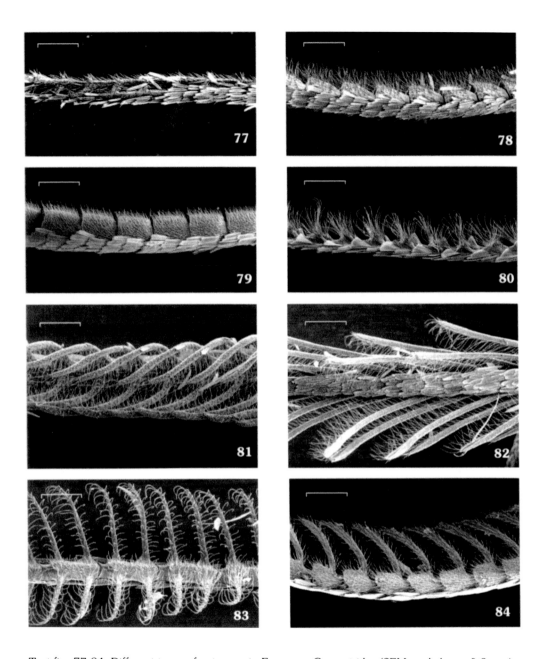

Text-figs 77-84: Different types of antennae in European Geometridae (SEM; scale bar = 0.2 mm)
Text-fig. 77: Filiform: *Thera variata* ([DENIS & SCHIFFERMÜLLER], 1775), ♂.
Text-fig. 78: (Slightly) dentate: *Thera britannica* (TURNER, 1925), ♂.
Text-fig. 79: Ciliate-setose: *Euphyia frustata* (TREITSCHKE, 1828), ♂.
Text-fig. 80: Ciliate-fasciculate: *Idaea cervantaria* (MILLIÈRE, 1869), ♂.
Text-fig. 81: Bipectinate, normal type: *Thetidia smaragdaria* (FABRICIUS, 1787), ♂.
Text-fig. 82: Bipectinate, appressed, untidy type: *Jodis lactearia* (LINNAEUS, 1758), ♂.
Text-fig. 83: Quadripectinate: *Rhodostrophia calabra* (PETAGNA, 1787), ♂.
Text-fig. 84: Unipectinate: *Amygdaloptera testaria* (FABRICIUS, 1794), ♂.

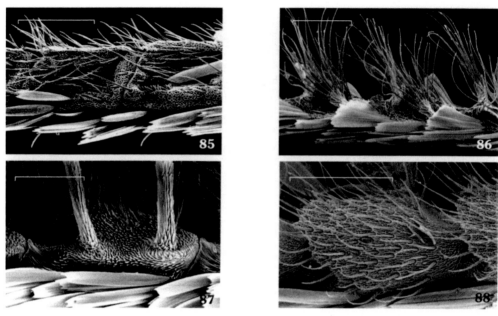

Text-figs 85-88: ♂ antenna, details (SEM; scale bar = 0.1 mm). Text-fig. 85: *Thera variata* ([DENIS & SCHIFFERMÜLLER], 1775). Text-fig. 86: *Idaea cervantaria* (MILLIÈRE, 1869). Text-fig. 87: *Rhodostrophia calabra* (PETAGNA, 1787). Text-fig. 88: *Amygdaloptera testaria* (FABRICIUS, 1794).

Legs

Each insect leg can typically be subdivided into coxa, trochanter, femur, tibia, tarsus, praetarsus (Text-figs 89-90). An 'epiphysis' on the foreleg tibia serves for the cleaning of the antennae and the proboscis. The number of spurs of the fore- (=prothoracic), mid- (=mesothoracic) and hindlegs (=metathoracic legs) is expressed with the 'spur formula'. The typical formula for Geometridae is 0-2-4. Spurs are sometimes vestigial or absent in the hindleg. The hindtibia of some species bears a pencil, usually belonging to a male scent organ (Text-fig. 92) and it can be dilated, e.g. in the males of various *Idaea* species. The tarsus consists of five segments, the tarsomeres. It is typically shortened in the males of the genus *Idaea*. The praetarsus ends in a pair of claws.

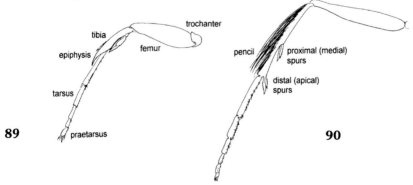

Text-figs 89-90: Legs of geometrid moths. Text-fig. 89: Foreleg. Text-fig. 90: Hindleg (after SCOBLE 1995, modified).

56

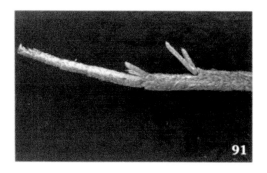

Text-figs. 91-92: Hindlegs of geometrid moths. Text-fig. 91: *Cabera pusaria* (LINNAEUS, 1758), ♂. Text-fig. 92: *Idaea biselata* (HUFNAGEL, 1767), ♂.

Abdomen and tympanal organ

The abdomen consists of 10 segments, parts of the last two being modified to form the genital organs (see below). Nearly all the segments have a sclerotized ventral plate, the **sternum** (=sternite) and a dorsal plate, the **tergum** (=tergite). On the pleura (sing. 'pleurum'), i.e. the membranous lateral parts between sternum and tergum there are openings of the respiratory ('tracheal') system, the spiracles or **stigmata** (sing.: 'stigma').

Narrow appendages, called 'apodemes', can arise from different segments, e.g. laterally from sternum A2 (many Geometrinae) or from the anterior margin of sternum A2 extending into the thoracic lumen. The latter is characteristic for ditrysian Lepidoptera. In many Geometridae those apodemes are only slightly sclerotized, but may be stronger, e.g. in the genera *Xenochlorodes, Eupithecia* and *Euchrognophos*. Paired setal patches on sternum A3 and external tufts ('crests') of hair-like scales on the tergites are characteristic for many Geometrinae species. 'Bolte's pockets' in the females of some *Eupithecia* species correspond exactly to the forked processes of the male sternum A8 and may play a role in copulation (MIKKOLA 1994).

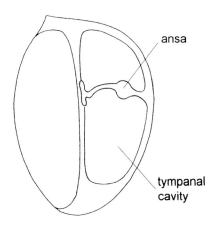

Text-fig. 93: Left half of the tympanal organ in geometrid moth (after SCOBLE 1995, modified)

The sternum A1, and parts of A2, are modified to form the paired **tympanal organ**. This auditory organ enables the moth to perceive sounds, mainly those of the ultra-sound frequencies used by hunting bats. In contrast to the Pyralidae, which also bear a tympanal organ in the first abdominal segment, in geometrid moths the two halves of the tympanal organ are divided from each other. They are furthermore characterized by an invaginated sclerite, the 'ansa', which is lacking in all the other families (Text-fig. 93). An 'accessory tympanum' in segment T3 (meta-thorax) is usually present in Geometridae (synapomorphy), but is lacking in the Archiearinae.

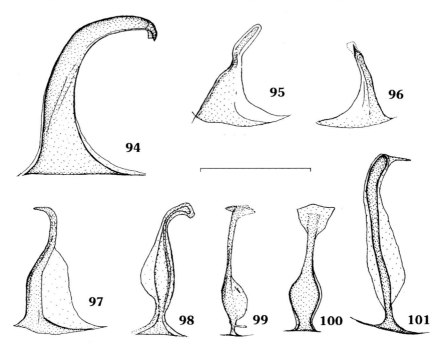

Text-figs 94-101: Shape of ansa in various subfamilies of Geometridae. Scale bar = 0.5 mm.
Text-fig. 94: *Archiearis parthenias* (LINNAEUS, 1761) (Archiearinae)
Text-fig. 95: *Orthostixis cribraria* (HÜBNER, [1799]) (Orthostixinae)
Text-fig. 96: *Gypsochroa renitidata* (HÜBNER, [1817]) (Desmobathrinae)
Text-fig. 97: *Alsophila aescularia* ([DENIS & SCHIFFERMÜLLER], 1775) (Alsophilinae)
Text-fig. 98: *Pseudoterpna coronillaria* (HÜBNER, [1813]) (Geometrinae)
Text-fig. 99: *Scopula marginepunctata* (GOEZE, 1781) (Sterrhinae)
Text-fig. 100: *Xanthorhoe fluctuata* (LINNAEUS, 1758) (Larentiinae)
Text-fig. 101: *Peribatodes correptaria* (ZELLER, 1847) (Ennominae)

Genitalia

The genitalia of Geometridae, and other Lepidoptera, are largely modifications of segments A9 and parts of A10. Both male and female genitalia usually bear important diagnostic features. A confusing quantity of terms has been introduced for the different structural details. In this book series we will largely follow the terminology proposed by KLOTS (1970) and applied in exemplary modern revisions such as SCOBLE (1994), COOK & SCOBLE (1995) or PITKIN (1993; 1996) (Text-figs 102-104).

Male genitalia

The male genitalia mainly consist of a sclerotized ring (tegumen, vinculum, saccus), two movable lateral appendages, the valvae, and at the posterior end forcipated 'claws', the uncus and the gnathos. The shape of the **uncus** is diverse and often diagnostic, in some species and genera, such as *Scopula*, it is absent. The **valvae** usually bear the most important differential features, sometimes they are reduced to short projections, as in *Scopula* and *Rhodometra* species. The shape of the **juxta** may be diagnostic, too. It is often characterized by paired dorsal projections. The '**aed(o)eagus**' contains the membranous vesica, which is everted in copulation. The vesica often bears sclerotized spines ('**cornuti**').

In many European Geometridae the male sternum 'A8', i.e. that of the eighth abdominal segment, bears particular processes, called 'cerata' (sing.: 'ceras') or 'octavals'. They evidently play a functional role in copulation, which however is poorly studied to date. The modified sternum A8 is usually regarded as belonging to the genitalia.

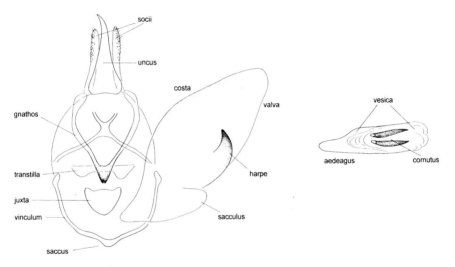

Text-fig. 102: Terminology of ♂ genitalia in Geometridae (without left valva); aedeagus separate. Semi-diagrammatic, ventral view.

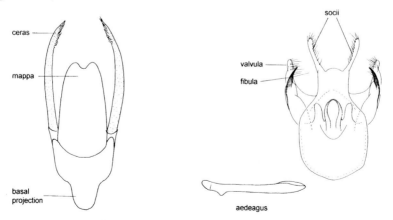

Text-fig. 103: Male genitalia of *Scopula luridata* (ZELLER, 1847), with sternum A8 and aedeagus. Ventral view.

The different structural details usually show little variation, although some cases of considerable plasticity have been reported (e.g. AMSEL 1951; BOURGOGNE 1963; ITÄMIES & TABELL 1997). With regard to the special case of polymorphism in the genitalia of the genera *Scopula* and *Glossotrophia* see chapter 5.

Female genitalia

In the 'Ditrysia', i.e. in 95% of all lepidopterans, there are two different openings for egg positioning and fertilization. The opening for egg positioning is at the end of the abdomen and bears laterally two membranous and setose lobes, the '**papillae anales**', also called the ovipositor. The opening for fertilization, the '**ostium bursae**', is situated on the ventral side of the abdomen on sternum A8 or A7. It leads to the **bursa-copulatrix** system, which stores the spermatophore(s) after fertilization. The ostium bursae is encircled by the '**sterigma**', also called the 'vaginal plate', which is often subdivided into the 'lamella antevaginalis' and 'lamella postvaginalis'.

The posterior part of the **ductus bursae**, the '**antrum**', is often dilated and strongly sclerotized. A **colliculum**, i.e. the subterminal sclerotization of the ductus bursae according to the definition in SCOBLE (1995) (Text-fig. 104), is typical for the Asthenini and the genus *Catarhoe*. The shape of the ductus bursae is diverse in European Geometridae, often straight, sometimes elongate, curved or helical, the latter e.g. in *Idaea deversaria* (HERRICH-SCHÄFFER, 1847).

The **corpus bursae** often bears a **signum** (bursae), which in many representatives of the genera *Eupithecia* and *Scopula* consists of innumerable small spines (signa) situated on the inner surface. Some species, e.g. *Larentia clavaria* (HAWORTH, 1809) and many Ennominae have very large signa ('lamina dentalis'). The shape of the corpus bursae may be variable, e.g. in many *Eupithecia* species, depending on whether the female has already been fertilized or not. Sometimes there is also an appendix bursae (=digitabulum), e.g. in the genera *Hydriomena, Lithostege* and in many Gnophini.

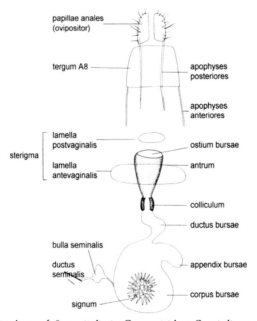

Text-fig. 104: Terminology of ♀ genitalia in Geometridae. Semi-diagrammatic, ventral view.

The membranous oviduct is situated dorsad of the bursa copulatrix and leads to the 'ovipositor' opening. Both oviduct and bursa copulatrix are connected by the tender, membranous ductus seminalis.

The papillae anales sometimes (genera *Aleucis* and *Theria*, Ennominae) bear modified, broad-headed scales, called 'floricomus' by PIERCE (1914) and described by PELLMYR (1980), sometimes it is densely covered by an anal tuft of hair-like scales, as e.g. in the *Alsophila* species. In both cases these hairs serve to cover the eggs and protect them. It is not known, if they have further olfactory functions. In some species the papillae anales can be telescoped (e.g. various Ennominae species).

In many cases the shape of the female genitalia seems to be correlated to that of the male genitalia. In species with long male aedeagi (or better: long vesicae) the females often have a long ductus bursae. An extreme example is the Sterrhinae species *Idaea straminata* (BORKHAUSEN, 1794), where the length of both the aedeagus and the bursa copulatrix exceeds 20 mm (=wingspan of the moth). Nevertheless the 'key-lock principle' of the sclerotized structures of male and female genitalia as an exclusive mechanism capable of obstructing interspecific fertilization is hardly acceptable. In copulation the opening of the spermatophore of the ♂ must be placed as near as possible to the outlet of the ductus seminalis in the corpus bursae. Only the inner key-lock (vesica versus ductus bursae) seems well substantiated (PETERSEN 1909; MIKKOLA 1993; 1994).

Morphology of the egg

As in other insects the life cycle of the geometrid moth begins with the egg (or 'ovum'). As a separate volume is planned for the early stages of European Geometridae, the egg morphology is only discussed briefly in the present book series. The phylogenetic and taxonomic aspects of the morphology of the egg are poorly studied and much remains to be discovered. In TONGE (1932) the eggs of many British Geometridae are figured and described. There is further information on the morphology of Geometridae eggs in SALKELD (1983) and DÖRING (1955).

In European Geometridae the **shape** of the eggs is very diverse, there are e.g. round, oval, flat, disc-shaped, furrowed and cylindrical eggs. The **sculpture** of the surface is also rather diverse (see Text-figs 105-108), but usually rough compared to the smoother eggs of many other moth families, e.g. Notodontidae, Lymantriidae, Arctiidae or Sphingidae. Smooth eggs sometimes occur also in geometrid moths, e.g. in the genera *Chloroclysta, Solitanea, Eupithecia* and *Ennomos*, but usually there is still faint sculpturation. The most common types of sculpturation in eggs of European Geometridae are the ribbed type with longitudinal ribs and smaller horizontal ridges between (Text-figs 107-108), and a reticulate network, usually with hexagonal facets nearly all over the surface (Text-figs 105-106). The coloration and change of **colour** during maturation of the egg is often characteristic for the species.

On the surface of the egg there is a particular area, where fertilization takes place, the so-called **micropylar area**. This zone usually consists of a few very small holes and it is encircled by a particular, rosette-like sculpturation, often widely extended in Geometridae. The eggs of most geometrid moth species are of the '**flat type**', i.e. with the longitudinal axis lying parallel to the substrate and the micropylar area situated laterally. Exceptionally there are, however, also eggs of the '**upright type**', as e.g. in some species of the genera *Biston* and *Idaea*. Many other Lepidoptera, e.g. most Rhopalocera and Noctuoidea, have eggs of the upright type.

Eggs of European Geometridae are usually well attached (glued) to the surface of the host-plant. Various habits and strategies of egg positioning are presented in chapter 3 (behaviour).

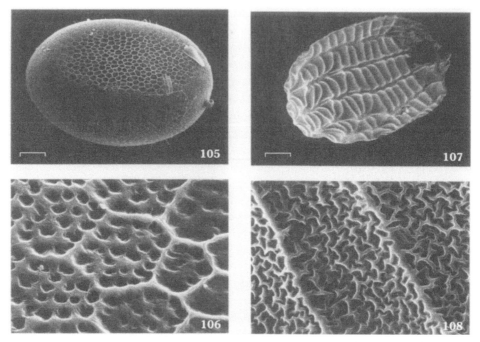

Text-figs 105-108: Geometrid moth eggs and sculpturation of surface. Text-figs 105-106: *Pseudo-terpna coronillaria flamignii* HAUSMANN, 1997. Text-figs 107-108: *Idaea elongaria* (RAMBUR, 1833), after hatching of larva. SEM, scale bar = 0.1 mm.

Morphology of the larva

Structure

As a separate volume is planned for the early stages of European Geometridae, the larval morphology is only discussed briefly in the present book series. Though larval morphology is important for the taxonomy of Geometridae, unfortunately little serious research has been done in this sector, to date. Comparative studies and descriptions of large numbers of species are given in SINGH (1951; India) and MCGUFFIN (1967-1990; Canada).

The so-called '**chaetotaxy**' is based on the assumption that there is a homology of '**primary setae**' throughout the Lepidoptera. '**Setal maps**' are worked out on the basis of a defined setal nomenclature. Modified positions of certain setae are often diagnostic for species or higher taxa. Just a few Geometridae species, such as the Orthostixinae, *Aplasta ononaria* (FUESSLY, 1783), *Athroolopha pennigeraria* (HÜBNER, [1813]) and *Minoa murinata* (SCOPOLI, 1763)) have 'hairy' larvae, i.e. they are covered by many secondary setae. Urticating 'hairs' are unknown in Geometridae. The abdominal segments of the Bistonini larvae are densely covered by scale-like structures ('ctenidia').

Caterpillars of Lepidoptera typically bear four pairs of 'ventral prolegs' on segments A3-A6, and one pair of 'anal prolegs' on A10, but in Geometridae the ventral prolegs of segments A3-A5 are absent or vestigial (synapomorphy). This absence of legs causes the typical 'looping' in the movement of the larvae (Text-figs 109-110). In the subfamily Archiearinae all the prolegs of the larva are present, but those of the segments A3-A4 become more reduced and less functional with each moult from L1 to L4 (Text-fig. 138). The adult Archiearinae larva walks 'semilooping', as also some Noctuidae species do. In the Orthostixinae the legs of segment A5, sometimes also of A4 are present, but reduced and without function. In the Alsophilinae and Campaeini (Ennominae) vestigial prolegs are present on segment A5. Geometridae prolegs are characterized by their grasping structures, the 'crochets', which are biordinal, and arranged in a mesoseries, a character shared by all the 'Macrolepidoptera' (SCOBLE 1995; SINGH 1951).

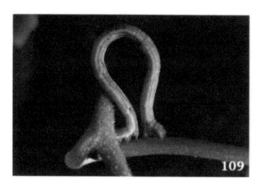

Text-figs 109-110: 'Looping' of Geometrid moth larvae as a consequence of the reduction of ventral prolegs. Text-fig. 109: *Scopula imitaria* (HÜBNER, [1799]). Text-fig. 110: *Biston betularia* (LINNAEUS, 1758) (photo 110 NIPPE).

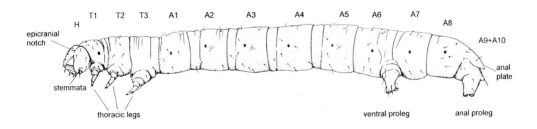

Text-fig. 111: Geometrid moth larva, terminology. H = head; T = thoracic segment(s); A = abdominal segment(s).

The **head capsule** of the geometrid moth larva is usually clearly 'hypognathous', i.e. the mouth parts (mandibulae) are vertically positioned to the axis of the body. The 'spinneret' between the mandibulae enables the larva to spin silk. On the basal part of the head near the mouth parts there are on each side six single eyes, the 'stemmata' (sometimes called 'ocelli', but without being homologous to the ocelli of the adult). These stemmata allow visual orientation, but no good optical perception. The head capsule typically shows a dorsal ('epicranial') notch, which can sometimes be very deep (subfamily Geometrinae). The caterpillars of Geometridae have no auditory hairs such as occur in some species in other Lepidoptera families.

The thoracic (T1-T3) and abdominal segments (A1-A10) often bear projections or warts, which are usually diagnostic, such as the paired horns on the prothorax (T1) and the elongation of the dorsally situated 'anal plate' (A10) in most representatives of the subfamily Geometrinae, or the hump on segment A8 in many Bistonini (Ennominae). The last segments, particularly A9-A10, are largely fused.

Coloration

In Geometridae the larval coloration may vary greatly within the same species, population or even within the same brood. Dichroic green or brown larvae occur in various species. Often each larval 'instar' (stage) has its own characteristic pattern and coloration. Slight differences sometimes occur between male and female as a consequence of different blood colour (VIIDALEPP pers. comm.). A correlation between the host-plant and the coloration is reported from flower-feeding *Phaiogramma faustinata* (MILLIÈRE, 1868), see p. 206. The pattern of the full-grown caterpillar, however, in most species bears typical specific characters. There are often longitudinal lines on the dorsum ('dorsalis'), laterally ('lateralis') and on the ventrum ('ventralis') with additional lines between ('subdorsalis' and 'sublateralis'). Caterpillars of 296 (British) geometrid moth species are figured in PORTER (1997).

Life cycle

Each caterpillar usually sloughs four times (with exceptions), followed by the fifth sloughing, which is the 'pupation'. The different larval stages are termed 'L1' (egg larva) to 'L5' (full-grown larva). With regard to various defence strategies and host-plant interactions of European Geometridae caterpillars see chapter 3.

Preservation

The preservation of caterpillars is problematic. To avoid decomposition due to micro-organisms in the digestive tract, larvae are usually 'blown out'. They can also be stored in alcohol after boiling in water or dried with special methods, such as freeze-drying or critical-point drying. The latter method is the only one to preserve the colours from fading, particularly in green caterpillars.

Morphology of the pupa

The Lepidoptera are 'holometabolic' insecta, i.e. the metamorphosis is complete with a pupal stage interpolated between larval and adult stage. Some primitive Lepidoptera families have a 'pupa exarata' with free appendages. In the pupae of Geometridae, and many other families, all the appendages are fused to the body ('pupa obtecta'). Exceptionally, the proboscis may be free and looped, e.g. in the genus *Glossotrophia* (Text-fig. 113).

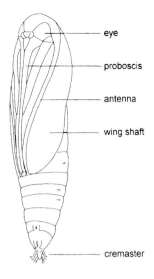

eye

proboscis

antenna

wing shaft

cremaster

Text-fig. 112: Geometrid moth pupa.
Semi-diagrammatic, semi-lateral view.

Text-figs 113-114: Geometrid moth pupae. Text-fig. 113: *Glossotrophia mentzeri* HAUSMANN, 1993.
Text-fig. 114: Girdle pupa of *Cyclophora puppillaria* (HÜBNER, [1799]) (photos LEIPNITZ).

The pupal morphology of European Geometridae has been published in detail by
PATOCKA (1978-1995). The publications of MCGUFFIN (1967-1990) on Canadian
Geometridae are recommended for additional information. In PATOCKA (1995) a
valuable key for the pupae of 161 European Geometridae genera is presented. For some
further remarks concerning pupation behaviour, see chapter 3 (defence strategies).

The **cremaster** region at the anal end of the pupa often bears diagnostic characters. It
is therefore urgently recommended not to throw away the 'exuvia' (= the empty pupal
integument after emergence of the moth) after the rearing of a species. These can be easily fixed
under the specimens or on separate pins. The cremaster is sometimes forked and usually bears
a characteristic number of hook-shaped hairs or bristles (Text-figs 115-123).

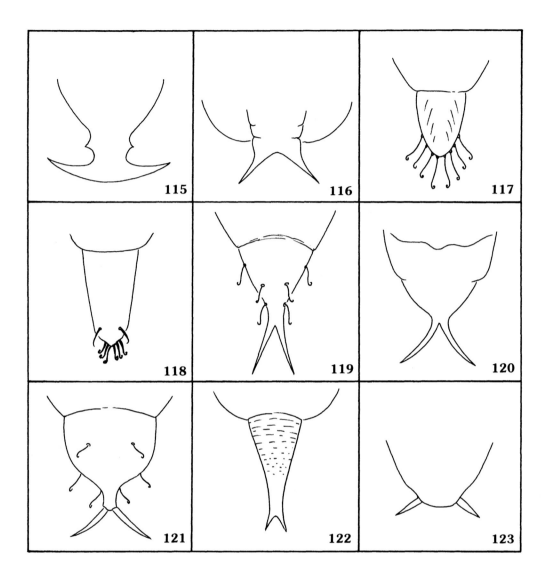

Text-figs 115-123: Shape of cremaster in some geometrid moth pupae (after PATOCKA 1995; modified).

Text-fig. 115: *Archiearis notha* (HÜBNER, [1803]) (Archiearinae)

Text-fig. 116: *Alsophila aceraria* ([DENIS & SCHIFFERMÜLLER], 1775) (Alsophilinae)

Text-fig. 117: *Pseudoterpna pruinata* (HUFNAGEL, 1767) (Geometrinae)

Text-fig. 118: *Cyclophora linearia* (HÜBNER, [1799]) (Sterrhinae)

Text-fig. 119: *Lampropteryx otregiata* (METCALFE, 1917) (Larentiinae)

Text-fig. 120: *Pelurga comitata* (LINNAEUS, 1758) (Larentiinae)

Text-fig. 121: *Nothocasis sertata* (HÜBNER, [1817]) (Larentiinae)

Text-fig. 122: *Tephrina arenacearia* ([DENIS & SCHIFFERMÜLLER], 1775) (Ennominae)

Text-fig. 123: *Theria rupicapraria* ([DENIS & SCHIFFERMÜLLER], 1775) (Ennominae)

5. The natural system of Geometridae

Taxonomy and phylogeny

Search for an order

The Geometridae are one of the most species-rich Lepidoptera-families. More than 21,000 species have been described worldwide (MUNROE 1982; SCOBLE et al. 1995; SCOBLE 1999), and we do not know how many species still await discovery. Most Geometridae species live in the tropics, e.g. about 6,500 described species in central and South America (Neotropical region), 4,300 species in southern and south-eastern Asia (Indo-Pacific or Oriental region) and about 3,200 species in central and southern Africa (Afrotropical or Ethiopian region) (SCOBLE et al. 1995; slightly modified).

Obviously, there is a need to find an order for such huge quantities of species. To arrange them alphabetically would not be satisfactory, even from a practical point of view. Biologists are essentially more interested in a 'natural system', i.e. an order which represents the natural (i.e. evolutionary) relationships between the various species, genera, or higher categories.

The classical phylogenetic analysis usually follows the principles and methods of **HENNIG** (1966): As the speciation processes cause 'dichotomous' or rarely 'polytomous' divisions of the evolutionary lineages starting from an ancestor and leading to two (or more) descendants, the following terms are introduced:

Plesiomorphic feature: 'old' feature, inherited from the ancestor. This term is relative and depends on which 'ancestor' one compares, i.e. how far one goes back in the time scale.

Apomorphic feature: 'new' feature (mutation), absent in the ancestor. Apomorphic characters that are present in several taxa as common heritage from the same ancestor are called **synapomorphic**.

Any groups (e.g. genera, tribes, subfamilies) are **monophyletic**, if they represent the whole recent descendants of one single ancestor. Monophyly can be recognized by common synapomorphic features.

They are however **polyphyletic**, if they descend from more than one ancestor. Such erroneous assumption of relationship is often the consequence of misinterpretation of similar, but not homologous features, which in evolution have been developed independently several times (**convergence**), because such mutations bear advantages for the species in its environment. Finally there are **paraphyletic** groups (e.g. the homoneurous Lepidoptera) that descend from one single ancestor without being monophyletic: Their members split from the main genealogical lineage in different periods, whilst other, younger (out-)groups, in the above example the Heteroneura, derive over the same lineage from the same ancestor.

Recently many attempts have been made to base phylogeny on algorhythmic computer programs (computer **cladistics**). After the input of full character sets the computer calculates various genealogical trees ('**clusters**'), from which the most simple, i.e. the shortest one is supposed to be the most probable tree (parsimony principle). In practice however the results may not be convincing, especially when the research is done by zoologists who are not specialists in the investigated group. The input and the weighting of the features has to be done very carefully and the character sets have to be published together with the results.

Modern systematics is based more and more on the comparison of sequences of amino-acids within proteins and on RNA and DNA sequences. These methods, however, seem to have their shortcomings in macrosystematics, where results are often speculative and do not coincide with morphological character sets. In the investigation of the phylogeny of (sister) species complexes **molecular taxonomy** will probably prove to be more important.

Speculations about the phylogeny of the Geometridae

The subtitle may appear provocative, but in fact we know little about the phylogeny of the higher systematic categories within the Geometridae. The monophyly of most of the subfamilies is in doubt, the relationships between the subfamilies and the best way to arrange them in a systematic order is still controversial.

The Lepidoptera most probably had their origin in the early Jurassic, about 150 Million years ago (WHALLEY 1986; SCOBLE 1995; KRISTENSEN & SKALSKI 1999). The first Geometridae may have appeared in the late Jurassic or in the early Cretaceous, 130-140 Million years ago, with the evolutionary 'explosion' of the Angiosperm plants (see POWELL & al. 1999). Unfortunately we have only a very few fossils of Geometridae (KERNBACH 1967; KRISTENSEN & SKALSKI 1999).

Where did the first geometrid moths come from? We know little about this descent, however Geometridae are '**Glossata**', having a proboscis in their ground plan in contrast to some primitive Lepidoptera-families. Within the Glossata, the Geometridae belong to the '**Heteroneura**' having a different venation in the fore- and the hindwing in contrast to some primitive families of the Glossata. They are '**Ditrysia**', i.e. the females have two separate openings for egg positioning and fertilization in contrast to some of the more 'primitive' heteroneurous Lepidoptera. Geometridae are furthermore attributed to the '**Macrolepidoptera**' (higher Ditrysia), a group of Ditrysian families, whose members characteristically are large-sized and usually have caterpillars that live exposed. The grasping structures ('crochets') of the larval prolegs are formed and arranged in a typical way (see chapter 4). The monophyly of the 'Macrolepidoptera' is controversial (see KRISTENSEN & SKALSKI 1999), but, actually, rejected by the majority of zoologists.

The **monophyly of the Geometridae** is supported by some synapomorphic characters such as the position and the structure of the tympanal organ (see 6.1.5.) and the reduction of the larval prolegs, which causes the typical walking as 'loopers'. The hindwing vein Sc+R1 is strongly bent at its base in most species. Geometridae furthermore usually have broad and delicate wings and are typically slender bodied. The families Uraniidae, Sematuridae, Epicopeiidae and Drepanidae are considered as sister groups of the Geometridae (MINET 1983; 1991; SCOBLE 1995; MINET & SCOBLE 1999). These sister groups differ from the Geometridae in having the complete number of larval prolegs (except for some Drepanidae) and different structure of the tympanal organ. Actually the superfamily Geometroidea is retained to include also the 'sister families' Uraniidae and Sematuridae (MINET & SCOBLE 1999), while the Drepanidae and Epicopeiidae belong to Drepanoidea.

The ('macrosystematic') relationships of the various **subfamilies** within the family are still unclear and controversial. No generally accepted system exists. Usually the subdivision into 6-8 subfamilies is postulated for the geometrid moths worldwide. As a good basis for discussion the 'tentative phylogeny for the Geometridae' in HOLLOWAY (1997) is largely accepted in the following presentation of subfamilies.

The Archiearinae may be the sister group of all the other Geometridae (the latter having an accessory tympanum in segment T3 as synapomorphy) and thus are placed at the beginning of the system. All the other subfamilies are arranged according to conventional and practical aspects. Whilst MÜLLER (1996) places the Ennominae at the beginning of the system immediately after the Archiearinae, in this book series the 'classical' order is maintained, following well known publications such as PROUT (in SEITZ 1912-1916; 1930-1938; 1934-1938; 1920-1941), FORSTER & WOHLFAHRT (1981), SKOU (1986) and LERAUT (1997). The future will show whether more arguments (see below) will be found for the basal position of Ennominae or not.

Also the 'Oenochrominae' have often been regarded as a 'primitive' subfamily. Many authors emphasize the heterogeneity and the polyphyletic origin of this 'subfamily' as it was conceived until recently. There is now increasing acceptance of the division of the group into Oenochrominae sensu stricto, Desmobathrinae, Orthostixinae and Alsophilinae (see SCOBLE & EDWARDS 1990; HOLLOWAY 1996; 1997; HAUSMANN 1996a; 1996b; SCOBLE 1999:999; Table 1).

Archiearinae
16 species in 6 genera worldwide, 4 species in Europe. Colourful diurnal moths. Important differential features: Absence of the accessory tympanum, presence of larval prolegs on the segments A3-A6, presence of two anal veins A2-A3 in the hind-wing with A1 present but not tubular. Pupa with exposed labial palps (NAKAMURA 1987). All these characters very probably plesiomorphic features. Monophyletic.

Oenochrominae
Strictly conceived (SCOBLE & EDWARDS 1990; HOLLOWAY 1996; 1997) including more than 300 species, nearly exclusively in Australia, one genus (*Sarcinodes*) distributed also in south-eastern Asia, not represented in Europe. Very robust bodied Geometridae with the lacinia of the tympanal organ lying parallel to the tympanon, a synapomorphy of the subfamily (COOK & SCOBLE 1992). Larval prolegs present on the segments A5 and A6, those of A5 being reduced or rarely absent. Possibly monophyletic.

Ennominae
About 9,800 species, distributed worldwide, 300 species in Europe. Important common features: Vein M2 weak or absent in the hindwing, never tubular. Wing pattern of fore- and hindwing usually similar. Structure of tympanal organ diverse, but ansa usually narrow at base. Phylogenetically basal groups possibly the Campaeini (larva with vestigial prolegs on segment A5) and Theriini (archaic pupal characters). Numerous genera from outside Europe (e.g. *Mnesampela* and *Thalaina* from Australia, *Gonodontis* from India) with additional, usually vestigial, larval prolegs on segments A4-A5 (SINGH 1951; MCQUILLAN 1981; MCFARLAND 1988; SCOBLE 1995; MINET & SCOBLE 1999). Such additional larval prolegs and weak M2 vein of hindwing

reminiscent of subfamily Archiearinae. Tribal subdivision controversial and still insufficiently resolved. Monophyly of subfamily somewhat in doubt, M2 of hindwing possibly lost several times independently.

Alsophilinae
26 species in four genera in the Holarctic, two species in Europe. Winter moths, females wingless, eggs positioned in ribbons around twigs. Forewings without areoles. Tympanal organ with ansa broad at base and tapering towards end, reminiscent to that of Desmobathrinae. Vestigial larval prolegs on segment A5. Some characters similar to equivalents in both Desmobathrinae and certain Larentiinae. Monophyletic. Previously included in 'Oenochrominae s. lat.'.

Orthostixinae
17 species in four genera in the Palaearctic and Indo-Pacific, one species in Europe. Larval prolegs vestigial on segment A5, sometimes even on A4. Larvae setose. Tympanal organ with ansa broad at base and tapering towards end, reminiscent to that of Desmobathrinae. Venation resembling that of the Desmobathrinae genus *Derambila*. Other characters similar to their equivalents in the Ennominae (HOLLOWAY 1996). Monophyletic.

Desmobathrinae
About 230 species in 25 genera worldwide, four species in Europe. Without unambigous synapomorphic features (HOLLOWAY 1996). Rather slender bodied with long legs and antennae. Larvae not setose, usually without prolegs on segments A4-A5. Tympanal organ with ansa broad at base and tapering towards end. External appearance and genitalia of some members similar to those in the Larentiinae tribus Chesiadini. Sister group to the Geometrinae according to the tentative phylogeny of HOLLOWAY (1997).

Geometrinae
About 2,350 species, distributed worldwide, 32 species in Europe. Subfamily monophyletic (MINET & SCOBLE 1999), perhaps except for some tribes or genera. Synapomorphies: Predominance of green pigment (COOK et al. 1994), some larval characters, origin of vein M2 from cell, female genitalia with oblique, papillate ovipositor lobes (HOLLOWAY 1996). Usually without areole in the forewing. Sternum A3 often with setose patches. In male genitalia socii usually present. Phylogenetically basal groups: Dysphaniini and the other robust-bodied genera (e.g. Pseudoterpnini). Genera *Heliothea* and *Aplasta* with some characters reminiscent of equivalents in Desmobathrinae. Arrangement of the Geometrinae as sister group of the Desmobathrinae in HOLLOWAY (1997).

Larentiinae
About 5,800 species, distributed worldwide, 390 species in Europe, 127 belonging to the genus *Eupithecia*. Important common features: Forewing with 1-2 areoles. Hindwing: Sc+R1 and Rs with long anastomosis. Colour of hindwing often pale, without the pattern of the forewing. Phylogenetically basal groups: Lythriini, Asthenini and Chesiadini perhaps with relationships to Sterrhinae tribes Cyclophorini, Rhodometrini and/or Desmobathrinae. Possibly polyphyletic.

70

Sterrhinae

About 2,800 species, worldwide distributed, 188 species in Europe, 106 belonging to the genus *Idaea*. Important common features: Forewing with 1-2 areoles. Hindwing: Sc+R1 and Rs without long anastomosis, except for a very few basal taxa. Adult moths usually small, whitish or sand coloured. Food plants low herbs, except for Cyclophorini. Phylogenetically basal groups: Cyclophorini, Timandrini and Rhodometrini, in various features reminiscent to some basal Larentiinae genera (e.g. *Lythria*). Probably poly- or paraphyletic.

The European species numbers are taken, with slight modifications, from MÜLLER (1996). The total species numbers based on SCOBLE et al. (1995; slightly modified).

Table 2: Geometridae subfamilies of the world, matrix of character sets. hw = hindwing; fw = forewing; v = vestigial; (+), (v) = usually absent, but present/vestigial in some species or groups; -/+* absent/present with a very few exceptions; anast. = anastomosis.

	larval prolegs A3 A4 A5			vein M2 of hw tubular	anast. Sc+R1 & Rs long	areole in fw	tubular hw-anal veins	predominant geoverdin	accessory tympanum	basis of ansa broad	larvae arboreal
Archiearinae	v	v	+	+	-	-	2	-	-	+	+
Oenochrominae	-	-	v*	+	-	(+)	2	-	+	-*	+*
Ennominae	-*	-*	(+/v)	-	-*	(+)	(1-)2	-	+	-*	+*
Alsophilinae	-	-	v	+	+*	(+)	2	-	+	+	+
Orthostixinae	-	-*	v	+	-	+	2	-	+	+	-/+
Desmobathrinae	-	-*	-*	+	-	+	1-2	-	+	+	-/+
Geometrinae	-	-	-	+	-*	-*	1(-2)	+	+	-*	+*
Larentiinae	-	-	-	+	+	+	1*	-	+	-	-*
Sterrhinae	-	-	-	+	-*	+	1*	-	+	-	-*

Speciation

Speciation processes in Europe can often be explained by isolation of populations into different glacial refugia with subsequent inter- and postglacial expansion to the actual distribution area. When discussing the Mediterranean species pair *Idaea distinctaria* (BOISDUVAL, 1840) and *Idaea predotaria* (HARTIG, 1951), REZBANYAI-RESER (1987) dates the speciation of these comparatively young species in the Riss-glaciation (200,000-100,000 years ago). Today the distribution areas of both species are overlapping, i.e. they '**occur sympatrically**', in some regions (Italy, southern France, Corsica, Sardinia). Species pairs that remain geographically separated are called 'allopatric species' or '**allopatric vicariants**' (see Text-fig. 124)

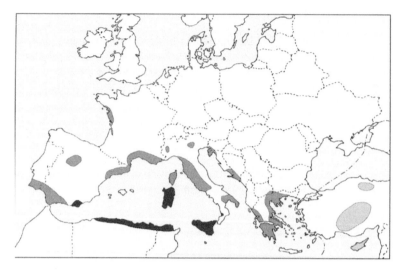

Text-fig. 124: Recent distribution area of the three allopatric vicariants *Idaea circuitaria* (HÜBNER, [1819]) (dark grey), *Idaea rainerii* (HAUSMANN, 1994) (black), and *Idaea mimosaria* (GUENÉE, [1858]) (light grey).

Glacial climatic changes and the expansion from glacial refugia are probably also responsible for a peculiar distribution pattern in the Sterrhinae species *Glossotrophia asellaria* (HERRICH-SCHÄFFER, 1847). The different populations form an almost closed ring of subspecies with gradually modified morphology, whilst the extremes, i.e. the two most northerly distributed subspecies, *isabellaria* (MILLIÈRE, 1868) (south-western France) and *romanaria* (MILLIÈRE, 1869) (north-western Italy) are structurally fairly different and have probably lost the capacity of mutual reproduction (Text-fig. 125). Taxonomically and nomenclaturally such '**ring species**' cannot be expressed in a satisfactory way.

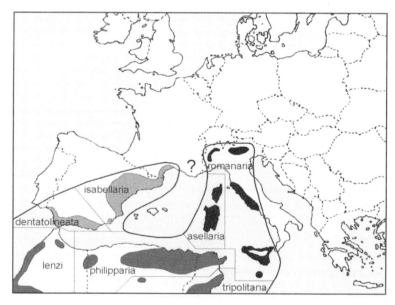

Text-fig. 125: 'Ring species': Distribution of *Glossotrophia asellaria* (HERRICH-SCHÄFFER, 1847) with seven subspecies in the western Mediterranean.

Certain species pairs, though usually well separated from each other in their reproduction may exceptionally produce fertile **hybrids** under natural conditions: *Lycia graecaria* (STAUDINGER, 1861) and *Lycia florentina* (STEFANELLI, 1882) share a 'common' intermediate population named ssp. *istriana* (STAUDINGER, 1901). In the southern Alps intermediate populations between *Isturgia limbaria* (FABRICIUS, 1775) and *Isturgia roraria* (FABRICIUS, 1776) are known (Text-figs 126-131). *Pseudoterpna pruinata* (HUFNAGEL, 1767) and *Pseudoterpna coronillaria* (HÜBNER, [1817]) have common hybridization areas in the Pyrenee mountains, western France, central Italy and western Turkey. Often, spreading out of such 'buffer zones' into the range of one of the genetically 'pure' species is disadvantageous, as many hybrids have smaller reproductive success.

Natural, accidental hybrids are reported from the genus *Cyclophora* (GELBRECHT pers. comm.) and *Euphyia* (GERSTBERGER 1979). In captivity hybridizations are easy to obtain between species of the tribe Bistonini (HARRISON 1913; BRETSCHNEIDER 1939; 1953; WEHRLI 1941a) and in *Cyclophora* (URBAHN 1939:481).

Text-figs 126-131: Hybridization in nature. Text-figs 126, 127 (underside): *Isturgia limbaria* (FABRICIUS, 1775), western Germany. Text-figs 128, 129 (underside): Specimen from intermediate population, southern Alps. Text-figs 130, 131 (underside): *Isturgia roraria* (FABRICIUS, 1776), southeastern Germany.

Text-figs 132-134: Hybridization in captivity. Text-fig. 132: *Lycia hirtaria* (CLERCK, 1759) ♀. Text-fig. 133: hybrid *'pilzii'* ♀ (*hirtaria* ♂ x *pomonaria* ♀). Text-fig. 134: *Lycia pomonaria* (HÜBNER, 1790) ♀.

Other species show in their distribution area a continuous and slight modification of certain features along a geographical gradient ('**cline**'). *Pseudoterpna pruinata* (HUFNAGEL, 1767), for example, gets continuously darker with the wing pattern more contrasted from central Europe towards the coasts of the North Sea.

Intermediate (hybrid) populations and clines are not satisfactorily describable with the tools of taxonomy and nomenclature.

Exaggerated applications of the **subspecies concept** as in the extreme case of *Parnassius apollo* with innumerable subspecies are neither helpful nor practicable. In certain cases, however, subspecific rank is appropriate to express constant differences of isolated populations, which can be interpreted as potentially initiating the process of speciation. 'Substrate races' have evolved in correlation with the particular coloration of the substrate on which the moth usually rests ('crypsis'; see chapter 3, defence strategies). Usually, such forms do not merit a separate subspecific name, as they are linked with the nominate form by clinous transitions or mixed up with it sympatrically. Sometimes, however, such populations show the particular coloration to 100% and seem to indicate very limited genetic exchange with neighbouring populations, such as the dark brown *Glossotrophia confinaria prouti* HAUS-MANN, 1993 from the Porphyr Alps in northern Italy or the dark grey *Idaea camparia cossurata* (MILLIÈRE, 1875) from the lava slopes of the Etna region in Sicily.

Coloration and morphology vary within species and populations with characteristic frequency and pattern for each feature. Slight '**variation**', with all kinds of forms transitional to the typical form can be distinguished from mutations, which sometimes cause totally different '**individual aberrations**' (examples see chapter 4, wings; Text-figs. 55-62). Variability can be regarded as the experimental laboratory of evolution, a silent and continual trial to create something new, that could be useful.

Sometimes variants of totally different structure or coloration occur, genetically caused, not singly, but with a certain percentage in the population over long periods. Such '**polymorphism**' (structure) or '**polychroism**' (colour) differs from variation by the absence of transitions to the typical forms. They often lead to misinterpretations in taxonomy. In European Geometridae particular caution should be given to the genitalia of the Sterrhinae genera *Scopula* and *Glossotrophia*. Many species show polymorphisms in the structure of sternum A8 (Text-figs 135-137; HAUSMANN 1999b).

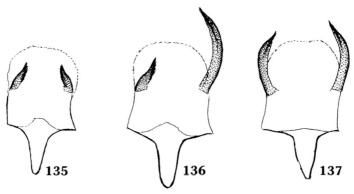

Text-figs 135-137: Polymorphism in *Glossotrophia alba* HAUSMANN, 1993: Sternum A8 of three males reared from the eggs of the same mother.

Nomenclature

The tenth edition of the 'Systema Naturae', published by LINNAEUS (Carl von LINNÉ) in 1758, is considered the starting point for the '**binominal**' denomination of animals, i.e. names have to be composed of both a scientific genus and species name. As many species were in fact described several times, increasingly arguments arose on the validity of certain names. Thus zoologists and botanists began (separately) to establish detailed rules for a correct nomenclature. These are given in the **International Code of Zoological Nomenclature** (ICZN 1999), now in its fourth edition.

The most important rules and principles of the Code are briefly outlined to present some fundamental nomenclatural terms. The author recommends reading and studying the detailed rulings of the Code itself.

Stability as basic principle

'The object of the Code is to promote stability ...' (Preamble). For this reason, the International Commission reserves the right to suspend in certain cases the strict application of the Code to validate and thus maintain names, that have been in common use over a long period of time.

Principle of priority (synonymy)

For each taxon the first published name is valid, if it has been introduced in formal accordance with the rules ('available'). All subsequent names that have been correctly introduced for specimens of the same (sub-)species are '(junior) synonyms'. Names in prevailing use, however, can take precedence over earlier disused synonyms according to the new, fourth edition of the Code.

When two generic names are based upon the same type-species, and thus synonymy is unambiguous, the younger one is an 'objective synonym'. Two generic names, whose type-species are supposed to be 'congeneric', i.e. to belong to the same genus, are 'subjective synonyms'.

Principle of homonymy

The Code does not allow the creation or current existence of two species with exactly the same binomen (genus and species name). Therefore such 'homonyms' (= '**preoccupied names**') are declared as **invalid**, even if they have been introduced according to the rules of availability (see below). Two kinds of homonyms can be distinguished:

When a newly introduced name (binomen) is identical with an already existing binomen of a previous description, it is a '(junior) **primary homonym**'.

Sometimes in revisions species names have to be transferred to another genus ('comb. nov.'), because they had been previously placed in the wrong genus. If in this new genus the same species name already exists we get the same binomen twice. The younger name is called '(junior) **secondary homonym**'.

Such preoccupied names have to be replaced by the oldest available synonym. If there is no available synonym, a new '**replacement name**' has to be introduced ('nom. nov.' = nomen novum).

Type-designation

Types of species and subspecies are the specimens mentioned in the original description. They are either collectively name-bearing (syntypes) or one **name-bearing type** specimen is established (holo-, neo- or lectotype). The type is, like the original metre in Paris, a definition, standing on the one hand as an exemplary specimen for its biological reproduction unit, the (sub-)species, constituting on the other hand the unequivocal link to a zoological name. In museums, name-bearing types are usually labelled with red labels.

Holotype: One single specimen upon which the original author explicitly based his species.

Syntypes: Two or more specimens upon which the original author based his species without choosing a particular specimen as name-bearing type.

Lectotype: One single specimen from a syntype series, which has been designated subsequently as the name-bearing type. Lectotype-designation is important as syntype series may contain material from different regions or even different species.

Neotype: When the type (series) of a taxon is known to be lost, and when it is necessary to avoid nomenclatural confusions with other, similar species, any single specimen can be designated as the name-bearing type in a revisional work. It has to be deposited in a public museum.

Paratype: All the other 'types' (typoids) except for the holotype, in older publications sometimes called 'cotypes'. In museums they are often marked with a yellow label.

Allotype: Particular 'type' of the opposite sex to the holotype. It is not more than a paratype and without taxonomic importance.

Locus typicus: Locality where the name-bearing type came from.

Availability

Under certain conditions, (new) names are '**unavailable**', i.e. nomenclaturally not existing. Such formal incorrectness can be:

Incorrect subsequent spelling: Typographic error or 'lapsus calami' in a subsequent citation of the original description.

Nomen nudum: A new name that has been introduced without description.

Misidentification: Erroneously used name for a misidentified species, without the intention of the author to introduce a new name. Unavailable even when accompanied by a description.

Non binominal: Names of the species group that were introduced in the original description 'uninominal', i.e. without generic name or reference thereto.

Infrasubspecific: Names that have been introduced by the describer explicitly in a category lower than the subspecific level. Under certain conditions old infrasubspecific names can be raised to an available rank ('stat. nov.').

Synonyms and homonyms as such are available, i.e. 'existent' names, unless unavailable for other reasons!

Misidentifications in historical publications are often a source of confusion: '*Phalaena vernaria*' (originally = *Jodis lactearia*), for instance, was attributed to five different species by various authors in the period between 1761-1809.

Problematic cases

In spite of the Code, some cases still remain, where strict acceptance of the rules would seriously obstruct stability of nomenclature: For example, the validity and treatment of the names introduced by [DENIS & SCHIFFERMÜLLER] (1775) (see chapter 1) have been widely discussed in recent years. As nearly all the 'SCHIFFERMÜLLER names' have been in continuous use for more than 200 years, and some of the species (e.g. *Alsophila aescularia*) have economic importance, the replacement of these names would lead to nomenclatural 'chaos'.

The first problem is, that nearly all the names were introduced with short descriptions of characters for species groups in keys, but without adequate description of the individual species. For this reason KOCAK (1982-1986) and MENTZER (1984) proposed rejection of these names as nomina nuda according to the Code sensu stricto. Unfortunately the types were lost by fire. The interpretation of the species identity is usually possible with the help of subsequent publications, e.g. those of BORKHAUSEN (1794), TREITSCHKE (1827-1828), CHARPENTIER (1821) and FABRICIUS (1775-1798), who had studied the SCHIFFERMÜLLER collection and largely based their publications on this material. This first problem is resolved by 'Opinion 516' of the International Commission for Zoological Nomenclature saying that such interpretations can validate the content of DENIS & SCHIFFERMÜLLER's book. Furthermore the recommendations of SATTLER & TREMEWAN (1984) and WOLF (1988) furnish reasonable and logical arguments which help to validate and maintain the SCHIFFERMÜLLER names.

Another problem is the question whether those names refer to the generic name *Geometra* or to *Phalaena*. The first is originally abbreviated as 'G.' before the species name in most cases, the latter usually appears only in the head of a chapter in the form 'Phalaenae Geometrae'. This question is quite important, as some names become homonyms with the generic name *Geometra*, others with *Phalaena*. While this debate is still controversial (e.g. KOCAK 1982-1984; MENTZER 1984; SATTLER & TREMEWAN 1984), there is contradictory usage in important monographs, such as DE FREINA & WITT (1987), who validate *Phalaena* as genus (*Bombyx* as subgenus), whereas FLETCHER (1979), HACKER (1989; 1996) and FIBIGER (1990) reject this usage combining the species names with '*Geometra*' and '*Noctua*' respectively. SATTLER & TREMEWAN (1984) recommend applying the latter solution which favours nomenclatural stability.

PART II: Systematic account

Abbreviations and conventions

Nomenclatural formalities

The following formal scheme is followed throughout the book series:

Scientific names of genera and species of animals and plants are written in *italics*, author's names in CAPITALS. Author and date are enclosed in parentheses, e.g. '(LINNAEUS, 1758)', when the taxon has been transferred into another genus than the original one. Square brackets are used, when the name of the author or the date of the publication was inferred from a source outside the original description. Author and date are separated by a comma. In some recent publications this comma is deliberately omitted.

The **gender** of the species name is given in the original form, without changing it to the greek or latin gender of the genus name. In this detail we do not follow the Code but the justified and substantiated practice in many modern publications such as SCOBLE (1995; 1999), HOLLOWAY (1997), KARSHOLT & RAZOWSKI (1996) or NIELSEN et al. (1996).

The systematic **category** is the genus (single nomen), species (binomen) or subspecies (trinomen) unless stated otherwise.

In the **synonymic lists** the taxa are arranged in chronological order. When citing the original description, the terms 'ab.' (i.e. aberratio), 'var.' (i.e. varietas) or 'f.' (i.e. forma) used there are quoted here notwithstanding the actual taxonomic qualification. The appositions 'ssp.(n.)' (i.e. subspecies (nova)) or 'sp.(n.)' (i.e. species (nova)) in an original description are omitted in the citation given here. Old unavailable names, mainly homonyms, are usually fully cited in the synonymic list, but prefixed with a double dagger ('‡').

Other **unavailable names** are (alphabetically) arranged at the end of the synonymic lists and grouped in the following categories: 'misidentification(s)', 'infrasubspecific' names and 'incorrect subsequent spellings'. As far as possible, the year of the original description is also indicated for the unavailable names quoted. But that was not a priority, and more than one case had to (and will in the other volumes) be left for subsequent updating.

The **nomenclature of the genera** is not fully presented here. The necessary information may be taken from FLETCHER (1979), and subsequent amendments. For further valuable informations on the nomenclature of the European Geometridae see MÜLLER (1996) and SCOBLE (1999).

Diagnosis

The 'wingspan' is measured including the fringe, in traditionally set specimens (inner margin of forewing right-angled to body). For the terminology of the imaginal features, e.g. venation or genitalia, and for the morphology of the early stages see introduction, chapter 4. The descriptions of the morphology are based on the examination of large numbers of specimens: for example, for the 42 species of Vol. 1 more than 1,000 genitalia slides have been examined.

Distribution

In the distribution maps a hypothetical resident distribution area (grey) is combined with localities from which specimens have been examined (black dots ●; see introduction, chapter 3, zoogeography). Old records are marked with a ring (O), when the species has probably disappeared from that locality. Important unverified records (e.g. outside the residence area, see *Aplasta ononaria*, p. 115 in this Volume) are marked with an asterisk (∗). Doubtful records are indicated with a question mark. For zoogeographical terms see chapter 3. For published faunistic information on European Geometridae see chapter 1. The geographical names in the text are taken from the current Times Atlas.

Ecology, phenology

In the **host-plant indications** the term 'found on ...' refers to authentic field observations of feeding caterpillars (with communicator and country). When a species is 'recorded (also) on' certain food-plants, the data are taken from literature and may need verification. The term 'reared on ...' indicates the host-plant under the artificial conditions of captivity; such a plant is not necessarily the preferred food in the natural habitat of the species. Just the generic names of the food-plants are mentioned, when the species feeds on several species of the plant genus. Unless otherwise stated the caterpillars feed on the leaves of the mentioned plants. The latin plant names are mainly taken from RAUH & SENGHAS (1976). The **phenology** is described indicating the 'main flight period'. It is important to note, that the flight period of many species can vary greatly. Therefore single outlying records are omitted. The phenology of the 'larval stages' indicates the life cycle of the caterpillar from leaving the egg until pupation, furthermore the overwintering stage is presented.
The indications of the **habitat** refer to the habitat of the adult moths (see chapter 3).

Figures, illustrations, plates

The colour plates and the genitalia figures are arranged at the end of the volume. This facilitates the comparison of many different species by turning just a few pages. The illustrations of the ♂ genitalia, ♀ genitalia and those of the colour plates bear the same numbers for each species. The specimens on the colour plates are reproduced at natural size throughout this volume. In the text some specimens are figured (black and white) at a larger scale, mainly in closely related species complexes or in very small species (e.g. genera *Idaea, Eupithecia*). Diagnostic characters are marked with arrows.

References

The bibliography is arranged at the end of each volume. For the convenience of the reader repeated citation of literature in separate volumes of the book series recurs.

Abbreviations

Acronyms (=abbreviations) of the museums see introduction, chapter 2.
'pers. comm.' = personal communication
'var. cit.' = various citations, used only in the host plant records: the complete references are substituted by 'var. cit.', when at least three identical references for the host-plant were found. Usually such an indication refers to books such as SPULER (1903-1910), REBEL (1910), CULOT (1917-1919), ALLAN (1949), BERGMANN (1955), BLASCHKE (1955), FORSTER & WOHLFAHRT (1981), FRIEDRICH (1984; 1986), SKINNER (1984), SKOU (1986), CARTER & HARGREAVES (1987), PORTER (1997), etc.

Subfamily **Archiearinae** FLETCHER, 1953
=Brephinae sensu auct. nec HÜBNER, 1826

Small subfamily with 6 genera and 16 species. Represented also by some species in the southern Andes and Tasmania (SCOBLE 1992; 1999; SCOBLE & MINET 1999). Two genera only in the Holarctic. Many features supposedly plesiomorphic, e.g. full number of larval prolegs and venation. Thus Archiearinae generally placed at the beginning as presumably the most basal ('primitive') recent group of Geometridae.

DIAGNOSIS: Venation see Text-fig. 46: R3 of forewing usually lacking (R3+4). Sc+R1 of hindwing touching Rs but not fused. M2 of hindwing usually developed, even in cell, but weak. A1 developed, but not tubular, A2 fully developed, even A3 rather long. Accessory tympanum reduced (COOK & SCOBLE 1992; SCOBLE 1992). Head hairy. Compound eyes small, oval.

Eggs of the upright type. Larva with legs present on abdominal segments A3-A6, the first two abdominal pairs (A3, A4) however weakly developed, larva therefore walking similarly to other Geometridae ('semilooping'). Skin covered with setose warts.

Archiearis HÜBNER, [1823]
=*Brephos* OCHSENHEIMER, 1816 nec HÜBNER [1813]

TYPE-SPECIES: *Phalaena parthenias* LINNAEUS, 1761. Four species in the Palaearctic, one other in North America. Discussion of larval morphology in SINGH (1960), of pupal morphology in PATOCKA & ZACH (1994).

DIAGNOSIS: See diagnosis of subfamily. Head, thorax, abdomen and legs strongly hairy. Eyes small. Palpi short. Frenulum present. Proboscis developed. Hindtibia of both ♂ and ♀ with two pairs of very short spurs.

REMARKS: 'Exclusively' diurnal, though both *Archiearis parthenias* and *Archiearis notha* are exceptionally attracted to light (BUSSE 1989; VIIDALEPP, GELBRECHT pers. comm.). Resting position with wings rolled around small twigs. Adults emerging in early spring. Hibernation as pre-imago.

Text-fig. 138: Caterpillar of *Archiearis parthenias* (LINNAEUS, 1761), with complete number of ventral prolegs (photo LEIPNITZ).

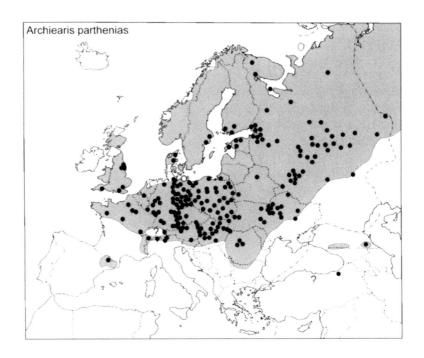

1. *Archiearis parthenias* (LINNAEUS, 1761)

Phalaena (Noctua) parthenias LINNAEUS, 1761: Fauna Suecica (Edn 2): 308 ([Sweden]). Type(s) in LSL.
Phalaena plebeja LINNAEUS, 1761: Faun. Suec. (Edn 2): 320 (Sweden: 'Uplandia'). Type(s) in LSL.
Junior synonym.
Phalaena fulvulata PALLAS, 1773: Reise Russ. Reich, 2: 732 ('Siberia'). Type(s) not traced. Junior synonym.
‡ *Phalaena vidua* FABRICIUS, 1775: Syst. Ent.: 580, nec PODA, 1763 (identity: *Arctia villica*, Arctiidae). Junior primary homonym.
Phalaena (Noctua) glaucescens GOEZE, 1781: Ent. Beitr. 3 (3): 206 (type locality not stated). Type(s) not traced. Junior synonym.
Phalaena (Geometra) glaucofasciata GOEZE, 1781: Ent. Beitr. 3 (3): 387 (type locality not stated). Type(s) not traced. Junior synonym.
Brephos parthenias lapponica RANGNOW, 1935: Ent. Rdsch. 53 (2): 22, pl. 4, fig. 11 ([Sweden/Finland]: Lappland). Type(s) not traced. Junior synonym (but doubtfully conspecific; see remarks).
Unavailable names (misidentification): *notha:* sensu HAWORTH (1809) nec HÜBNER, [1803]. - (infrasubspecific): *albofasciata:* COCKAYNE (1952); *brunnea:* CLOSS; *contrasta:* LEMPKE (1949); *cuprea:* COCKAYNE (1952); *dealbata:* KLEMENSIEWICZ (1913); *dilutior:* HEINRICH (1916); *extrema:* REBEL; *fasciata:* LEMPKE (1949); *flava:* WOOD; *indigena:* HAVERKAMPF (LAMBILLON); *intermedia:* LEMPKE (1949); *luteata:* HENNIN (1910); *muliercula:* STEPHAN (1923); *nigra:* TUTT; *nigrobasalis:* SPULER; *obscura:* PROUT; *passetii:* THIERRY-MIEG; *szymanskii:* ISAAK (1920); *unicolor:* HEINRICH (1916); *variegata:* LEMPKE (1949). - (incorrect subsequent spellings): *extremata:* BURESCH (1924); *fulvata:* WOLF (1988); *passeti:* LHOMME (1935); *plebeia:* PROUT (1912b).

DIAGNOSIS: Wingspan ♂ 31-36, ♀ 33-38 mm. Female larger than male. Forewings brownish with faint darker grey suffusion. Usually three white costal spots present, exceptionally absent. Ante- and postmedial lines more distinct in ♀ than in ♂. Hindwings orange with brown border along apex and termen. Discal spot large, brown. Fringe chequered black and white. Venation (Text-fig. 46): Sc+R1 and Rs of hindwing approximate for nearly 1/2 length of cell; Rs and M1 distinctly stalked. Antennae slightly dentate in ♂, branches with velvety covering of extremely short cilia. Antennae simple in ♀. Further features under diagnosis of genus and subfamily.

MALE GENITALIA: Uncus and valvae very long and slender, without projections. Uncus curved dorsoventrally. Juxta pyriform. Central plate of gnathos rounded, spinulose. Aedeagus long and narrow, with long membranous 'cornutus' and round terminal lobe.

FEMALE GENITALIA: Apophyses anteriores and posteriores comparatively short and stout. Tergum A8 narrow, ribbon-shaped. Sternum A7 (lamella antevaginalis) large, triangular. Ductus bursae laterally projecting. Signum bursae long, tongue-shaped, laterally with small teeth, slightly variable.

DISTRIBUTION: Palaearctic. Usually common all over central, northern and eastern Europe except Iceland, Ireland, southern Jutland (Denmark) and northern Schleswig-Holstein (Germany). In western Europe, in Great Britain and in the colder parts of France southwards to the Alps and north-western Bulgaria. Also an isolated population in the Pyrenees. In central Europe decreasing in the 20th century.
    Outside Europe in northern Caucasus (TIKHONOV 1993), reputedly in 'north-eastern Turkey' (1♀ in ZSM, not completely labelled), in the east across Siberia and eastern Asia (subsp. *sajana* PROUT, 1912) to Sakhalin (subsp. *hilara* SAWAMOTO, 1937) and Japan (subsp. *bella* INOUE, 1955, and subsp. *elegans* INOUE, 1955).

PHENOLOGY: Univoltine. Main flight period: early March to late April, sometimes even in mid-February; in northern Europe early April to late May. Protandrous (OSTHELDER 1929; URBAHN 1939). Larval stages early May to mid-July, sometimes in late April, in northern Europe later, June to July (SEPPÄNEN 1970). Overwintering as pupa, sometimes twice (PROUT 1912a; FORSTER & WOHLFAHRT 1981; etc.).

BIOLOGY: Larva oligophagous. Found on various species of *Betula* (Germany: WEGNER, LEPINITZ pers. comm.; SCHNEIDER 1934; Finland: SEPPÄNEN 1954; 1970), preferring young trees and sitting on top of branches (OSTHELDER 1929; KOCH 1984; BERGMANN 1955). Also found on *Sorbus aucuparia* (Rosaceae) (Great Britain: SKINNER pers. comm.; SCOBLE 1999). Recorded also on *Fagus* (var. cit.). Early instars on catkins, later between loosely interwoven leaves (REID 1986). Pupation in rotting wood or bark. Sonic hearing discussed in SURLYKKE et al. (1998).

HABITAT: Birch forests and birch marshes, mainly glades and forest rides. Most abundant around 10-20 year old birches (WEGNER pers. comm.). Also on heaths, near birches. Adults attracted to blossom of *Salix* (BROOKS 1991), sap runs and to excrement of horses (LEIPNITZ pers. comm.). Exclusively dayflying, mainly from 9:00-12:00, in sunshine, high around birch trees. From 0 m up to 1,000 m above sea-level (e.g. northern Italy: Mte. Baldo, Brenta).

PARASITOIDS: Diptera, Tachinidae: *Lydella nigripes* (THOMPSON 1944).

SIMILAR SPECIES: *Archiearis notha* can easily be distinguished by the bipectinate ♂ antennae and the much smaller ♀. Further features are the absence of white costal spots on the forewings, smaller hindwing discal spot, unchequered fringe and the hindwing venation. The ♀ of *A. notha* is characterized by a very dark oblique antemedial line, distally bordered by a dark grey medial band, which is narrower than in *A. parthenias*.

REMARKS: The northern European populations are said to constitute a separate subspecies (RANGNOW 1935). There are, however, no structural differences and the slight difference in habitus shows clinous transitions to the central European populations. The Finnish specimen upon which the description of *lapponica* RANGNOW, 1935 (fig. 11) is based, seems to represent *Archiearis notha* ♀. The type should be traced and examined.

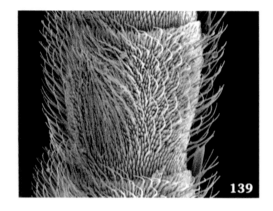

Text-fig. 139. Male antenna of *Archiearis parthenias* (LINNAEUS, 1761).

Text-fig. 140. Male antenna of *Archiearis notha* (HÜBNER, [1803]).

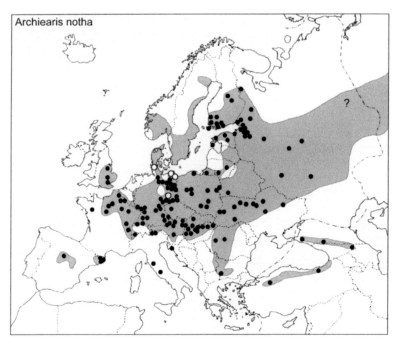

2. *Archiearis notha* (HÜBNER, [1803])

Noctua notha HÜBNER, [1803]: Natur. eur. Schmett. 4, Noctuae 3: pl. 74, figs 343, 344 (Europe). Type(s) lost.
‡ *Phalaena hyemalis* GIORNA, 1791: Calend. Ent.: 18, nec RETZIUS, 1783. Junior primary homonym.
Brephos nothum ab. *laeta* REBEL, 1910: Berge's Schmetterlingsbuch (9): 306 (type locality not stated).
  Type(s) not traced in NHMW. Available, not explicitly infrasubspecific.
Unavailable names (misidentifications): *parthenias*: sensu ESPER (1786) nec LINNAEUS, 1761; *puella*:
  sensu LHOMME ([1935]) nec ESPER, 1787; *vidua*: sensu BILLBERG (1820) nec FABRICIUS, 1775 (identi-
  ty: *Archiearis parthenias*). - (infrasubspecific): *cuprina*: COCKAYNE (1952); *diluta*: COCKAYNE (1952);
  intermedia: COCKAYNE (1952); *obscura*: COCKAYNE (1952). - (incorrect subsequent spellings): *nota*:
  FORSTER & WOHLFAHRT (1981); *nothoa*: BURESCH (1915).

DIAGNOSIS: Wingspan ♂ 29-34, ♀ 27-31 mm. Female smaller than male. Wing pattern
similar to that of *Archiearis parthenias*, but forewings brownish with pale grey admixtu-
re. Females sometimes with slightly bluish tinge, wing pattern more contrasted, antemedial line black, oblique, distally bordered by dark grey (medial-)field, rest of medial area and
basal area usually light grey. Hindwings with discal spot small and sharp. Fringe dark, not
chequered. Venation: Sc+R1 and Rs of hindwing approximate over nearly 1/2 length of
cell; Rs and M1 usually connate, rarely shortly stalked. Antennae shortly bipectinate in ♂,
longest branches 2-3 times width of flagellum. Antennae of ♀ filiform. Further features
under diagnosis of genus and subfamily.

MALE GENITALIA: Uncus and valvae short and broad, without projections. One examined
central Italian ♂ with uncus slightly narrower. Valva distally tapered. Juxta small, caudally
often tapered. Central plate of gnathos caudally angled, strongly spinulose. Aedeagus
short, basally narrow, with a short rounded, membranous 'cornutus', bearing a small
terminal spine; terminal lobe laterally tapered.

FEMALE GENITALIA: Apophyses anteriores and posteriores long and slender. Tergum A8 rectangular. Sternum A7 (lamella antevaginalis) slender, slightly narrower still in one examined ♀ from central Italy. Ductus bursae narrow, longitudinally wrinkled, enlarged and strongly sclerotized near ostium bursae. Signum bursae short, lanceolate, laterally, rarely sublaterally, with small teeth, terminally with two stout teeth, deeply concave between.

DISTRIBUTION: Palaearctic. Temperate areas of Europe, absent in great parts of northern Europe. In southern Europe absent except for the eastern Pyrenees, central Spain, central Italy (Toscana and Lazio; ZILLI pers. comm.), Krk Island (HABELER pers. comm.) and the Northern Balkans. In the east to the Urals, but absent in the south (i.e. southern Ukraine). Endangered in various areas, with abundance drastically decreasing e.g. in northern Germany (WEGNER pers. comm.). However, recently reported as abundant from some localities in eastern and southern Germany (GELBRECHT, WEIGERT pers. comm.), perhaps often overlooked.

Outside Europe in northern Caucasus (TIKHONOV 1993) and northern Turkey (KOCAK 1990), eastwards to Ussuri (subsp. *suifunensis* KARDAKOFF, 1928) and Japan (subsp. *okanoi* INOUE, 1958).

PHENOLOGY: Univoltine. Main flight period: early March to late April, in northern Europe early April to mid-May. Larval stages early May to early July, in northern Europe later, June to July (SEPPÄNEN 1970). Overwintering as pupa, sometimes two or three times (REBEL 1910; PROUT 1912a; FORSTER & WOHLFAHRT 1981; etc.).

BIOLOGY: Larva oligophagous. Found on *Populus tremula* (Germany: WEGNER pers. comm.; Poland: URBAHN 1939; Finland: SEPPÄNEN 1954; 1970; Great Britain: SKINNER pers. comm.) and *Betula* (Germany: LEIPNITZ pers. comm.). Recorded also on other species of *Populus*, e.g. *P. nigra* (Japan: NAKAJIMA & SATO 1979), in central Europe exceptionally even on *Salix* (var. cit.). Reared on *Salix alba* and *Salix viminalis* (Hungary: RONKAY pers. comm.). Caterpillars living between spun leaves (REID 1986). Pupation in a strong cocoon.

HABITAT: Wind-protected areas, such as glades, valleys, forest-rides usually on wet, peaty or sandy soils near ground water (WEGNER, GELBRECHT pers. comm.). Often near pioneer trees of *Populus tremula*, but also around old trees. Adults attracted to blossom of *Salix* and sap runs. Exclusively diurnal, mainly from 9:00-12:00, in sunshine. In the afternoon flying high around *Populus tremula* trees or resting on their twigs (GELBRECHT pers. comm.; BURGERMEISTER 1955). From 0 m up to 1,000 m above sea-level.

PARASITOIDS: Hymenoptera, Ichneumonidae: *Agrypon anxium*, *A. flaveolatum*, *A. stenostigma*, *Eulimneria alkae*. - Diptera, Tachinidae: *Zenillia libatrix* (THOMPSON 1944).

SIMILAR SPECIES: *Archiearis parthenias* can easily be distinguished by the filiform ♂ antennae and the larger ♀. Furthermore it is usually characterized by whitish costal spots on the forewings, large hindwing discal spots, chequered fringe and different hindwing venation. The ♀ of *A. parthenias* is characterized by the narrower dark oblique antemedial fascia. *A. puella* differs in the narrower forewing shape, coloration of the forewing more greyish, hindwing yellow, postmedial line of hindwing complete, dark markings more extensive and the fringe of the hindwings with one white spot near tornus. The medial band of the forewing is less contrasted in the ♀ of *A. puella*.

Text-figs 141-146: Diagnostic characters (indicated) of European *Archiearis* species. Text-figs 141 (♂) and 142 (♀): *A. parthenias* (LINNAEUS, 1761). Text-figs 143 (♂) and 144 (♀): *A. notha* (HÜBNER, [1803]). Text-figs 145 (♂) and 146 (♀): *A. puella* (ESPER, 1787).

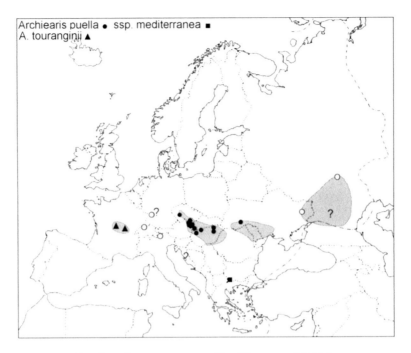

3. *Archiearis puella* (ESPER, 1787)

Phalaena (Noctua) puella ESPER, 1787: Schmett. in Abbild. 4: 1, pl. 106, figs 2, 3 (Germany: Frankfurt a. M.). Lectotype ♂ (LMW, examined: HACKER 1999).
Phalaena (Noctua) caelebs HÜBNER, 1789: Beitr. 1 (4): 21, pl. 3, fig. Q (Austria: Vienna district). Type(s) lost. Junior synonym.
Noctua spuria HÜBNER, [1803]: Natur. Eur. Schmett. 4, Noctuae 3: pl. 74, fig 345 ('Europe'). Type(s) lost. Junior synonym.
Unavailable names (infrasubspecific): *inversa*: NITSCHE (1932); *latevirgata*: KITT (1920); *treitschkei*: PROUT (1912).

DIAGNOSIS: Wingspan ♂ 29-34, ♀ 27-30 mm. Female smaller than male. Forewings slender, in ♂ yellowish grey, wing pattern diffuse. In ♀ antemedial line of forewing black, very sharp, medial band only slightly contrasted to rest of wing. Hindwings yellow, with black postmedial line complete, yellow parts comparatively small. Dark border along termen very broad. Fringe unicolorous, hindwing fringe with whitish spot near tornus. Hindwing venation as described for *Archiearis notha*, Rs and M1 however more often shortly stalked. Antennae shortly bipectinate in ♂, longest branches about twice width of flagellum. Antennae simple in ♀. Further features under description of genus and subfamily.

MALE GENITALIA: Uncus comparatively short and slender. Valvae short and broad, costa more convex than in *Archiearis notha*. Juxta large, with two spinulose fields. Central plate of gnathos caudally rounded, shortly spinulose. Aedeagus very short and basally broad with membranous 'cornutus' weak, but elongate, terminally bearing a sclerotized spine (sometimes split into 3-4 single needles). 'Terminal lobe' of aedeagus with one stout, strongly sclerotized lateral spine.

FEMALE GENITALIA: Apophyses anteriores and posteriores long and slender. Sternum A7 (lamella antevaginalis) extremely narrow. Ductus bursae comparatively broad, longitudinally wrinkled, near ostium bursae with two sclerotized lateral lobes. Signum bursae lanceolate, (sub-)laterally with small teeth, terminally with two stout teeth, concave between.

DISTRIBUTION: (East-)European. In rather isolated populations from central Europe and, perhaps, Dalmatia eastwards to southern European Russia. Some old records from Germany (Hessen, Württemberg, Thüringen) discussed in GELBRECHT (1999). Not recently confirmed for northern Switzerland, northern Italy, Dalmatia. Everywhere highly endangered due to local distribution and habitat destructions (loss of gravel banks and flood plain forests caused by the straightening of rivers).

PHENOLOGY: Univoltine. Main flight period: Early February to early April. Larval stages early May to late June. Overwintering as pupa.

BIOLOGY: Larva recorded on *Populus tremula* and *P. alba* (var.cit.). Reared on *P. alba* and *P. nigra* (Hungary: RONKAY pers. comm.).

HABITAT: Xerothermic localities, with natural occurrence of larval host-plants, near rivers and streams, i.e. 'floodplain forest, Salicion albae' (LASTUVKA 1994). Adults exclusively diurnal. From 0 m up to about 300 m above sea-level.

SIMILAR SPECIES: *Archiearis notha* differs in the broader forewing shape, forewing less greyish, hindwing orange, postmedial line of the hindwing not complete, dark terminal markings narrow, fringe unicolorous and in the strongly contrasted medial band of the ♀ forewing.

Archiearis puella mediterranea GANEV, 1984

Archiearis puella mediterranea GANEV, 1984: Z. Arb.Gem. öst. Ent. 36 (1/2): 15, figs 3-5 (southwestern Bulgaria: Kozuch). Holotype ♂ (coll. Ganev, Sofia), paratype ♂ (NHMW, examined).

DIAGNOSIS: Wingspan ♂ 28-31, ♀ 26-27 mm. Smaller than the nominate subspecies. External structure, wing pattern and colour as described for *Archiearis puella*, but hindwing light red, underside of forewing yellowish red.

MALE GENITALIA: As described for *Archiearis puella*. According to the original description valva straighter than in nominate subspecies, juxta truncate on caudal margin, its lateral margins parallel. One male dissected by the author, however, with comparatively broad valva.

FEMALE GENITALIA: As described for the nominate subspecies, but signum shorter, laterally more convex. Lamella antevaginalis slightly broader, granulose.

DISTRIBUTION: Endemic to south-western Bulgaria. Known from the type locality only. Highly endangered (see nominate subspecies).

PHENOLOGY: Univoltine. Main flight period: early March to late March.

BIOLOGY: Larval host-plant and ecology unknown.

HABITAT: Type locality: A small river valley at about 200 m above sea-level.

SIMILAR SPECIES: See nominate subspecies. This subspecies is easily confused with (allopatric) *Archiearis notha*, due to the same hindwing colour.

3A. *Archiearis touranginii* (BERCE, 1870)

Brephos notha var. *touranginii* BERCE, 1870: Faune Entom. Franc. Lép. 4: 169 (France, Cher: Tourangin). Type(s) not traced.
Unavailable names (incorrect subsequent spelling): *tourangini*: SAND (1880)

DIAGNOSIS: Wingspan ♂♀ 22-27 mm. Smaller than *Archiearis notha* and *A. puella*. Ground colour of hindwing red. External structure, wing pattern and colour similar to *Archiearis notha*, even on the underside, but transverse lines of forewing upperside dark brown, fine and sharp, postmedial line of hindwing complete.

MALE GENITALIA: As in *A. puella*, but valva straighter, less convex at costa, juxta smaller, its lateral margins parallel.

FEMALE GENITALIA : Apophyses short, signum very small (BÉRARD 2000).

DISTRIBUTION: Endemic to central France (dept. Cher, Loire, Saône-et-Loire). Rare and probably endangered (see *A. puella*).

PHENOLOGY: Univoltine. Main flight period: March.

BIOLOGY: Larva recorded on *Salix purpurea* (BÉRARD 2000).

SIMILAR SPECIES: This species is easily confused with *Archiearis notha*, due to the same hindwing colour. Identification is possible based on size, wing pattern and genitalia.

REMARKS: *A. touranginii* is closely related to *Archiearis puella*, but considered a separate species by BÉRARD (2000).

Subfamily **Orthostixinae** MEYRICK, 1892

Palaearctic and Indo-Pacific. 17 species in four genera, until recently placed in the subfamily 'Oenochrominae' (see SCOBLE & EDWARDS 1990; HOLLOWAY 1996; 1997 and HAUSMANN 1996). Common ('archaic') features with Archiearinae: larval prolegs on segments A5 (A4) and hindwing venation. Certain similarities however also to Ennominae (HOLLOWAY 1996) and Desmobathrinae (HAUSMANN 1996). The latter hypothesis supported by structural (e.g. tegumen loops in ♂ genitalia) and ecological arguments, main food-plants of *Orthostixis* (Labiatae) being closely related to those preferred by many Desmobathrinae (Verbenaceae). *Naxa* species however feeding on Oleaceae (SINGH 1951; NAKAJIMA & SATO 1979; HOLLOWAY 1996) and ferns (HOLLOWAY 1996).

DIAGNOSIS: Venation (Text-fig. 47) of hindwing: Sc+R1 and RR not fused, but with short crossvein (R1) between Sc and Rs in the type-species of *Orthostixis*. M2 present, tubular. A1 present, but not tubular, A2 long, A3 about 1/3 of hindwing length. Ansa of tympanal organ broad at base, apically tapered. Frons slightly convex.

IMMATURE STAGES: Larva setose, with rudimentary ventral prolegs on segment A5, sometimes even on A4. Larva and pupa often whitish with conspicuous black and yellow markings. Chaetotaxy described for *Naxa* in SINGH (1951).

REMARKS: The caterpillars of *Naxa* live colonially, the pupae hanging freely exposed in the common web, fixed with silken threads (HOLLOWAY 1997, pl. 4).

Orthostixis HÜBNER, [1823]

TYPE-SPECIES: *Geometra cribraria* HÜBNER, [1799]. Three species in the eastern Mediterranean (HAUSMANN 1996).

DIAGNOSIS: Venation (Text-fig. 47): Forewing with only one (distal) accessory cell (compare Desmobathrinae). Hindwing: Rs and M1 separate. M3 and CuA1 separate. Frenulum in ♂ present, replaced by tuft of stiff hair-scales in ♀. Frons comparatively flat, only ventral part strongly convex. Hindtibia with two short spurs in both sexes.

MALE GENITALIA: Uncus broad, not tapered towards end. Small basal lobes near uncus ('socii' or 'tegumen loops') present. Gnathos present. Costa of valva strongly sclerotized, spinulose. Juxta with long flask-shaped posterior projection. Aedeagus without stout cornuti.

FEMALE GENITALIA: Apophyses anteriores 1/3-1/2 length of apophyses posteriores. Antrum long, caudally deeply notched. Ductus bursae fairly long, longitudinally wrinkled in the oral part. Signum well developed, elliptical, with serrate margin.

IMMATURE STAGES: Larva thick, cylindric, squat, wrinkled transversely.

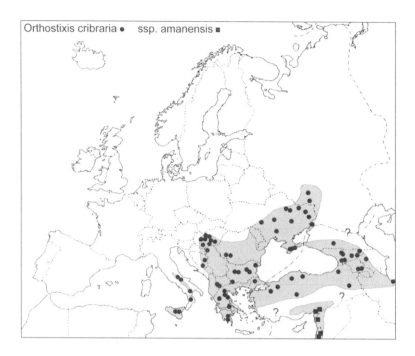

Orthostixis cribraria ● ssp. amanensis ■

4. *Orthostixis cribraria* (HÜBNER, [1799])

Geometra cribraria HÜBNER, [1799]: Samml. Eur. Schmett. 5: pl. 16, fig. 83 ('Europe'). Type(s) lost. Subjective replacement name.
‡ *Phalaena laetata* FABRICIUS, 1798: Syst. Ent. (Suppl.): 456 ('southern Russia') nec FABRICIUS, 1794 (identity: *Agathia laetata*, Geometrinae, India). Junior primary homonym.
Zerene cribrata TREITSCHKE, 1825: Schmett. Eur. 5 (2): 445. No separate types. Unjustified emendation.

DIAGNOSIS: Wingspan ♂ usually 28-32 mm, ♀ 29-34 mm, occasionally smaller, mainly in the second generation (sometimes 24 mm only). Both forewings and hindwings clear white, postmedial lines replaced by rows of black dots. Frons and vertex whitish. Venation with short cross-vein (R1) between Sc and Rs in the hindwing (absent in *Orthostixis calcularia* LEDERER, 1853 from Turkey). Antennae shortly ciliate in both sexes, dentate in ♂, slightly dentate in ♀. Flagellum with pale brown scales on upperside. Proboscis long. Palpi of medium length, with outstanding scales, tip pale brown. Further features under generic diagnosis.

MALE GENITALIA: Uncus comparatively short. Spinulose area in the central part of the valva large. Costa of valva less spinulose than in *Orthostixis calcularia*. Aedeagus shorter than in *Orthostixis cinerea* REBEL, 1916 from Cyprus.

FEMALE GENITALIA: Corpus bursae comparatively large. Signum bursae lanceolate or elliptical, margins with small teeth.

DISTRIBUTION: SE-European species, extending to Asia minor and Italy ('Pontic'): Sicily, central and southern Italy, from central Hungary all over the Balkans including Peloponnes and European Turkey, Ukraine with Crimea to south-western European Russia. Absent on the Greek islands.

Outside Europe in Turkey, Caucasus, Transcaucasus, northern Iran. In southern Turkey, northern and central Lebanon replaced by subsp. *amanensis* WEHRLI, 1932.

PHENOLOGY: Usually univoltine at higher altitudes: early June to late July, exceptionally late August (data of about 100 examined specimens). In lowlands bivoltine with flight periods from May to June and July to September, e.g. in Bulgaria (NESTOROVA 1998). Probably overwintering as egg (FORSTER & WOHLFAHRT 1981).

BIOLOGY: Larva found on *Scutellaria rubicunda* (REBEL 1903, as *S. peregrina*), recorded also on *Scutellaria altissima* (FORSTER & WOHLFAHRT 1981). Adults mainly day-flying, sometimes attracted to light (THURNER 1967).

HABITAT: In southern Italy mountainous woodland with *Quercus* and *Acer* (PARENZAN 1988). From 0 m up to about 1,800 m above sea-level, preferably from 500 m to 1,000 m. In Bulgaria and Ukraine however regularly reported from 0 m to 400 m (NESTOROVA 1998; GELBRECHT, KOSTJUK pers. comm.). Subsp. *amanensis* exclusively from 800 m to 1,600 m.

SIMILAR SPECIES: *Orthostixis calcularia* from Turkey is larger, the wings less transparent, more creamy white, the terminal spots larger, and the position of the antemedial spots more oblique. *Orthostixis cinerea*, endemic to Cyprus, differs in its darker wing colour.

Subfamily **Desmobathrinae** MEYRICK, 1886

Cosmopolitan. According to the present concept about 25 genera with some 230 species. Defined as 'delicately built Oenochrominae' and separated from both Oenochrominae s. str. and Orthostixinae by HOLLOWAY (1996; 1997). Certain relationships to Orthostixinae suggested by intermediate features of some genera such as *Derambila* WALKER, [1863] (HAUSMANN 1996b). Further structural (e.g. tegumen loops in ♂ genitalia) and ecological arguments under preceding subfamily.

DIAGNOSIS: Thorax and abdomen slender in ♂. Females often smaller than males, with large abdomen (Palaearctic genera). Legs and antennae usually elongated and slender. Venation (Text-fig. 48): Forewing with one or two accessory cells. Hindwing with Sc+R1 and Rs not fused. M2 tubular. A1 weak, A2 long, A3 about half length of wing. Ansa of tympanal organ broad at base, apically tapered. Frons sometimes strongly convex.

MALE GENITALIA: Costa of valva often strongly sclerotized.

IMMATURE STAGES: Larva not setose (as in Orthostixinae), segments A3-A5 without ventral prolegs, as far as known. Chaetotaxy of genus *Ozola* described in SINGH (1951).

REMARKS: Representatives of many genera day-flying or easily flushed up during daytime.

Gypsochroa HÜBNER, [1825]

TYPE-SPECIES: *Geometra renitidata* HÜBNER, [1817]. One species only. Considered by various authors as a Larentiinae genus, but to be transferred into Desmobathrinae because sharing some common features with genera *Myinodes* and *Derambila* (HAUSMANN 1996b). Antennae and legs much longer than in *Orthostixis*, similar to equivalents in *Myinodes*, *Derambila* etc. Perhaps a phylogenetically basal genus.

DIAGNOSIS: Forewings slender, apically pointed. Venation (Text-fig. 48) very similar to that of *Derambila*: Forewing with one accessory cell only. Accessory cell at its costa bordered by R1 and R2, on the opposite side by R3-R5. Both cells and accessory cell fairly long. Hindwing: Sc+R1 and Rs not fused, quite distant from each other. Frenulum in ♂ present (retinaculum as in *Myinodes* very close to forewing costa), in ♀ replaced by some rather stout hairs. Frons convex. Proboscis very long, with shield-shaped black sclerite at base. Palpi of medium length, slender. Antennae (♂♀) long, filiform and shortly ciliate, dentate in ♂. Flagellum with white scales on upperside. Legs very long. Foreleg with tibia and femur long and slender, without grasping structure at tibia (compare *Lithostege*). Hindtibia (♂♀) with very short terminal spurs only. Abdomen (♂♀) very long and slender. Sternum A5 in ♀ with two posterior projections, A6 small and quadrate, A7 small and lens-shaped.

GENITALIA: see *Gypsochroa renitidata*.

IMMATURE STAGES: Larva with deep intersegmental wrinkles and two (additional) pairs of vestigial prolegs on segments A4 and A5. Pupa with wings terminally free.

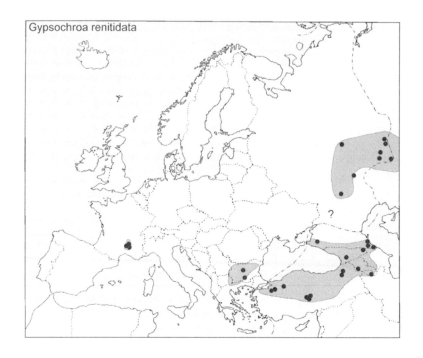

Gypsochroa renitidata

5. *Gypsochroa renitidata* (HÜBNER, [1817])

Geometra renitidata HÜBNER, [1817]: Samml. Eur. Schmett. 5: pl. 94, figs 485, 486 ('Europe').
  Type(s) lost.
Siona renitidaria BOISDUVAL, 1840: Genera Index Meth. eur. Lepid.: 228. No separate types. Unjustified
  emendation.
Gypsochroa renitidata gallica LE MOULT, 1948: Miscell. ent. 45: 47 (France, Ardèche: La Voulte-
  sur-Rhône). Holotype ♂ (coll. HERBULOT, ZSM). Junior synonym.
Unavailable names (infrasubspecific): *postgallica* LE MOULT, 1948.

DIAGNOSIS: Wingspan (♂ ♀) 21-27 mm. Fore- and hindwings white, without pattern, scales smooth and shining, no terminal line or spots, fringe white. Frons and vertex white. Palpi white, third segment pale brown. Further features under generic diagnosis.

MALE GENITALIA: Uncus long, somewhat tapered. Gnathos weak. Costa of valva prominent. Harpe forked, with stout terminal spines, varying in number from one to three on each arm.

FEMALE GENITALIA: Lamella antevaginalis with two strongly sclerotized, rounded lobes. Ductus bursae short. Corpus bursae without signum.

DISTRIBUTION: Disjunct area of distribution: In France only recorded from the departments Ardèche and Drôme on the slopes of the Rhône valley. Western and northern Bulgaria, northern Greece, southern European Russia, southern Urals.

  Outside Europe in northern and central Turkey, Caucasus, Transcaucasus, south-western Siberia, northern Kazakhstan, Touva Region, Mongolia (VIIDALEPP 1996). Endangered in the European parts of the distribution area due to local distribution and habitat destruction. The last habitats in France highly endangered, meriting special conservation.

PHENOLOGY: Bivoltine. Main flight period mid-May to mid-June and early July to mid-August. First generation later in Turkish mountains, specimens emerging in June. From European Russia recorded in the second half of June (univoltine?). Larval phenology unclear, in 'summer and autumn' according to CULOT (1919). Overwintering stage unknown.

BIOLOGY: Eggs found on *Linaria repens* (CHRÉTIEN 1910; as '*Linaria striata* var. *monspeliensis*'). Larva recorded on the same plant (CULOT 1919). Reputedly feeding also on lichens of herbaceous plants (LHOMME 1935). Adults resting on larval host-plants with wings crossed or slightly rolled around twigs or leaves. Moths diurnal (Turkey: GELBRECHT pers. comm.), but also attracted to light (France: GOATER, SKOU pers. comm.). Pupa in a cocoon on the host-plant.

HABITAT: Species of dry, hot and karstic open habitats with ruderal (or pioneer) flora. In Turkey on flower-rich mountainous steppes (GELBRECHT pers. comm.). K-strategy. Moths not moving far from host-plant. In Europe from 0 m up to 300 m above sea-level (limited data). In Turkey up to 2,200 m.

SIMILAR SPECIES: *Lithostege farinata* (HUFNAGEL, 1767) and other, similar *Lithostege* species differ in having a shorter and more robust abdomen, shorter legs and antennae, forelegs modified, wings broader, coloration less shiny.

REMARKS: No remarkable and constant difference in structure or habitus between the various isolated populations from south-eastern France to the Urals.

Myinodes MEYRICK, 1892
=*Pseudotagma* STAUDINGER, 1892

TYPE-SPECIES: *Sterrha interpunctaria* HERRICH-SCHÄFFER, 1839. Three species in the western Palaearctic. Venation closely matching that of North African genera *Eumegethes*, *Drepanopterula* and even of *Epirranthis*. Generic revision by HAUSMANN (1994a). Frequent misspellings in subsequent literature: *Myinos* (KRÜGER 1939); *Myidodes* (ELLISON & WILTSHIRE 1939); *Myinoides* (MARIANI 1943); *Myrinodes* (SCHMIDLIN 1964); *Myiniodes* (RUNGS 1981).

DIAGNOSIS: Venation of forewing: R1 and R2 arising separately from cell, forming with Sc two accessory cells. Frenulum strongly developed in ♂ (retinaculum very near forewing costa), in ♀ replaced by some stout hairs. Forewings comparatively slender, veins pale with white intervenal streaks. ♀ smaller, with large abdomen. Abdomen of ♂ slender. Proboscis long. Frons strongly convex. Palpi long. ♂ antennae ciliate with two rows of cilia, ♀ antennae simple, finely ciliate beneath. ♂ and ♀ hindlegs long and slender, with two pairs of long spurs.

MALE GENITALIA: Uncus tapered, without basal lobes ('socii'). Costa of valva strongly sclerotized. Ventral part of valva medially with a stout inwardly directed projection. Lateral sclerotization of aedeagus and shape of cornutus diagnostic.

FEMALE GENITALIA: Ductus bursae strongly sclerotized. Bursa copulatrix usually longitudinally ribbed, without signa, joining ductus bursae laterally on the latter.

REMARKS: Early flying 'winter moths'. Larval morphology unknown.

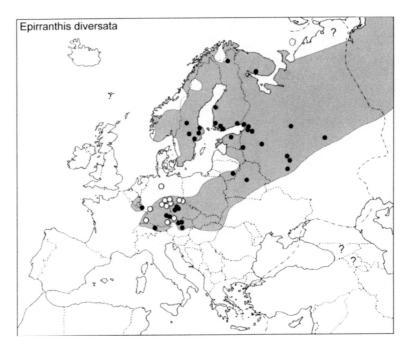

Epirranthis diversata

6. *Myinodes interpunctaria* (HERRICH-SCHÄFFER, [1839])

Sterrha interpunctaria HERRICH-SCHÄFFER, [1839]: in PANZER, Faunae Ins. Germ.: pl. 6 and wrapper (Sicily). Type(s) not traced.
Myinodes interpunctaria atlantica HAUSMANN, 1994: Nota lepidopterologica 17 (1/2): 36, figs 2, 3, 11, 15 (Spain, Andalusia: Alcolea). Holotype ♂ (ZSM, examined). Valid at subspecific rank.

DIAGNOSIS: Wingspan ♂ 25-28 mm, ♀ 21-25 mm. Forewing sand coloured. Postmedial line and terminal area brown. Postmedial line only slightly dentate. White (intervenal) line just below forewing apex crossing postmedial fascia. Terminal spots black, triangular, surrounded by forked white intervenal line. Hindwing pattern similar to that of forewing, but paler, mainly in basal area. Palpi dark brown, upperside white. Frons white, with two semilateral projections. Cilia of ♂ antennae slightly exceeding width of flagellum, but somewhat shorter in populations from Spain and Morocco. Further features under generic diagnosis.

MALE GENITALIA: Uncus short, with stout subapical process. Medial notch of juxta very deep, base of juxta forked. Costal sclerotization of valva broad, smoothly edged. Caudad directed spine (harpe) prominent, pointed, slightly curved. Basal lobe of harpe strongly convex, with numerous small spines. In specimens from Spain and Morocco (ssp. *atlantica*) harpe somewhat more curved, prominent and pointed, basal lobe of harpe slightly convex, less heavily rounded. Aedeagus slender and long. Cornutus weakly sclerotized, situated subterminally. Aedeagus terminally bearing a longitudinal row of about four to six sharp, stoutly sclerotized teeth. Subsp. *atlantica* with aedeagus terminally more heavily sclerotized, with just 3 or 4 sharp, lateral teeth.

FEMALE GENITALIA: Ductus bursae comparatively long and narrow in females examined from eastern Algeria, shorter and broader in subsp. *atlantica*. Left lateral margin concave.

96

Corpus bursae broad and large, but narrower than in subsp. *atlantica*. Caudal margin of lamella postvaginalis convex in nominate subspecies, slightly concave in subsp. *atlantica*.

DISTRIBUTION: West-Mediterranean (Circumtirrhenic), but lacking in the northern parts of the Mediterranean. Nominate subspecies in isolated populations in the central and southern parts of the Mediterranean basin: Sicily and southern Italy (Basilicata and Puglia). One old record from Marche (SPADA 1893) awaits confirmation. Subspecies *atlantica* in central and southern Spain, Portugal.

   Outside Europe in northern Tunisia, north-eastern Algeria. Subspecies *atlantica* in western and northern Morocco. Records for the European part of Turkey (GRAVES 1926; SEVEN 1991) probably refer to *M. shohami* (see below).

PHENOLOGY: Univoltine. Main flight period: Late February to mid April. One very late catch at Toledo (May) perhaps indicating a later flight period in central Spain. Larval phenology unknown, probably overwintering as pupa.

BIOLOGY: Larval host-plant *Rhamnus cathartica* according to SPADA (1893). To be verified.

HABITAT: In southern Italy abundant in a xerothermic locality (Mte. Camplo) with remnants of Mediterranean macchia. Ecological niche not precisely known. ♀ seemingly inactive flyers, probably K-strategy. ♀ ratio at light very low, due to low flight activity. From 0 m up to 900 m above sea-level, nominate subspecies usually below 300 m.

SIMILAR SPECIES: *Myinodes shohami* from the eastern Mediterranean and *M. constantina* HAUSMANN, 1994 from North Africa differ in their single-bulbed frons (Text-figs 147-148), white basal scales of palpi and in the punctiform black terminal spots (Text-figs 149-150). The intervenal line in the forewing apex touches the postmedial line (*M. shohami*) or is much shorter (*M. constantina*), furthermore the palpi are very long in *M. constantina*.

Text-fig. 147: Head parts of *Myinodes inter-punctaria* (HERRICH-SCHÄFFER, [1839]).

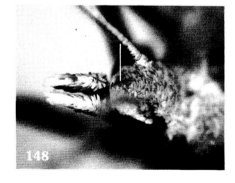

Text-fig. 148: Head parts of *Myinodes M. shohami* HAUSMANN, 1994.

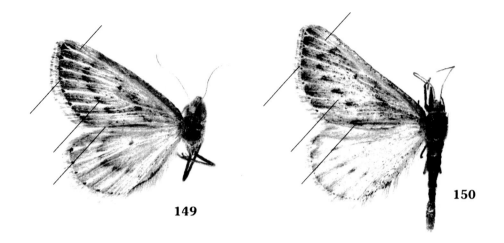

149　　**150**

Text-figs 149-150: Diagnostic characters (indicated) of European *Myinodes* species. Text-fig. 149: *M. interpunctaria* (HERRICH-SCHÄFFER, [1839]). Text-fig. 150: *M. shohami* HAUSMANN, 1994.

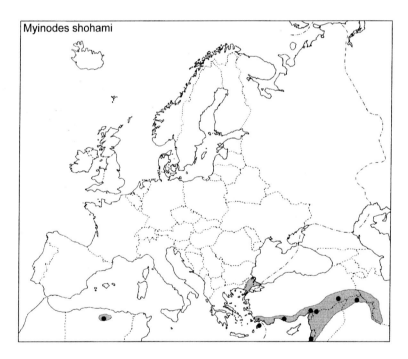

Myinodes shohami

7. *Myinodes shohami* HAUSMANN, 1994

Myinodes shohamı HAUSMANN, 1994· Nota lepidopterologica 17 (1/2): 38, figs 6, 7, 13, 17 (Jordan: Qasr el Hallabad). Holotype ♂ (ZSM, examined).
Unavailable names (misidentification): *interpunctaria*: sensu STAUDINGER (1892) nec HERRICH-SCHÄF-FER, [1839]. (incorrect subsequent spelling): *interpunctata*: AMSEL (1933).

DIAGNOSIS: Wingspan ♂ 22-26 mm, ♀ 19-21 mm. Forewing sand-coloured. Postmedial line and terminal area brown. Postmedial line strongly dentate, more outwardly curved than in congeneric species, white intervenal line just below forewing apex not crossing postmedial fascia. Small terminal spots black, punctiform and small, thinly encircled by white forked intervenal line. Hindwing pattern similar to that of forewing, but paler, mainly in basal area, postmedial line dotted. Palpi dark brown, upperside and basal scales near proboscis white. Frons white, with only one central projection. Cilia of ♂ antennae about 1.3 times width of flagellum. Further features under generic diagnosis.

MALE GENITALIA: Uncus long, subapical process lacking. Caudal excavation of juxta U-shaped, less deep than in *Myinodes interpunctaria*, base convex. Costal sclerotization of valva comparatively narrow, distally pointed. Harpe prominent, S-shaped, without spines, basal lobe lacking. Aedeagus comparatively broad and short, with a terminally situated, very stout cornutus, and a heavily sclerotized, distally pointed plate with one or two lateral teeth.

FEMALE GENITALIA: Ductus bursae broad and short. Corpus bursae small and narrow, caudal half more sclerotized than in congeneric species.

DISTRIBUTION: (East-)Mediterranean-Turanian. Rhodos (GERSTBERGER pers. comm.). Records for European Turkey as '*Myinodes interpunctaria*' (GRAVES 1926; SEVEN 1991) await verification and confirmation.
    Outside Europe in southern Turkey, Syria, Lebanon, Israel, Jordan, northern Iraq, western Iran. In BMNH and ZMUC material from Libya and Algeria.

PHENOLOGY: Univoltine. Main flight period: Mid-February to late March, in Iraq until April. Larval phenology unknown.

BIOLOGY: Larval host-plant unknown.

HABITAT: Ecological niche not precisely known. In Israel probably K-strategy, females seemingly inactive flyers (HAUSMANN 1997a). ♀ ratio at light very low, due to low flight activity. From -390 m (Israel) up to 900 m (southern Turkey) above sea-level.

SIMILAR SPECIES: *Myinodes interpunctaria* differs in the double-bulbed frons, dark brown basal scales of palpi, triangular black terminal spots, and in the longer intervenal line in the forewing apex (Text-figs 147-150). The North African *Myinodes constantina* HAUSMANN, 1994 differs in very long palpi and the shorter intervenal line in the forewing apex. There are also strong differences in ♂ ♀ genitalia.

Epirranthis HÜBNER, [1823]

TYPE-SPECIES: *Geometra diversata* [DENIS & SCHIFFERMÜLLER], 1775. Two species in the Holarctic. Systematic position controversial, some authors (e.g. McGUFFIN 1981; SKOU 1986; WOLF 1988; VIIDALEPP 1996) regard *Epirranthis* as an Ennominae genus, others (e.g. MÜLLER 1996; LERAUT 1997) as belonging to 'Oenochrominae s.lat.'. Relationship to tribe Ennomini supported by pupal morphology (PATOCKA 1992), pupal morphology of Desmobathrinae however not studied and compared. Detailed study of morphology, especially that of venation and tympanal organ reveals many similarities with the Mediterranean Desmobathrinae genera *Myinodes, Eumegethes* and, particularily, *Drepanopterula*. Frequently misspelled as 'Epirrhantis' or 'Epirrhanthis'.

DIAGNOSIS: Forewings broad, costa strongly convex, termen slightly angled at M3. ♀ smaller with forewing apex pointed. Venation: Forewing: R1 and R2 arising separately from cell, forming with Sc two accessory cells. Hindwing with Sc+R1 and Rs approximate for a short distance, but not touching. M2 tubular, but weak in the Palaearctic species, reduced and not tubular in the Nearctic sister species. Two anal veins present (A2, A3). Frenulum strongly developed and long in ♂ (retinaculum near forewing costa), in ♀ replaced by a large tuft of stout hairs. Proboscis developed. Frons strongly convex, smoothly scaled. Palpi short. ♂ antennae ciliate, ♀ antennae simple, without cilia. Abdomen and thorax very slender in ♂, abdomen shorter and thick in ♀. Tympanal organ with ansa broad at base, tapered towards end, terminally bent. ♂ and ♀ hindlegs of medium length, with four spurs. ♂ and ♀ genitalia under diagnosis of *Epirranthis diversata*.

IMMATURE STAGES: Larva slender, not setose. Ventral prolegs absent on segments A3-A5. Segment A8 with transverse hump. Pupal morphology see PATOCKA (1992).

REMARKS: Diurnal springtime moths.

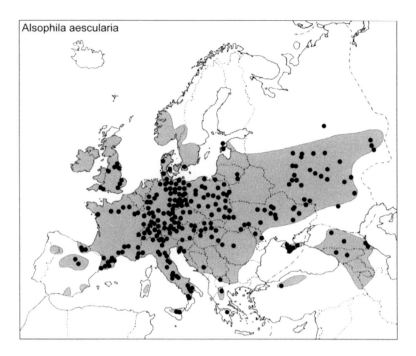

Alsophila aescularia

8. *Epirranthis diversata* ([DENIS & SCHIFFERMÜLLER], 1775)

Geometra diversata [DENIS & SCHIFFERMÜLLER], 1775: Ank. syst. Werkes Schmett. Wienergegend: 315 (Germany: Frankfurt am Main). Type(s) lost (? 'ex coll. Gernings').
Geometra pulverata THUNBERG, 1784: Diss. ent. sistens Ins. Suec. 1: 9, fig. ('Sweden'). Lectotype ♀ (UZIU, Uppsala). Junior synonym.
Phalaena aurantiata FABRICIUS, 1787: Mantissa Insect. 2: 201 (Germany). Type(s) lost. Junior synonym.
Geometra diversaria HÜBNER, [1799]: Samml. Eur. Schmett. 5: pl. 39, fig. 202 ('Europe'). Unjustified emendation, available.
Unavailable names (infrasubspecific): *fasciata* SÄLZL; *obscurior* NORDSTRÖM, 1941; *pallidaria* WEND-LANDT, 1909.

DIAGNOSIS: Wingspan ♂ usually 37-42 mm, ♀ much smaller, 27-34 mm. Northern European populations slightly smaller with males 35-40 mm only. Ground colour of forewing pale yellowish brown, often with tinge of violet. ♀ darker than ♂. Northern European males darker than central European ones, mainly on hindwings. Antemedial and postmedial lines dark brown, near costa with whitish scales beside. Postmedial line usually dotted, but in northern European populations often uninterrupted. Hindwing yellowish in ♂, orange in ♀. Both wings mottled with many dark scales. Cell spots large, dark brown. Underside of wings yellowish to orange (♀), that of hindwing strongly, that of forewing slightly mottled with dark scales. Palpi small in both sexes, slightly exceeding diameter of eye, in ♂ sometimes shorter. Frons dark brown, convex. Antennae of ♂ segmental, with intersegmental incisions, shortly, but densely ciliate. Antennae of ♀ slender, filiform, without cilia. Further features under generic description.

MALE GENITALIA: Uncus slender; without basal lobes ('socii'). Gnathos developed. Valva slender, strongly sclerotized on costa, ventral part membranous. Costa of valva basally elongated into inwardly directed sclerites. Juxta large, bottle-shaped. Aedeagus short and broad, with many long cornuti, varying in number.

FEMALE GENITALIA: Apophyses long and slender. Antrum (or better 'colliculum') short, cylindrical, well sclerotized. Lamella ante- and postvaginalis large, membranous. Ductus bursae (=upper half of corpus bursae) strongly ribbed longitudinally. Signum well sclerotized, elliptical, surface corrugated.

DISTRIBUTION: Eurasiatic: From northern Scandinavia southwards to central Europe and the northern parts of western Europe. In the east to the Ural mountains. Strongly decreasing over the whole distribution area in 20th century, endangered. In Germany not far from extinction, recently recorded for some Bavarian localities. In eastern Germany no record since 1938 (MÜLLER & GELBRECHT 1992), in south-western Germany since 1928 (EBERT pers. comm.). Even in Byelorussia very rare (KULAK 1999 pers. comm.). Special programs for research and conservation of habitats are urgently needed.

Outside Europe known from western and central Siberia, in the east to Irkutsk and Shantar Island in the Okhotskoe Sea (MIRONOV pers. comm.). Doubtful records from 'Tiflis', 'Armenia' and 'Kazakhstan' (four specimens in ZFMK and coll. SKOU, mislabelled?).

PHENOLOGY: Univoltine. Main flight period in central Europe: mid-March to early May. In northern Europe later, from late April to early June. Larval stages early May to mid-June, in northern Europe late May to mid-July. Overwintering as pupa.

BIOLOGY: Larva monophagous. Found on *Populus tremula* (Germany: BERGMANN 1955; Austria: specimen in ZSM), recorded on this plant by many other authors. Caterpillars chiefly on suckers of the host-plant (OSTHELDER 1929). Reared also from other *Populus* species (var. cit.). Pupation in a loose spinning between leaves. Adult moths diurnal, flying rapidly near food-plant, often together with *Archiearis* species. In Estonia mainly active at dusk, during daytime easily flushed up near young, 1-1.5 m high trees of food-plant (VIIDALEPP pers. comm.). Also, occasionally, attracted to light. Nocturnal visits to flowers of *Salix caprea* recorded by BERGMANN (1955).

HABITAT: Forests, glades, forest fringes, hedges, often in wet localities. ♀ seemingly inactive flyers. K-strategy. Usually from 0 m up to 500 m above sea-level.

SIMILAR SPECIES: The wing shape is diagnostic for the genus *Epirranthis*. *Pungeleria capreolaria* ([DENIS & SCHIFFERMÜLLER], 1775) clearly differs in the bright hindwing without cell spot, different phenology, etc.

REMARKS: Some authors, such as SHELJUZHKO (1955), regard the northern European populations as a distinct subspecies (*pulverata* THUNBERG, 1784) as there are some differences in coloration and wing pattern (see above). There are however no structural differences and the differences in habitus underlie clinous transitions to the nominate subspecies in the Baltic states and northern Russia (NORDSTRÖM et al. 1941).

Subfamily **Alsophilinae** HERBULOT, 1962

Subfamily with 26 species in four genera in the Palaearctic, one further species in the Nearctic. Actual subfamily concept mainly based on venation (see below), but with some exceptions in the eastern Palaearctic (HAUSMANN 1996b). In literature often placed in 'Oenochrominae s.lat.' (INOUE 1977; SCOBLE 1992; MÜLLER 1996), but surely not monophyletic with robust-bodied Oenochrominae sensu stricto. Pupal morphology suggests 'relationships' to Theriini (Ennominae) and Operophterini (Larentiinae) according to PATOCKA (1978) and PATOCKA & ZACH (1994). Larval morphology (additional ventral prolegs) supports basal position of Alsophilinae in phylogenesis of Geometridae.

DIAGNOSIS: Venation (Text-fig. 49) of forewing: R3-R5 distinctly stalked. R3-R5 and M1 typically, but not always, separate. Hindwing with Sc+R1 and Rs anastomosing for a long distance (as in Larentiinae), except for some East Asiatic species. Rs and M1 usually shortly stalked. M2 developed on all wings, tubular but weak, even inside cell. Two anal veins present. ♀ 'apterous', i.e. completely lacking visible wings, some rudiments of wings being present. Tympanal organ secondarily lost in ♀, interesting parallelism to reduction of tympanal organ in Operophterini.

IMMATURE STAGES: Larva characterized by presence of vestigial prolegs on segment A5, skin very smooth, head flat. Chaetotaxy of larva described in NAKAJIMA (1998). Detailed discussion of pupal morphology in PATOCKA & ZACH (1994), with statement of 'striking similarity' between Alsophilinae and Theriini (Ennominae)

REMARKS: Adults emerging in late autumn or in early spring.

Alsophila HÜBNER, [1825]
=*Anisopteryx* auct. nec STEPHENS, 1827

TYPE-SPECIES: *Geometra aescularia* [DENIS & SCHIFFERMÜLLER], 1775. Important generic revisions and publications by INOUE (1943), VIIDALEPP (1986) and NAKAJIMA (1998). Genus with 10 species in the Holarctic. Discussion of pupal morphology in PATOCKA & ZACH (1994).

DIAGNOSIS: Venation (Text-fig. 49) with R1 and R2 of forewing both arising free from cell, hindwing discocellular angled, S-shaped. Palpi very short. ♂ antennae slightly dentate, ciliate with tufts of long cilia. Proboscis vestigial. Frenulum present. Hindtibia (♂♀) in European species with four spurs, in ♀ very short, in eastern Palaearctic species sometimes only two spurs. Further features and larval characters under diagnosis of subfamily.

REMARKS: Resting position of adults with crossed forewings and folded hindwings. Eggs glued in ribbons around twigs of host-plant in large numbers, covered with hair scales from anal tuft. Larvae arboreal, often as pest. Pupation in a cocoon in the ground near surface, or rarely free.

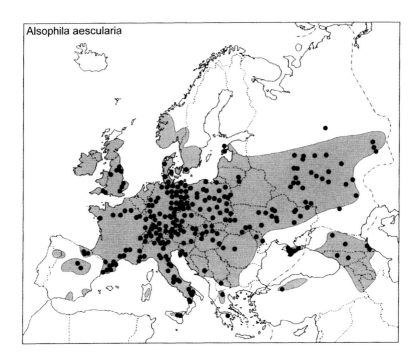

Alsophila aescularia

9. *Alsophila aescularia* ([Denis & Schiffermüller], 1775)

Geometra aescularia [Denis & Schiffermüller], 1775: Ank. syst. Werkes Schmett. Wienergegend: 102 (Austria: Vienna distr.). Type(s) lost. For the treatment of the Schiffermüller names see chapter 5. Name in continuous use for 200 years, also in pest control.

‡ *Phalaena hirtaria* Fabricius, [1776]: Genera Insect.: 286, nec Clerck, 1759 (identity: *Lycia hirtaria*). Junior primary homonym.

Phalaena hirsutaria Fabricius, 1781: Spec. Ins. 2: 508 ('Europe'). Type(s) (of *hirtaria*) probably lost. Junior synonym. Introduced as replacement name for *hirtaria* Fabricius, 1777.

‡ *Phalaena (Geometra) murinaria*: sensu Borkhausen, 1794: Eur. Schmett. 5: 210, nec [Denis & Schiffermüller], 1775 (identity: *Tephrina murinaria*, Ennominae). Misidentification. Junior primary homonym nec [Denis & Schiffermüller], 1775, if *Phalaena* is retained valid as original generic name (see chapter 5).

‡ *Phalaena (Geometra) ligustriaria*: sensu Borkhausen, 1794: Eur. Schmett. 5: 212, nec Lang, 1789 (identity *Agriopis bajaria*). Misidentification.

Geometra apteraria Haworth, 1809: Lep. Brit. (2): 306, (♀) ([England]). Type(s) not traced. Junior synonym.

Further names (availability to be checked): *Anisopteryx fagaria*: sensu Wallengren, 1873: Bih. Svensk. Ak. Handl. 2 (4): 28, nec Thunberg (identity: *Dyscia fagaria,* Ennominae)

Unavailable names (misidentification): *capreolaria*: sensu Wood (1834), partim ♀; Wood's ♂ = *capreolaria* sensu Esper, 1801 (identity: *Agriopis marginaria*) nec [Denis & Schiffermüller], 1775 (identity: *Pungeleria capreolaria*, Ennominae). - (infrasubspecific): *albina*: Lucas (1912); *astrigaria*: Rebel (1913); *brunnea*: Hannemann (1917); *fasciata*: Lempke (1949); *tangens*: Lempke (1949). - (incorrect subsequent spellings): *aesculi*: Duponchel (1829); *rescularia*: Haan (1830); *ligustriariae*: Huemer & Tarmann (1993). - (nomen nudum): *cineraria*: Haworth (1809).

DIAGNOSIS: Wingspan ♂ 28-34 mm, exceptionally only 25 mm. ♀ completely lacking visible wings, body brownish grey, length 8-10 mm. ♀ abdomen with grey brown anal tuft, about 2 mm wide, i.e. narrower than abdomen. Forewing of ♂ pale grey brown. Slight tendency to melanism, usually less than 1%. Postmedial line strongly dentate. Apical streak distinct, dark grey. Hindwing paler than forewing. Dark grey discal spots and dark brown terminal line on all wings. Frons flat in both sexes. Antennae with cilia about 4 times width of flagellum in ♂. Hindtibiae of both ♂ and ♀ with four spurs. Further features under diagnosis of genus and subfamily.

MALE GENITALIA: Uncus broad, terminally tapered. Juxta Y-shaped. Valva with long and narrow costal process, on ventral part terminally bifurcate. Aedeagus comparatively narrow, straight, without cornuti.

FEMALE GENITALIA: Apophyses anteriores and posteriores comparatively stout and short (compare PELLMYR 1980). Ostium bursae triangular, cup-shaped, surrounded by strongly sclerotized vaginal plate. Ductus bursae short and narrow. Corpus bursae long, weakly sclerotized, without signa.

DISTRIBUTION: West-Palaearctic. Widely distributed in Europe except for the most northern parts of Europe, northern Scotland, Corsica, Malta, Albania and Greek islands. Very isolated populations in the mountains of the Iberian peninsula, Sardinia, Sicily and Greece.

Outside Europe distributed in northern and eastern Turkey, Caucasus, Cis- and Transcaucasus, in the east an isolated population in the Kopet Dagh mountains (Turkmeniya, Iran). In the eastern Palaearctic replaced by several rather closely related species, such as *Alsophila acroama* INOUE, 1943 and *A. japonensis* (WARREN, 1894).

PHENOLOGY: Univoltine. Main flight period: Early February to mid-April, in northern Europe and at higher altitudes later, until early May. Larval stages late April to early July, in northern Europe ten days later. Overwintering as pupa.

BIOLOGY: Larva polyphagous, mainly on deciduous trees (var. cit.). Found on many plants, e.g. *Prunus spinosa, P. padus, P. domestica, Rosa, Malus domestica, Carpinus, Corylus, Tilia, Ulmus, Quercus robur, Q. petraea, Acer campestre, Crataegus, Rham-*

Text-fig. 151: Eggs of *Alsophila aescularia* ([DENIS & SCHIFFERMÜLLER], 1775), covered with hair scales from anal tuft.

nus cathartica, *Betula pendula* (Denmark: SKOU pers. comm.; Poland: BUSZKO pers. comm.; Germany: HAUSMANN, LEIPNITZ, WEGNER pers. comm.; SCHNEIDER 1934; BERGMANN 1955; Great Britain: PORTER 1997), also ornamental trees and shrubs (CARTER 1984). Reared from *Fraxinus excelsior* (Germany: KRAUS 1993). Sometimes abundant enough to become a pest (KUDLER 1978). Active at dusk and by night, the ♂♂ being well attracted to light.

HABITAT: Forests glades, forest fringes, hedges, gardens, orchards. Width of host-plant spectrum and mobility of males allowing ubiquitous occurrence in most parts of the distribution area and high level of gene-flow (HAUSMANN 1990). From 0 m up to 1,600 m above sea-level.

PARASITOIDS: Hymenoptera, Braconidae: *Acampsis alternipes, Earinus nitidulus, Meteorus pulchricornis, Microgaster tiro.* - Hymenoptera, Ichneumonidae: *Dasona leptogaster, Agrypon stenostigma, Aphanistes xanthopus, Cryptus viduatorius, Cymodusa cruentata, Labrorychus tenuicornis, Pimpla arundinator, P. examinator, Sagaritis incissa.* - Diptera, Tachinidae: *Blepharomyia amplicornis* (KUDLER 1978; THOMPSON 1944).

SIMILAR SPECIES: Females of *Alsophila aceraria* are smaller, anal tuft of hairs broader and brighter, frons convex (see below).

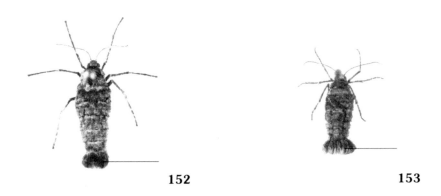

152   **153**

Text-figs 152-153: Diagnostic female characters (indicated) of European *Alsophila* species. Text-fig. 152: *A. aescularia* ([DENIS & SCHIFFERMÜLLER], 1775). Text-fig. 153: *A. aceraria* ([DENIS & SCHIFFERMÜLLER], 1775).

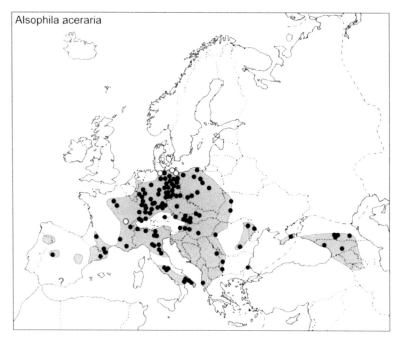

Alsophila aceraria

10. *Alsophila aceraria* ([DENIS & SCHIFFERMÜLLER], 1775)

Geometra aceraria [DENIS & SCHIFFERMÜLLER], 1775: Ank. syst. Werkes Schmett. Wienergegend: 102
  (Austria: Vienna distr.). Type(s) lost. For the treatment of the SCHIFFERMÜLLER names see chapter 5.
  Name in general use for 200 years, also in pest control. Junior primary homonym nec HUFNAGEL, 1767
  (identity: *Ematurga atomaria*, Ennominae), if '*Phalaena*' is retained valid as original generic name.
‡ *Phalaena (Geometra) quadripunctaria* ESPER, 1801: Schmett. in Abbild. 5 (8): 205, pl. 36, figs 10,11,
  nec PODA, 1761 (identity: *Euplagia quadripunctaria*, Arctiidae). Syntype ♂ (ZSM, examined), syntypes 3♂2♀ (LMW). Junior primary homonym.
Phalaena (Geometra) mellearia SCHARFENBERG, 1805: Vollst. Naturgesch. d. schädl. Forstinsekten: 676
  (Germany: Hannover). Type(s) not traced. Synonym of *aceraria* according to WOLF (1988), synonym
  of *Agriopis aurantiaria* (Ennominae) according to SCOBLE (1999). Introduced as replacement name
  with redescription, *quadripunctaria* being said to be 'preoccupied by GOEZE'; correct spelling of the
  latter however '*quadripunctata*' GOEZE, 1781 (identity: *Aspitates gilvaria*, Ennominae).
Unavailable names (misidentification): *progemmaria*: sensu TREITSCHKE (1827; partim ♀) nec HÜBNER,
  [1799] (identity: *Agriopis marginaria*). - (infrasubspecific): *obscura*: LEMPKE (1949); *obsoleta*: LEMPKE
  (1949); *umbrata*: HEINRICH. - (incorrect subsequent spellings): *aceris*: DUPONCHEL (1829); *accuraria*:
  HÜBNER ([1799]); *acararia*: MACK (1985); *quadripuuctaria*: INOUE (1943); *quadripunctata*: PROUT
  (1912); *quadripuncta*: FORSTER & WOHLFAHRT (1981).

DIAGNOSIS: Wingspan ♂ 26-34 mm, ♀ completely lacking visible wings, body brownish
grey, length 6-9 mm. ♀ abdomen with grey brown anal tuft, about 4 mm wide, i.e. as
broad as abdomen. Forewings (♂) light ochreous. Postmedial line slightly dentate, both
ante- and postmedial lines diffuse. Apical streak nearly absent. Hindwings much paler than
forewings. Grey discal spots small. Brown terminal line very thin. Venation with R3-R5
and M1 of forewing slightly stalked. Frons convex in both sexes. ♂ antennae with cilia
about twice width of flagellum. Hindtibiae of both ♂ and ♀ with four spurs. Further features under description of genus and subfamily.

MALE GENITALIA: Uncus broad, terminally rounded. Juxta Y-shaped. Valva with long and
stout costal process, on ventral part terminally with one stout and tapered process. Aede-

agus comparatively broad, with one central patch of numerous very small 'cornuti' and with stout external thorn near tip.

FEMALE GENITALIA: Apophyses narrow, apophyses anteriores very short and weak. Ostium bursae small, semicircular. Corpus bursae more rounded than that of *Alsophila aescularia*.

DISTRIBUTION: Euro-Caucasian: Central and southern Europe except for the most southern parts. In many areas, e.g. in Spain in isolated populations only (DANTART 1991). Doubtfully recorded from southern Spain in 'September' (RIBBE 1912), an unusually early record. Local populations in western Ukraine, eastern Balkans and southern Crimea. Absent from wide parts of the Alps and all Mediterranean islands. Abundance apparently decreasing in many areas such as Austria (DESCHKA & WIMMER 1996), western Germany (KRAUS 1993) and most northern parts of Germany, without records since 1980 (GELBRECHT, WEGNER 1998 pers. comm.).
Outside Europe recorded from Caucasus and Transcaucasus.

PHENOLOGY: Univoltine. Main flight period: mid-October to mid-December, rarely emerging in late September. In southern Europe late November to late January. Exceptionally single specimens in February and March from overwintering pupae (FORSTER & WOHLFAHRT 1981; BERGMANN 1955). Larval stages early May to mid-July. Overwintering as egg, exceptionally as pupa.

BIOLOGY: Larva oligophagous. Found on *Quercus robur, Q. petraea* (Germany: WEGNER, LEIPNITZ, GELBRECHT pers. comm.; OSTHELDER 1929; Poland: URBAHN 1939; Austria: SCHWINGENSCHUSS 1953) and on young shrubs of *Acer campestre* (south-western Germany: SCHNEIDER 1934; LEIPNITZ pers. comm.). Recorded also on *Fagus, Carpinus, Prunus* and many other deciduous trees (var. cit.). Preferred food-plant *Quercus*, 'probably on old trees' (DESCHKA & WIMMER 1996), but found also on young shrubs (LEIPNITZ pers. comm.). Caterpillars living between loosely spun leaves of the host-plant. Sometimes parasitized to 90% (SCHNEIDER 1934). Males nocturnal, readily attracted to light. Easily beaten from lower twigs of old oaks in the main flight season (GELBRECHT pers. comm.).

HABITAT: Clearly preferring sun-exposed areas inside oak forests, but, occasionally, also in gardens, tree-lined walks, forest fringes. In central Europe tends to be thermophilous. K-strategy. From 0 m up to 1,000 m above sea-level.

SIMILAR SPECIES: Females of *Alsophila aescularia* are larger, anal tuft of hairs narrower and darker, frons flat (see above).

Subfamily **Geometrinae** STEPHENS, 1829

About 2,350 described species worldwide, well established in the Neotropical region with 470 species (PITKIN 1996), Afrotropical region with nearly 570 species (HERBULOT 1992), Indo-Pacific with 584 species and Australia with 404 species (SCOBLE et al. 1995). Predominance of green pigment ('geoverdin') synapomorphic for subfamily (COOK et al. 1994).

Other features, e.g. shape of ansa of the tympanal organ support monophyly (see below). Presence of second anal vein (A3) in hindwing perhaps valuable as plesiomorphic character indicating phylogenetically basal groups within subfamily.

Systematic concept of HOLLOWAY (1996) accepts only two tribes, Dysphaniini and Geometrini, the latter including nearly all the Geometrinae species (> 2,300). Though there are strong arguments that the Dysphaniini are the sister group of all the other Geometrinae, the present volume follows the systematic concept of PITKIN (1996) in validating HOLLOWAY's subtribes at tribe level to obtain a more lucid and balanced grouping. At least some of these tribes show very characteristic homogeneous feature patterns. Important papers on the subfamily systematics worldwide: VIIDALEPP (1981), STEKOLNIKOV & KUZNETZOV (1981), FERGUSON (1985), HOLLOWAY (1996), PITKIN (1996), HAUSMANN (1996b).

DIAGNOSIS: Seven characteristic subfamily characters according to HOLLOWAY (1996): 'Green coloration, [tendency towards] reduction of the frenulum, paired setal patches on ♂ sternum A3, socii of ♂ genitalia well developed and often with parallel reduction of uncus, aedeagus with sclerotization reduced to a ventral strip along length, ♀ genitalia with oblique, papillate ovipositor lobes and bicornute signum'. Usually similar coloration on fore- and hindwings (PITKIN 1996). These 'ground plan' characters have some exceptions. Venation: Forewing: R1 and R2-R5 arising separately from cell and R2-R5 stalked, with a few exceptions. R1 usually free, rarely anastomosing with Sc or R2. Hindwing: Rs appressed or fused to Sc+R1 for a short distance only (exceptions e.g. many Microloxiini genera, *Xenochlorodes* and the neotropical genus *Pachycopsis*; see PITKIN 1996). M2 fully developed in both wings and arising from above middle of cell near origin of M1 (important differential feature from most Ennominae). Ansa of tympanal organ narrow at base, widening medially and then tapered towards apex in nearly all species occurring in the study area (compare COOK & SCOBLE 1992), with exceptions in the Neotropical region (PITKIN 1996). Antennae of ♂ usually bipectinate with the exception of only four genera in the study area: *Aplasta, Hemithea, Chlorissa, Phaiogramma*. Abdomen often with crests.

IMMATURE STAGES: Eggs usually lozenge-shaped (MINET & SCOBLE 1999). Larva of most species with head capsule deeply cleft dorsally. First thoracic segment T1 often with paired bifurcate projections and tergum of last segment (A10) extended to form anal process (see Text-figs 24, 25, 154). Surface of skin granular. Larval crochets arranged in two separate or nearly separate groups, especially those of anal prolegs (FERGUSON 1985). Description of chaetotaxy with subfamily characters in SINGH (1951). Pupae often speckled with blackish markings or with stripes. Cremaster furrowed radially (PATOCKA 1994).

REMARKS: The caterpillars feed mainly on trees and shrubs. They are cryptically coloured and when resting they resemble a rigid twig. The larvae usually pupate on the host-plant or on the ground among debris.

The pigment has been modified or lost secondarily in the evolution of a few genera only. As to the chemistry of this labile green pigment, which perhaps is a derivate of Chlorophyll, see COOK et al. (1994). It is very sensitive to light and acids (water). The chemistry of the yellow pigment in the Heliotheini is still unexamined, it is possible that it is, like that of the Asiatic Dysphaniini, just a derivate of the same molecule.

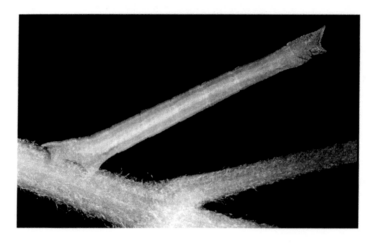

Text-fig. 154: Caterpillar of *Hemistola chrysoprasaria* (ESPER, 1795) with typical features for Geometrini: Note processes on head, segments T1 and A10 (photo NIPPE).

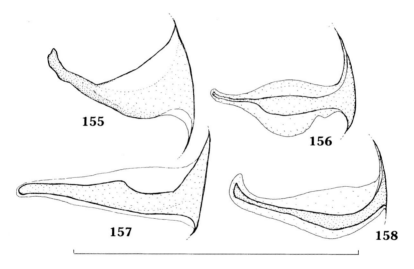

Text-figs 155-158: Ansa of tympanal organ in various tribes of the subfamily Geometrinae. Text-fig. 155: *Heliothea discoidaria* BOISDUVAL, 1840. Text-fig. 156: *Aplasta ononaria* (FUESSLY, 1783). Text-fig. 157: *Geometra papilionaria* (LINNAEUS, 1758). Text-fig. 158: *Thetidia smaragdaria* (FABRICIUS, 1787). Scale bar = 1 mm.

Tribus **Heliotheini** EXPOSITO, 1978

Described as subfamily, subfamily rank denied by VIVES MORENO (1994), MÜLLER (1996) and VIIDALEPP (1996). Tribe with one Palaearctic (*Heliothea*) and one Afrotropical genus (*Petovia* WALKER, 1854). Larval morphology (head and segment T1) indicating certain relationships to typical group of Geometrinae. Within Geometrinae tentatively better placed near Hemistolini, considering similar features in genitalia. Venation somewhat resembling that of genus *Aplasta*. Origin of M2 as in other Geometrinae. Tribe status of Heliotheini justified by shape of ansa in tympanal organ, being unique within Geometrinae (plesiomorphic?) and resembling that of Archiearinae and Desmobathrinae (Text-fig. 155), furthermore by yellow pigment of wing colour and some unusual larval features. The supposed isolated position in the phylogeny of the Geometrinae may be expressed by the placement of the tribe at the beginning of the subfamily. Interestingly, *Heliothea* and *Pseudoterpna* are hosts of two very closely related parasitoids of Braconidae (OLTRA et al. 1995).

DIAGNOSIS: Venation (Text-fig. 159) of forewing: Sc and R1 with short anastomosis. Hindwing: Sc+R1 and Rs closely approached but not fused for 1/2 length of cell. Frenulum lacking in both sexes. Tympanal organ with base of ansa very broad and tapered towards apex (Text-fig. 155).

IMMATURE STAGES: Larva cylindrical. Head small, dorsally cleft, slightly separated from segment T1, the latter bearing two short dorsal projections. Pupa similar to that of many other Geometrinae, with black dots and markings. Cremaster with 8 setae.

REMARKS: Adults diurnal.

Heliothea BOISDUVAL, 1840

TYPE-SPECIES: *Heliothea discoidaria* BOISDUVAL, 1840. Actual generic concept including central Asian *Apetovia* as subgenus (HAUSMANN 1996b).

DIAGNOSIS: Venation (Text-fig. 159) of forewing: Short fusion between R1 and R2-4 delimiting areole. R2 and R3 completely fused. R2-R5 and M1 usually connate, M2 arising from discocellular vein near M1. M3 and CuA1 separate. Hindwing: Rs and M1 shortly stalked. M3 and CuA1 separate. Only one anal vein (A2) present. Further features under diagnosis of tribe.

MALE GENITALIA: Socii present, diagnostic from *Petovia*.

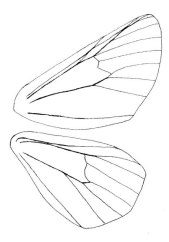

Text-fig. 159: Venation of *Heliothea discoidaria* BOISDUVAL, 1840.

Heliothea discoidaria

11. *Heliothea discoidaria* BOISDUVAL, 1840

Heliothea discoidaria BOISDUVAL, 1840: Genera Index Meth. eur. Lepid.: 178 (Spain, Andalusia). Syntypes 1♂2♀ (ZFMK). RAMBUR's manuscript name published by BOISDUVAL. Type series discussed in HERBULOT (1987).
Unavailable names (infrasubspecific): *albida* RIBBE, 1912.

DIAGNOSIS: Wingspan ♂♀ 20-27 mm. Wings deep yellow with black costa and black fringe. Hindwings often with stronger black suffusion. Cell spots on all wings black, round and strongly developed. Specimens from Sierra de Gredos (Castilia) with less black suffusion. Antennae totally black, bipectinate in both sexes, longest branches about four times width of flagellum in ♂, 1.5-2 times in ♀. Head, palpi and thorax with long black hair-like scales, similar to species of subfamily Archiearinae. Palpi of both sexes of medium size (length about 1.5 times diameter of eye). Proboscis reduced, length about 0.5 mm. Hindtibia of both sexes with two spurs. Hindtarsus not shortened. ♀ abdomen very large before oviposition.

MALE GENITALIA: Uncus short. Socii broad, length about ²/₃ uncus. Gnathos strongly developed. Juxta large with two basal lobes. Valva with strong costal projection and three ventral projections at ¹/₂. Aedeagus without cornuti, narrow basal part truncate at end. Sternum A8 simple, very broad.

FEMALE GENITALIA: Apophyses posteriores long, apophyses anteriores short and stout. Corpus bursae very long and slender, not differentiated from ductus bursae. Signum lacking.

DISTRIBUTION: West-Mediterranean. Mountains of central and southern Spain. Endangered due to local distribution, low density, strategy and diurnal activity. Habitats of this species merit protection.
    Outside Europe in Morocco (Middle Atlas).

PHENOLOGY: Univoltine. Main flight period: early June to mid-July, sometimes even in late May. Larval stages from July to May, overwintering. Larva slow growing.

BIOLOGY: Larva monophagous (3) on *Santolina* spp. (Compositae). Found on *Santolina rosmarinifolia* (GELBRECHT pers. comm.). Recorded on this plant also by CHRÉTIEN (1905) and GOMEZ DE AIZPURUA (1987), also on *S. chamaecyparissus* (SPULER 1904; RIBBE 1912; DOMÍNGUEZ & BAIXERAS 1995). Pupation in a silken cocoon on the ground near base of host-plant. Adults diurnal, usually not moving far from host-plants. Resting on flowers or twigs of larval food-plant with wings folded over abdomen as in Zygaenidae (tectiform). K-strategy.

PARASITOIDS: Hymenoptera, Braconidae: *Protapanteles santolinae* (OLTRA et al. 1995).

HABITAT: Montane species. On rocky slopes and scrub with host-plants. From 1,000 m up to 1,700 m above sea-level.

SIMILAR SPECIES: No similar species in Europe.

<h1 style="text-align:center">Tribus Pseudoterpnini WARREN, 1893</h1>
='Terpnini' sensu INOUE, 1961

Alternatively considered subtribe of Geometrini ('Pseudoterpniti') by HOLLOWAY (1996), compare taxonomic notes on subfamily. Distributed with many species and genera in Old World tropics, absent from Nearctic and Neotropical regions. Many features of *Holoterpna* and *Aplasta* anomalous within Pseudoterpnini, systematical position of *Aplasta* being particularily isolated, perhaps separated early from base of the genealogical tree.

DIAGNOSIS: Green colour reduced, 'mottled forewing ground with strong, crenulate fasciae' (HOLLOWAY, 1996). Venation (Text-fig. 50) of forewing: R1 arising from cell below apex. M3 and CuA1 unstalked. Hindwing: Rs + M1 connate or very shortly stalked (exception: *Aplasta*). M3 and CuA1 unstalked, often separate. Usually two anal veins (A2, A3) present. Frenulum of ♂ present (exception: *Holoterpna* and *Aplasta*). Hindtibia of both sexes usually with four spurs, rarely with two (*Holoterpna*). ♀ antennae filiform, with very short cilia. Robust-bodied. Abdominal crests developed (exception: *Holoterpna* and *Aplasta*). Setal patches on sternum A3 of ♂ usually present (absent in *Aplasta*). Tympanal organ as described for subfamily with base of ansa stalked, rather narrow (Text-fig. 156).

MALE GENITALIA: 'Uncus' deeply forked (synapomorphy of tribe). 'Socii' absent or vestigial (e.g. in *Pseudoterpna*), homology with socii of other Geometrinae doubtful. Gnathos well developed, often with teeth. Valvae usually with basal coremata. Costa of valva at base widely extended into inwardly directed sclerites. Aedeagus usually tubular, without stalk. Sternum A8 and tergum A8 simple.

FEMALE GENITALIA: Apophyses anteriores comparatively short. Lamella antevaginalis often band-shaped. Corpus bursae sac-shaped, usually without signum.

IMMATURE STAGES: Larva with typical characters of Geometrinae, i.e. head deeply cleft and paired dorsal projections on segment T1. Exceptions: *Aplasta* and *Holoterpna*. Pupal morphology with heterogeneous pattern and great differences between genera (PATOCKA 1994).

Aplasta HÜBNER, [1823]

TYPE-SPECIES: *Phalaena ononaria* FUESSLY, 1783. Genus with one species only. Isolated position within tribe supported by many features, such as larval morphology (see below), absence of ♂ frenulum, structure of antenna, number of hindtibial spurs etc.. Thus placed at the head of the tribe.

DIAGNOSIS: Venation: Forewing: Sc fused with R1, length of fusion very variable. R1 exceptionally re-anastomosing with R2-R4. R3 usually absent, with exceptions. R2-R5 and M1 connate or separate. Hindwing: Sc+R1 and Rs appressed for about half length of cell, sometimes even more. Rs and M1 distinctly stalked. M3 and CuA1 separate. One anal vein (A2) only. Frenulum in both sexes absent. Proboscis lacking, or sometimes vestigial. Antennae of ♂ with very thick flagellum and deep intersegmental incisions, shortly ciliate. Antennae of ♀ similar to that of ♂, somewhat less broad and segmental. Hindtibia (♂♀) with two pairs of spurs of nearly equal length. Setal patches on sternum A3 absent. Tympanal organ (Text-fig. 156) similar to that of other Geometrinae.

MALE GENITALIA: Valvae asymmetric, with small basal coremata. Aedeagus with heavily sclerotized basal 'stalk'.

FEMALE GENITALIA: Lamella antevaginalis poorly sclerotized.

IMMATURE STAGES: Larva (Text-fig. 160) setose, short and thick, Lycaenidae-like. Head small, flat, with shallow dorsal notch, but projections of segment T1 lacking. Abdominal legs absent on segments A3-A5. Pupa with warts.

Text-fig. 160: Caterpillar of *Aplasta ononaria* (FUESSLY, 1783) (photo: LEIPNITZ).

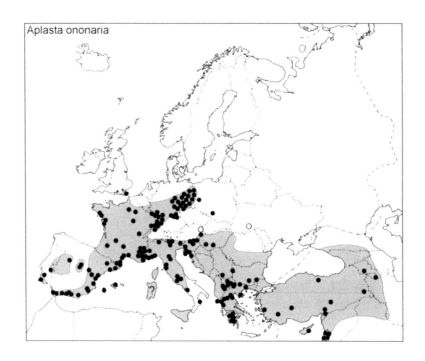

Aplasta ononaria

12. *Aplasta ononaria* (FUESSLY, 1783)

Phalaena (Geometra) ononaria FUESSLY, 1783: Arch. Insectengesch. 3: 1, pl. 17, figs 4-6 ([Europe]). Type(s) not traced.

Phalaena rubellata VILLERS, 1789: Linn. Ent. 2: 385 (southern France: 'Nemausum'). Type(s) (coll. Tissier, Lyon). Junior synonym.

Geometra sudataria HÜBNER, [1817]: Samml. Eur. Schmett. 5: pl. 95, fig. 492 ('Europe'). Type(s) lost. Junior synonym.

Aplasta ononaria ab.? (v.?) *berytaria* STAUDINGER, 1901: Cat. Lep. Palaearct. 1: 261 (Lebanon: Beirut). 5 syntypes (MNHU, examined). Junior synonym.

Geometra faecaturia HÜBNER, [1819]: Samml. Eur. Schmett. 5: pl. 97, fig. 503 ('Europe'). Type(s) lost. Junior synonym (seasonal form).

Aplasta faecataria HÜBNER, [1823]: Verz. Bek. Schmett.: 304. No separate types. Probably intentional change (correction of print error), unjustified emendation.

Aplasta ononata BELLIER, 1861: Ann. Soc. Ent. Fr. (3) 8 (4): 711. No separate types. Intentional change of final syllable because of filiform antennae. Unjustified emendation.

Aplasta ononaria spinosaria DANNEHL, 1926: Ent. Z. Frkf. 40 (17): 403 (Italia, South Tyrol: Etschtal). Lectotype ♀ (ZSM, examined). Junior synonym.

Unavailable names (misidentifications): *rubicapraria*: sensu HÜBNER (1793) nec [DENIS & SCHIFFERMÜLLER], 1775 (identity: *Theria rupicapraria*, Ennominae; misidentified and misspelled); *sciuraria*: sensu LERAUT (1997) nec '*sciurearia*' ESPER, [1806] (identity: *Pungeleria capreolaria*, Ennominae; misidentified and misspelled). - (nomen nudum): *rubraria*: SALAY (1910) - (infrasubspecific): *blanca*: RIBBE (1912); *monotonia*: STAUDER; *obscura*: PROUT; *rubraria*: PROUT (1912); *squamata*: STAUDER; *trifasciata*: STAUDER. - (incorrect subsequent spellings): *onoraria*: BERCE (1873); *falcaturia*: PROUT (1912); *falcataria*: FORSTER & WOHLFAHRT (1981).

DIAGNOSIS: Wingspan ♂ 20-28, ♀ 22-29 mm. Second generation in south-eastern Europe much smaller, often 15-18 mm only. Ground colour of wings ochreous, often with greyish and/or reddish tinge. Postmedial line and terminal area of forewing usually forming two broad reddish brown fasciae. Markings of hindwing less expressed. Rather variable in wing colour and wing pattern: Large, unicolorous greenish grey forms rare in Europe, commoner in the Levant ('f. *berytaria*'). Red-brown moths common in south-eastern Europe, mainly Greece ('f. *rubraria*'). In Spain usually with light grey tinge and slightly smaller, almost approaching subspecific value. Second generation in south-eastern Europe small, light ochreous ('f. *faecaturia*') with postmedial line sharper than in the first generation. Also in France and southern Alps second generation paler than first, but size and wing pattern similar. Frons rather flat, pale brown. Vertex ochreous, sometimes with greyish scales. Palpi brown, bushy-scaled. Length of palpi in ♂ 1.6-1.8 times diameter of eye, in south-eastern Europe and south-western Asia slightly longer, in ♀ 2.0-2.2 times diameter of eye. Last segment of palpi very slender, with short appressed scales, not elongate. Hindtarsus not shortened, hindtibia slightly dilated. Further features under generic diagnosis.

MALE GENITALIA: Spinose harpe present on right valva only. Considerable infrapopulational variability in size and shape of valva and harpe. Valvae of second-brood ♂ ♂ often more slender. Posterior margin of sternum A8 slightly notched medially. Tergum A8 simple.

FEMALE GENITALIA: Length of apophyses posteriores 3-4 times length of apophyses anteriores. Antrum sclerotized, shape rather variable, but usually dilating towards caudal end. Ductus bursae long and slender, Corpus bursae elongate. Signum elongate, smooth, with medial ridge, rather similar to that of Cosymbini, Timandrini and Rhodometrini (Sterrhinae).

DISTRIBUTION: Submediterranean. All over western and southern Europe, also on Balearics, Corsica, Sardinia and Sicily. As yet not recorded from Malta and Crete. In central Europe restricted to warm, southern localities and to the area between north-eastern France and north-western Poland (Pomerania). Very local in the southern Netherlands, the Channel islands and southern England (resident in western Kent only: REID and SKINNER 1998 pers. comm.).

Outside Europe in Turkey, Cyprus, Armenia, Caucasus, Syria, north-western Iran, northern Iraq and the Levant.

PHENOLOGY: Univoltine in central Europe and Great Britain, main flight period: early June to late July. Under warm conditions exceptionally a partial second brood in late July to early September (SCHNEIDER 1935). In eastern Germany second generation becoming more and more established in recent years (GELBRECHT pers. comm.). Obligatorily bivoltine in France, south-eastern Austria and southern Europe from mid-May to late July; mid-July to mid-September, first generation somewhat later in the mountains. Univoltine from mid-May to late July in mountainous habitats of central and southern Spain. Larva in univoltine populations from early July to late May, slow growing. In bivoltine populations from late June to late July; September to April. Overwintering as larva.

BIOLOGY: Larva monophagous (3). Found on *Ononis arvensis, O. spinosa* and *O. repens* (Great Britain: ALLAN 1949; PORTER 1997; SKINNER pers. comm.; Germany: ROMETSCH 1932; LEIPNITZ, GELBRECHT pers. comm.). Reputedly also on *Genista* (France: CHAPELON 1992) and on *Cytisus scoparius* ([France]: ALLAN 1949). Caterpillars bite through stems of food-plant. Thus, dried outer parts of food-plant clearly disclose presence of larvae in spring time. Pupation in a loose spinning at the base of the food-plant. Adults active by day and night. ♀ ratio at light high, usually near 50%, in Israel 30%.

HABITAT: In central Europe strongly xerothermophilous, in open habitats, mainly on lime-stone (BERGMANN 1955; KOCH 1984; KRAUS 1993) and basiphilous sandy soils (GELBRECHT pers. comm.). In Italy also reported as hygrophilous (ZELLER 1847b; DANNEHL 1927). Associated with Festuco-Brometea (CHAPELON 1992) and 'Artemisietea fragrantis anatolica' (KOCAK & SEVEN 1993). Expansion flights ('immigration') repeatedly recorded (LEMPKE 1949; REID, SKINNER pers. comm.). From 0 m up to 1,400 m above sea-level, in southern Spain more montane, up to 2,200 m.

SIMILAR SPECIES: No similar species in the area of distribution.

REMARKS: The occurrence of a possible second species in Turkey, Cyprus, Iraq and in the Levant has been discussed (LATTIN 1951). However, there is no constant correlation between structural details of the genitalia and flight season or habitus respectively (HAUS-MANN 1996b; 1997a). At least in Europe, small lightly coloured specimens always belong to subsequent generations of the same species.

Holoterpna PÜNGELER, 1900

TYPE-SPECIES: *Holoterpna diagrapharia* PÜNGELER, 1900. Genus with two species in the Palaearctic, one further species in South Africa. Erroneously synonymized with *Dyschloropsis* WARREN, 1895 (Thalerini) by VOJNITS (1976). Some features of *Holoterpna* reminiscent of Hemistolini (frenulum, venation, papillae anales in ♀ genitalia), but ♂ genitalia justify position in Pseudoterpnini. Certain unusual adult features and particular morphology of larva (see below) characterize *Holoterpna* as being quite isolated within phylogeny of Pseudoterpnini. Thus tentatively placed at the beginning.

DIAGNOSIS: Venation of forewing: R2-R5 distinctly stalked. R2-R5 and M1 connate (exceptionally on very short stalk). Hindwing: Sc+R1 and Rs appressed for short distance only. Rs and M1 shortly stalked, somewhat variable. M3 and CuA1 unstalked, connate (or exceptionally separate). Second anal vein (A3) present, short. Frenulum in both ♂ and ♀ absent. Proboscis lacking or extremely short (in the type-species). Antennae shortly bipectinate in ♂, filiform in ♀. Abdomen robust, abdominal crests absent. Hindtibia of both sexes with only one pair of terminal spurs. Male hindtarsus not shortened.

MALE GENITALIA: Uncus bifid. Valvae simple, basal coremata small.

FEMALE GENITALIA: With characteristic foldings between papillae anales.

IMMATURE STAGES: Larva colourful, yellow with red transverse ribbons. Projections on segments T1 and A10 absent (PROUT 1935a).

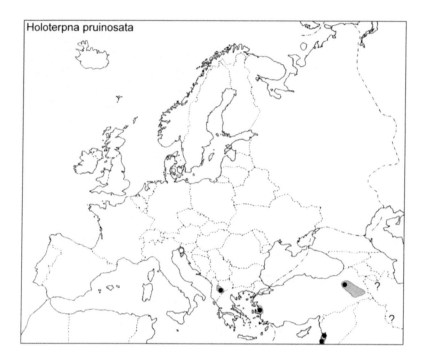

Holoterpna pruinosata

13. *Holoterpna pruinosata* (STAUDINGER, 1898)

Eucrostis pruinosata STAUDINGER, 1898: Dt. ent. Z. Iris 10: 303 (Israel: Jerusalem). Lectotype ♂ (MNHU, examined).
Further names (availability to be checked): *foulquieri* OBERTHÜR, 1910: Junior synonym (MÜLLER 1996).

DIAGNOSIS: Wingspan ♂ 22-26, ♀ 29-34 mm. Shape of wings diagnostic: Forewing costa straight, apex tapered, termen distinctly rounded, hindwing tornus rectangular. Ground colour pale green, in collections often faded to ochreous. Postmedial line hardly visible and indistinct, slightly paler than ground colour, no further markings. Frons flat, brown. Vertex white. Palpi white, small, in both sexes ca. 1.3 times diameter of eye. Last segment of palpi very slender, with short appressed scales, resembling that of *Pingasa*, but not elongate. Proboscis lacking. Antennae of ♂ shortly bipectinate, longest branches 1.5 times width of flagellum. Antennae of ♀ filiform, almost without cilia. Hindtibia of both sexes with two terminal spurs of nearly equal length. Length of tarsus somewhat exceeding length of tibia. Further features under generic diagnosis.

MALE GENITALIA: Valvae simple without any armature or projections. Basal coremata small. Aedeagus short, without cornuti.

FEMALE GENITALIA: Apophyses posteriores 3-4 times length of apophyses anteriores. Membrane between papillae anales characteristically folded. Ductus bursae short. Signum bursae lacking. Lamella antevaginalis only slightly sclerotized.

DISTRIBUTION: East-Mediterranean. No recent confirmation of old records from Istria around Trieste, perhaps extinct. Recorded also from Macedonia (♀ specimen in ZSM). In Europe highly endangered, perhaps near extinction.

Outside Europe in western and eastern Turkey, Levant and, reputedly, western Iran.

PHENOLOGY: Univoltine. Main flight period: mid-June to late August. In the Levant bivoltine (HAUSMANN 1997a) from early April to early June and early August to early November. Larval stages in August (PROUT 1935a). Overwintering as pupa, sometimes twice (Trieste: REBEL 1924).

BIOLOGY: Oligophagous (1) on Umbelliferae. Oviposition observed on *Ferulago galbanifera (=campestris)*, larva 'feeding on its flowers' (Trieste: REBEL 1924; PROUT 1935a). Recorded also on *Foeniculum* spp. (Israel: STAUDINGER 1898). ♀ ratio at light about 50%.

HABITAT: Open or bushy grassland, hot rocky slopes. From 0 m up to 700 m above sea-level.

SIMILAR SPECIES: Similar forms of *Pseudoterpna pruinata* (HUFNAGEL, 1767) differ in the presence of the proboscis, four hindtibial spurs in both sexes, presence of male frenulum etc. *Holoterpna diagrapharia* PÜNGELER, 1900 (allopatric) from Transcaspia, Turkmeniya and northern Iran is characterized by the white basal half of the hindwing, the frons is dirty green.

Pingasa MOORE, [1887]

TYPE-SPECIES: *Hypochroma ruginaria* GUENÉE, [1858] (northern India). About 45 species in the Old World. Five species only in the Palaearctic and its transitional zones towards Afrotropical and Indo-Pacific regions.

DIAGNOSIS: Large moths, wing pattern quite characteristic and rather uniform, with crenulate dark postmedial lines and reduced green colour. Hindwing with two transverse brushes of scales. Underside white, at base often yellow or orange, usually with two large black fasciae. Hindwing elongate, with long anal margin. Venation: Forewing: Sc and R1 free. R2-R5 stalked. R2-R5 and M1 connate. Hindwing: Sc+R1 and Rs appressed for short distance only. Both Rs/M1 and M3/CuA1 usually unstalked, sometimes on very short common stalk. Second anal vein (A3) present, short. Frenulum present in both sexes. Third segment of palpi elongate, 'naked' (JANSE 1935), i.e. slender, smooth, with short appressed scales. Antennae shortly bipectinate in ♂, filiform in ♀. Setal patches on ♂ sternum A3 strongly developed. Hindtibia of both sexes typically with four spurs.

MALE GENITALIA: Valvae with large coremata.

FEMALE GENITALIA: Ductus bursae short. Corpus bursae large.

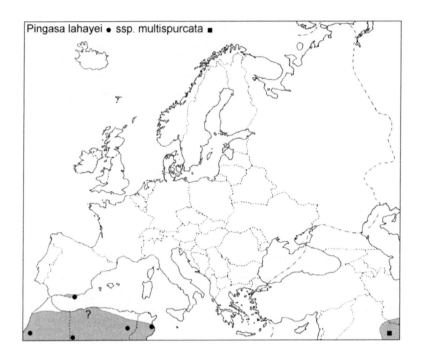

Pingasa lahayei ● ssp. multispurcata ■

14. *Pingasa lahayei* (OBERTHÜR, 1887)

Hypochroma lahayei OBERTHÜR, 1887: Bull. Soc. ent. Fr. 1887: 59 (Algeria, Oran: Ain-Sefra). Holotype ♂ (ZFMK).
Unavailable names (incorrect subsequent spellings): *lahayi*: PROUT (1930).

DIAGNOSIS: Wingspan ♂ 27-29, ♀ 35-36 mm. Ground colour light grey, terminal area somewhat darker. Ante- and postmedial lines very narrow, black and sharp, postmedial line crenulate. Cell spots of both wings elongated and narrow. Underside of wings snow-white, postmedial line with traces only, cell spots distinctly marked. Three dark marginal shades in forewing apex, hindwing tornus and around M2 of hindwing. Frons slightly convex, black, lower
third creamy white. Proboscis well developed, length 6-7 mm. Palpi (♂) 1.5 times diameter
of eye, 2.5 times in ♀, last segment very long (1.1 mm). Antennae bipectinate in ♂, with
longest branches 2.0-2.5 times width of flagellum. Antennae filiform in ♀. Hindtibia (♂ ♀)
with two pairs of spurs of unequal length. Further features under generic diagnosis.

MALE GENITALIA: Dorsal part of anellus ('superjuxta': INOUE 1961) strongly sclerotized.
Valvae tapered towards end, without projections. Aedeagus without cornuti.

FEMALE GENITALIA: Lamella antevaginalis transversely ribbed. Ductus bursae very short,
connection to corpus bursae broad. Corpus bursae large, without signa.

DISTRIBUTION: Saharo-Sindian. For Europe recorded only from southern Spain, Almeria
province as rare and very local, but probably resident (VALLHONRAT 1980a; KRAUS 1997;
1999).
    Outside Europe in North Africa, tropical Africa, e.g. Gambia, Nigeria, Zimbabwe,
Swaziland, etc. (subsp. *austrina* PROUT, 1917), in the east (subsp. *multispurcata* PROUT,

1913) from Yemen, central and southern Arabia across Iraq, Iran to Pakistan and north-western India. Absence of records from Egypt, Sudan, Ethiopia and Somalia suggests disjunct distribution area.

PHENOLOGY: Probably bivoltine. In Spain caught in early May and mid-August. In Morocco recorded from March to October with main flight periods in April and August (RUNGS 1981). Larval phenology unclear, overwintering as larva.

BIOLOGY: Larva oligophagous (2). Recorded on *Rhus tripartita* ('oxyacantha'; Anacardiaceae) and *Zizyphus lotus* (Rhamnaceae) (Algeria: PROUT 1934a). African subspecies found on *Sclerocarya birrea* (Anacardiaceae) (Swaziland: DUKE & DUKE 1998).

HABITAT: Hot open habitats, semideserts. From 0 m up to 1,000 m above sea-level. In Europe recorded from hot coastal lowlands only.

SIMILAR SPECIES: Pale grey forms of *Pseudoterpna coronillaria* can be easily distinguished by the more proximal position of the hindwing postmedial line and the greyish underside of wings (see Text-figs 161-162).

161   162

163   164

Text-figs 161-164: Diagnostic characters (indicated) of *Pingasa lahayei* (OBERTHÜR, 1887) (upperside of wings Text-fig. 161, underside 162) and *Pseudoterpna coronillaria* (HÜBNER, [1817]) (upperside of wings Text-fig. 163, underside 164).

Pseudoterpna HÜBNER, [1823]

TYPE-SPECIES: *Geometra cythisaria* [DENIS & SCHIFFERMÜLLER], 1775 (=*Pseudoterpna pruinata*). Small genus with six species in the western Palaearctic. Literature: LUTRAN (1989), NYST (1993), HAUSMANN (1996b; 1997b).

DIAGNOSIS: Venation (Text-fig. 50): Forewing without areole. R2-R5 and M1 usually connate. Hindwing: Sc+R1 and Rs appressed for about 1/3 length of cell. Rs and M1 unstalked. M3 and CuA1 unstalked. Second anal vein (A3) present, but more or less reduced. Frenulum present in ♂ but often weak, absent in ♀ (only some long hairs present). Proboscis well developed, length ca. 4-5 mm. Palpi of medium length, third segment short, with bushy scales. Antennae of ♂ bipectinate, branches terminally dilated. Flagellum dorsally with white scales. Antennae of ♀ filiform, ciliate. Abdominal crests usually present. Setal patches on ♂ sternum A3 developed. Hindtibia (♂ ♀) with two pairs of spurs of unequal length. ♂ tarsus somewhat shortened.

MALE GENITALIA: Socii vestigial, adherent to base of uncus. 'Superjuxta' (compare *Pingasa*) absent. Harpe and gnathos usually strongly spinulose.

FEMALE GENITALIA: Apophyses anteriores very short. Signum bursae lacking.

IMMATURE STAGES: Egg see Text-figs 105-106. Larva cylindrical, compact and thick. Skin granulose. Head and first segment (T1) with paired dorsal projections. Tergum of the last abdominal segment (A10) with tapered anal process.

REMARKS: The caterpillars feed exclusively on Papilionaceae. The behaviour of the larva is characterized by extreme immobility in the resting position, even when it is disturbed. Pupation in a loose spinning between leaves on the ground.

Text-figs 165-167: Diagnostic characters (indicated) of European *Pseudoterpna* species.
Text-fig. 165: *P. pruinata* (HUFNAGEL, 1767). Text-fig. 166: *P. coronillaria* (HÜBNER, [1817]).
Text-fig. 167: *P. corsicaria* (RAMBUR, 1833).

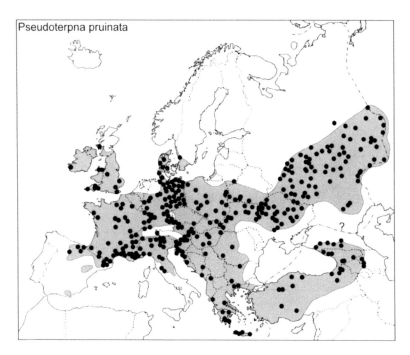

Pseudoterpna pruinata

15. *Pseudoterpna pruinata* (HUFNAGEL, 1767)

Phalaena pruinata HUFNAGEL, 1767: Berlin. Magazin 4 (5): 520 (Germany: Berlin region). Type(s) lost.

‡ *Phalaena prasinaria*: sensu FABRICIUS, 1775: Syst. Ent. 626, nec HUFNAGEL, 1767 (identity: *Geometra papilionaria* LINNAEUS, 1758). Misidentification (and homonym).

Geometra cythisaria [DENIS & SCHIFFERMÜLLER], 1775: Ank. syst. Werkes Schmett. Wienergegend: 97 (Austria: Vienna distr.). Type(s) lost. Junior synonym.

Phalaena pruinaria ROTTEMBURG, 1777: Naturf. 11: 72. No separate types. Intentional change of final syllable because of bipectinate antennae. Unjustified emendation.

‡ *Geometra thymiaria*: sensu THUNBERG, 1784: Ins. Suec. 1: 4, nec [DENIS & SCHIFFERMÜLLER], 1775 (identity: *Hemithea aestivaria* HÜBNER, 1789). Misidentification.

Phalaena (Geometra) genistaria VILLERS, 1789: Linn. ent. 2: 328 (Germany, Bavaria: by Regensburg). Type(s) not traced. Junior synonym.

Hemithea agrestaria DUPONCHEL, 1829: Hist. Nat. Lép. 7 (2): 257, pl. 152, fig. 4 (France: Montpeiller). Type(s) not traced. Junior synonym.

‡ *Hemithea porracearia* BOISDUVAL, 1840: Genera Index Meth. eur. Lepid.: 180. Nomen nudum, without description.

Phalaena (Geometra) viridisparsata ROQUETTE, 1857: Allg. dt. naturh. Ztg (N.S.) 3: 305 (type locality not stated). Type(s) not traced. Junior synonym.

Aspilates atropunctaria WALKER, 1863: List. Specimens lepid. Insects Colln Br. Mus. 26: 1673. ('East Florida' in the original description and on the label of type specimen; mislabelled [England]?). Holotype ♂ (BMNH, examined by M. PARSONS pers. comm.).

Pseudoterpna pruinata ab. *grisescens* REUTTI, 1898: Lep. Baden (ed. 2): 110 (south-western Germany: Gengenbach nr. Karlsruhe). Type(s) not traced. Not explicitly introduced at infrasubspecific category.

Pseudoterpna pruinata var. *virellata* KRULIKOVSKI, 1908: Soc. Ent. 23 (2): 11 ([Volga, central Russia]: Wiatka, Kasan). 3 syntypes (Museum Kiew, one ♂ examined). Junior synonym.

‡ *Pseudoterpna pruinata* f. *nigrolineata* SCHWINGENSCHUSS, 1918: Verh. zool.-bot. Ges. Wien 68 (6-8): 151 (northern Germany: Braunschweig). Explicitly introduced at infrasubspecific rank.

Pseudoterpna pruinata f. *candidata* STAUDER, 1920: Int. ent. Zt. Frkft. 14 (5): 36 (Italy: Trieste). Syntypes 2 ♀ (BMNH). Junior synonym.

Pseudoterpna pruinata var. *holsatica* WAGNER, 1922: Int. ent. Z. Guben 16 (1): 39 (northern Germany: Schleswig-Holstein, Hannover). Type(s) in NHMW. Junior synonym.

‡ *Pseudoterpna pruinata* ab. *syltica* PROUT, 1934: SEITZ Macrolep. Suppl. 4: 7. Introduced as junior synonym (manuscript name).

Further names (availability to be checked): *illibaria* SIEBOLD, 1841: Preuss. Prov.-Bl. 25: 431
Unavailable names (misidentification): *simplex*: sensu ZUKOWSKY (1937) nec ALPHERAKY, 1892 (identity: *Pseudoterpna simplex*). - (infrasubspecific): *albescens*: SCHWINGENSCHUSS; *albida*: KOLOSSOW (1936); *albolineata*: F. WAGNER (1922); *approximata*: LEMPKE (1949); *aurata*: BRETSCHNEIDER; *bilineata*: LEMPKE (1949); *cotangens*: COCKAYNE (1952); *extrema*: LEMPKE (1949); *fasciata*: PROUT (1912); *fuscomarginata*: LEMPKE (1949); *grisescens*: HANNEMANN (furthermore homonym to *grisescens* REUTTI, see above); *mixta*: LEMPKE (1949); *pallida*: ROCCI; *reducta*: LEMPKE (1949); *tangens*: LEMPKE (1949); *unilineata*: LEMPKE (1936); *viridimelaina*: HEYDEMANN (1938); *viridisquama*: HEYDEMANN. - (incorrect subsequent spellings): *agrestaria*: BURROWS (1940); *cithysaria*: ESPER (1803); *cytisaria*: HÜBNER ([1823]); *cythisiaria*: JUNG (1782); *cytisiaria*: TISCH.; *nigrilineata*: PROUT (1934a; pl.); *porracea*: WALKER (1861); *priunata*: SALAY (1910); *pruniata*: SAUER (1982); *viridisquamosa*: HEYDEMANN (1938); *viridissima*: PROUT (1912).

DIAGNOSIS: Wingspan ♂♀ 25-34 mm, in eastern Europe towards Urals larger, 32-35 mm. Ground colour pale green, in the north of the distribution area often with a faint blue tinge or greyish. Green colour quickly fading. Wing pattern rather variable, transverse lines and cell spot often weak or absent, mainly in southern and eastern Europe ('f. *agrestaria*', 'f. *virellata*'). Postmedial and antemedial lines darker green. Antemedial line usually strongly angled at CuA2 and M1. Wavy line pale grey or white. Fringe green at base, distally pale grey or white. In northern Germany, Denmark and Great Britain usually darker, with ante- and postmedial lines dark grey ('f. *atropunctaria*'). Frons convex, black. Vertex white. ♂♀ palpi white, but upperside with many black scales, length 1.5 times diameter of eye. Antennae bipectinate in ♂, length of branches ca. 2-2.5 times width of flagellum. Hindtarsus of ♂ ca. 3/4 length of hindtibia. Abdominal crests absent or very small and concolorous. Setal patches on sternum A3 (♂) paired, divided from each other. Further features under generic diagnosis.

MALE GENITALIA (see also Text-figs 171-178): Harpe short, length ca. 1.5-2 times breadth. Aedeagus short, ca. 2.0-2.2 mm. Distal cornutus long, narrow and pointed, often with small lateral or terminal teeth. Central cornutus terminally broad and truncate, laterally dentate at one side.

FEMALE GENITALIA: Lamella antevaginalis poorly sclerotized. Antrum rectangular, with deep caudal excision. Ductus bursae folded, sclerotized. Corpus bursae usually long and slender.

DISTRIBUTION: European-Westasiatic. Western Europe, from Ireland to the Pyrenees, central Europe except for northern Alps. Southwards to northern Italy, local populations in the Apennine Mountains, Balkan countries including Peloponnes and Crete. Not recorded on the other Greek Islands. In the north up to Denmark and southern Sweden, in the east across Lithuania, Byelorussia, and the Volga Plain to southern Urals. Records for central and southern Spain require confirmation and taxonomic study. In Sweden and northern Germany (WEIGT 1984; RETZLAFF et al. 1993) abundance drastically decreasing, endangered due to habitat destruction.

Outside Europe in Turkey, Caucasus and Transcaucasus. In the east to south-western Siberia.

PHENOLOGY: Usually univoltine, main flight period: late June to late July. Occasional specimens from late May until August. Under warm conditions exceptional 2nd-brood specimens in early August to late September. In southern Europe, except northern Spain, bivoltine late May to early July and late July to late September, exceptionally until mid-October. In Crete trivoltine mid-April to late May; early July to late July; late October to late November. Larval stages in central Europe late July to early June, in bivoltine populations mid-June to late July; mid-August to mid-May. Overwintering as small larva.

BIOLOGY: Larva oligophagous (1) on Papilionaceae. Found on *Cytisus (Sarothamnus) scoparius, Genista pilosa, G. anglica* and *Ulex europaeus* (Great Britain: PORTER 1997; SKINNER pers. comm.; Denmark: SKOU pers. comm.; Poland: BUSZKO pers. comm.; Germany: OSTHELDER 1929; KRAUS 1993; WEGNER, LEIPNITZ pers. comm.), on *Chamaecytisus ruthenicus* (European Russia: ANIKIN et al. 2000) and other 'Cytisus spp.' (Germany: LEIPNITZ pers. comm.). Recorded also on *Genista germanica* and *G. tinctoria* (var. cit.), reputedly also on *Coronilla* (CULOT 1917). Reared on *Laburnum anagyroides* (var. cit.). Egg positioning on the food-plant or on grass nearby (BURROWS 1940b). Adult moth nocturnal, but easily flushed up during daytime. ♀ ratio at light usually 30-50%.

PARASITOIDS: Hymenoptera, Braconidae: *Apanteles immunis, A. spurius, A. triangulator* - Hymenoptera, Ichneumonidae: *Canidiella tristis, Mesochorus brevipetiolatus, Ophion obscurus* (THOMPSON 1946).

HABITAT: Xerothermic heathlands and steppes with shrubs of the food-plant, pioneer vegetation, often on dams, sometimes forest edges, but usually in open habitats (WEGNER pers. comm.; BERGMANN 1955; FORSTER & WOHLFAHRT 1981; DESCHKA & WIMMER 1996). In northern Europe also on wet soils (SKOU 1986). Usually from 0 m up to 500 m above sea-level, sometimes 800 m, in the southern Alps, Bulgaria and northern Greece up to 1,500-1,800 m (FORSTER & WOHLFAHRT 1981; NESTOROVA 1998; SKOU, GOATER pers. comm.).

SIMILAR SPECIES: *Holoterpna pruinosata* differs in lacking medial hindspurs (♂♀), proboscis and ♂ frenulum. *Pseudoterpna simplex* ALPHÉRAKY, 1892 (allopatric) from central Asia can easily be distinguished by its greenish ochreous frons. *P. coronillaria* is rather similar to grey forms of *P. pruinata* from northern Germany ('f. *atropunctaria*'). The latter forms usually have straighter ante- and postmedial lines.

REMARKS: Some authors consider the populations from northern Germany and Great Britain as a separate subspecies ('atropunctaria WALKER, [1863]') but in these populations nominotypical forms are mixed in and no clear geographical separation can be found from the central European populations. Furthermore the type locality of the nominate subspecies (Berlin) is situated quite close to the transitional zone between the dark atlantic populations and the paler continental (and southern European) ones. Similarly the taxon 'virellata KRULIKOVSKI, 1908' is retained as valid at subspecies rank by some authors (PROUT 1912a; VIIDALEPP 1996). There are, however, clinous transitions to the features of the central European populations and no constant differences between 'virellata' and the pale southern European forms ('agrestaria').

Conspecificity of *Pseudoterpna coronillaria* and *P. pruinata* is suggested by NYST (1993) despite the above mentioned differential features. Furthermore there are differences in the larva, which in *P. pruinata* is characterized by a broader rosy lateral fascia and more strongly developed granulation of the skin (SPULER 1903; MILLIÈRE 1867; LEIPNITZ pers. comm.). Both *P. coronillaria* and *P. pruinata* occur sympatrically in various regions such as the Pyrenees, southern and western France, north-western and central Italy. In southern France and north-western Italy the former species is well separated by the grey wing colour and in the male genitalia by the longer harpe and the narrower basal cornutus. In the Pyrenees, north-western France, central Italy and Turkey, however, specimens exceptionally show intermediate features, probably due to hybridization (see chapter 5. and OBERTHÜR 1896; CULOT 1917). Hybrid-like specimens often have intermediate genitalia.

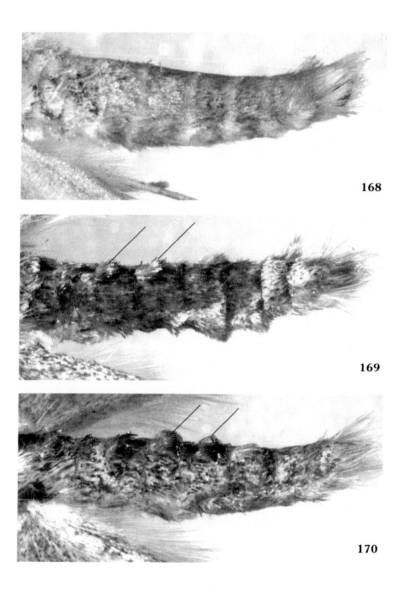

Text-figs 168-170: Abdomina of *Pseudoterpna* species. Text-fig. 168: *P. pruinata* (HUFNAGEL, 1767). Text-fig. 169: *P. coronillaria* (HÜBNER, [1817]). Text-fig. 170: *P. corsicaria* (RAMBUR, 1833).

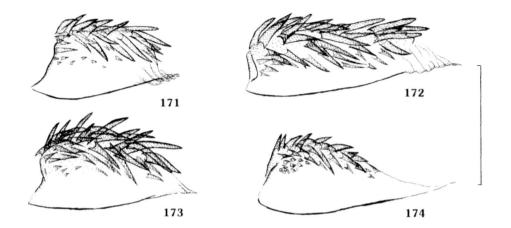

Text-figs 171-174: Differential features in ♂ genitalia of *Pseudoterpna* species: Shape of harpe (scale bar = 0,5 mm). Text-fig. 171: *P. pruinata* (HUFNAGEL, 1767). Text-fig. 172: *P. coronillaria* (HÜBNER, [1817]). Text-fig. 173: *P. coronillaria flamignii* HAUSMANN, 1997. Text-fig. 174: *P. corsicaria* (RAMBUR, 1833).

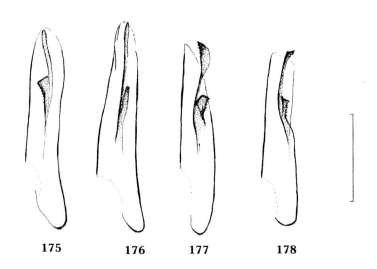

Text-figs 175-178: Differential features in ♂ genitalia of *Pseudoterpna* species: Shape of aedeagus and cornuti (scale bar = 1 mm). Text-fig. 175: *P. pruinata* (HUFNAGEL, 1767). Text-fig. 176: *P. coronillaria* (HÜBNER, [1817]). Text-fig. 177: *P. coronillaria flamignii* HAUSMANN, 1997. Text-fig. 178: *P. corsicaria* (RAMBUR, 1833).

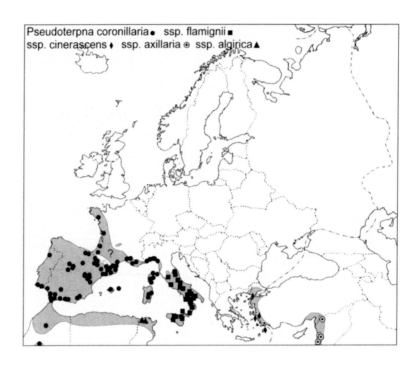

Pseudoterpna coronillaria● ssp. flamignii■
ssp. cinerascens◆ ssp. axillaria⊚ ssp. algirica▲

16. *Pseudoterpna coronillaria* (HÜBNER, [1817])

Geometra coronillaria HÜBNER, [1817]: Samml. Eur. Schmett. 5: pl. 93, fig. 479-482 ('Europe'; [France or Spain], *see* HAUSMANN 1997b). Type(s) lost.
‡ *Pseudoterpna coronillata* BELLIER, 1861: Ann. Soc. Ent. Fr. (3) 8 (4): 701. Incorrect subsequent spelling.
‡ *Pseudoterpna cytisaria*: sensu MILLIÈRE, 1867: Iconogr. Chen. Descr. Lép. inédits 2: pl. 91, fig. 10, nec [DENIS & SCHIFFERMÜLLER], 1775 (identity: *Pseudoterpna pruinata*). Misidentification and incorrect subsequent spelling.
Unavailable names (infrasubspecific): *armoraciaria*: OBERTHÜR (1896). - (incorrect subsequent spellings): *armoracearia*: SPULER (1910); *armoricana*: OBERTHÜR (1925); *armoricaria*: BUBACEK (1926); *coronilleria*: REBEL (1939)

DIAGNOSIS: Wingspan ♂♀ 24-33 mm, second generation smaller than first. Ground colour light grey, sometimes with strong dark suffusion all over ('f. *armoraciaria*'), mainly in western France and Portugal. Cell spot, postmedial and antemedial lines sharp, black or dark brown. Antemedial line rather straight, postmedial line strongly crenulate. In the populations along the Atlantic coast postmedial line of forewing distally with distinct brown shades. Wavy line whitish, undulated. Fringe grey at base, distal half pale grey or white. Frons slightly convex, black. Vertex whitish. Palpi whitish in both sexes, upperside black, length 1.5 times diameter of eye. Antennae bipectinate in ♂, longest branches ca. 2-2.5 times width of flagellum. Abdominal crests usually well developed on segments 1, 2 and 3, black with whitish tip. Paired setal patches of ♂ sternum A3 strongly developed, almost fused into one. Length of ♂ hindtarsus 2/3 length of hindtibia, tibia somewhat dilated. Further features under generic diagnosis.

MALE GENITALIA (see also Text-figs 171-178): Gnathos sometimes with long terminal spines, mainly in specimens from Spain and North Africa. Harpe elongate, length 2.5-3.5 times breadth. Length of aedeagus 2.3-2.7 mm. Two cornuti, both narrow and pointed, distal one rather long and without teeth, central one short, smooth, rarely with small teeth. Populations of north-western France with some features of ♂ genitalia (length of aedeagus and shape of harpe) becoming intermediate towards *Pseudoterpna pruinata*, but shape of cornuti matching that of *P. coronillaria*. Populations of Elba, Tuscany (Italy), Corsica and Sardinia with typical genitalia of the nominate subspecies. North African populations also corresponding well.

FEMALE GENITALIA: Lamella antevaginalis smooth, developed as elongated transverse ribbon. Antrum sclerotized with U-shaped caudal excision. Ductus bursae towards corpus bursae sclerotized and folded, obliquely 'sitting' on the latter. Corpus bursae usually broader and more rounded than in *P. pruinata*.

DISTRIBUTION: Species (Holo-)Mediterranean. Nominotypical subspecies West-Mediterranean: Portugal, Spain, Pyrenees, southern and western France (near coasts), north-western Italy (Liguria, western Tuscany), Elba, Corsica and Sardinia. Subsp. *flamignii* HAUSMANN, 1997 Adriato-Mediterranean: Sicily, central and southern Italy (see below). Subsp. *cinerascens* (ZELLER, 1847) East-Mediterranean: European Turkey (see below).

Outside Europe in North Africa (subsp. *algirica* WEHRLI, 1929), western Turkey (subsp. *cinerascens*), southern Turkey and Lebanon (subsp. *axillaria* GUENÉE, [1858]), Israel and northern Jordan (subsp. *halperini* HAUSMANN, 1996). Replaced by a sister species in Cyprus (*Pseudoterpna rectistrigaria* WILTSHIRE, 1948).

PHENOLOGY: Bivoltine. Main flight period: mid-May to mid-July; early August to early October. In southern Portugal emerging from mid-April (CORLEY pers. comm.). Larval stages September to early May; late June to early August. Overwintering as larva.

BIOLOGY: Larva oligophagous (1) on Papilionaceae: Found on *Spartium junceum* (Spain: LEIPNITZ pers. comm.). Recorded also on *Cytisus purgans* (MILLIÈRE 1867), *Adenocarpus complicatus* (Spain: RIBBE 1912 as 'intermedius'), *Lygos* (*Retama*) (Spain: GOMEZ DE AIZPURUA 1989), *Ulex, Genista* and *Cytisus* (GOMEZ DE AIZPURUA 1989; CULOT 1917) and *Calycotome villosa* (Corsica: MILLIÈRE, 1867; SPULER 1903; RIBBE 1912 as 'Cytisus lanigerus'). Reared from *Cytisus scoparius* and *Genista tinctoria* (LEIPNITZ pers. comm.). ♀ ratio at light low, 10-20%.

HABITAT: Mediterranean macchia, rocky slopes and forest fringes with shrubs of food-plant. Common in cork oak forest (*Quercus suber*) and pine forest (*Pinus pinaster, P. pinea*) with shrubby understorey (Portugal: CORLEY pers. comm.). From 0 m up to 1,000 m above sea-level, in Spain up to about 1,700 m, in Morocco even up to 2,600 m.

SIMILAR SPECIES: *Pingasa lahayei* can easily be distinguished by the white colour of the wing underside (Text-figs 161-164). Grey forms of *Pseudoterpna pruinata* differ in weakly developed abdominal crests, less crenulate ante- and postmedial lines, and in ♂ genitalia in the shape of cornuti and harpe. *P. lesuraria* LUCAS, 1933 from Morocco is more uniform coloured pale grey, genitalia strongly different.

REMARKS: Conspecificity of *P. coronillaria* and *P. pruinata* is suggested by NYST (1993) despite the above mentioned differential features. There are further differences in the larva, which in *P. coronillaria* has a weaker rosy lateral fascia and a less granulated skin (SPULER 1903; MILLIÈRE 1867; LEIPNITZ pers. comm.). In some regions natural hybridization is probable between both species (see remarks to *P. pruinata*).

Pseudoterpna coronillaria flamignii HAUSMANN, 1997

Pseudoterpna coronillaria flamignii HAUSMANN, 1997: Nachr. entomol. Ver. Apollo, N.F. 18 (2/3): 223 (Italy, Abruzzi: Montagna Grande). Holotype ♂ (ZSM, examined).
Unavailable names (misidentification): *cinerascens*: sensu HAUSMANN (1996) nec ZELLER, 1847 (identity: *P. c. cinerascens*, see below).

DIAGNOSIS: Wingspan ♂ ♀ 25-37 mm, second generation smaller than first. Ground colour light grey, in specimens from southern Italy and Sicily terminal area darker. In populations from Lazio (central Italy) often uniform dark grey. Wing pattern as described for nominate subspecies. Venation of forewing: R2-R5 and M1 sometimes shortly stalked, whilst never stalked in nominate subspecies. Abdominal crests usually present, but small and greyish. Further features under nominate subspecies.

MALE GENITALIA (see also Text-figs 171-178): Gnathos shortly spinulous. Harpe short, dorsally with long spines. Length of harpe 1.5-2 times breadth. Length of aedeagus 2.0-2.4 mm, with two short, broad and deeply dentate cornuti (Text-fig. 177).

FEMALE GENITALIA: Very similar to that of nominate subspecies, but antrum smaller.

DISTRIBUTION: Subspecies Adriato-Mediterranean: Central and southern Italy, Sicily. In the north up to provinces Umbria and Marche. Geographically isolated from populations of nominate subspecies in north-western Italy.

PHENOLOGY: Bivoltine. Main flight period: mid-May to mid-July, mid-August to late September, in southern Italy from late April, in Apennine mountains over 800 m mid-June to mid-September, here perhaps univoltine (PARENZAN 1994; ZAHM pers. comm.). Phenology of larval stages probably as in nominate subspecies. Overwintering as larva.

BIOLOGY: Adult larva found under *Genista* (southern Italy: HAUSMANN, no feeding observation). Probably also on *Cytisus scoparius* and *Genista aetnensis* (see under 'habitat'). ♀ ratio at light low, usually about 20-30%.

HABITAT: Mediterranean macchia near coasts and on hills. In the mountains sometimes abundant on rocky slopes with host-plants or near forest fringes with *Cytisus scoparius* (Sila Mts., southern Italy). In western Sicily on lava-slopes of Mt. Etna with endemic *Genista aetnensis* as only available food-plant. Usually from 0 m up to 1,300 m, sometimes up to 2,000 m above sea-level (Sila Mts., Mt. Etna).

SIMILAR SPECIES: No similar species in the area of distribution. *Pseudoterpna coronillaria flamignii* can easily be distinguished from the other subspecies by the shape of cornuti and harpe in the ♂ genitalia. These genitalia features match better the equivalents in *P. pruinata*. The external appearance of the populations from

Sicily and southern Italy is transitional to the North African subspecies, genitalically however they match well the populations from central Italy. Compare discussion in HAUSMANN (1997b).

REMARKS: With regard to the strong differences in ♂ genitalia, *flamignii* could be considered a separate species from *Pseudoterpna coronillaria*. The distribution borders and possible contact zones should be examined further (see remarks to *P. pruinata*). Breeding experiments might also be helpful.

Pseudoterpna coronillaria cinerascens (ZELLER, 1847)

Geometra cytisaria var. *cinerascens* ZELLER, 1847: Isis 1847: 18 (western Turkey: Mermeriza). Holotype ♂ not traced.
Unavailable names (incorrect subsequent spellings): *cinarescens*: KOCH (1854).

DIAGNOSIS: Wingspan ♂ 29, ♀ 32 mm. Forewings slender, apex pointed. Ground colour grey, transverse lines and cell spot black. Antemedial line of forewing and postmedial line of hindwing indistinct. Postmedial line of forewing sharp, slightly dentate. Wavy line missing or indistinct on all wings. Length of palpi (♂♀) ca. 1.8 times diameter of eye. Antennae bipectinate in ♂, longest branches ca. 2.5 times width of flagellum. Length of ♂ hindtarsus ca. 2/3 length of hindtibia, tibia comparatively slender. Abdominal crests developed, dark grey. Further features under nominate subspecies.

MALE GENITALIA: Gnathos shortly spinulous. Shape of harpe variable, length 1.7-2.6 times breadth. Harpe shorter and sub-triangular in subsp. *axillaria* (GUENÉE, [1858]) from southern Turkey and Lebanon. Distal cornutus long and slender as in nominate subspecies, central cornutus with short teeth. Length of aedeagus 2.2-2.5 mm.

FEMALE GENITALIA: Unknown. Corpus bursae in subsp. *axillaria* more elongate than in the other subspecies, tapered towards oral end. Apophyses posteriores slightly shorter than in European subspecies.

DISTRIBUTION: Subspecies East-Mediterranean. European Turkey, Samos, Rhodos. Very rare and probably near extinction.
   Outside Europe in western Turkey.

PHENOLOGY: Early May to mid-June (scarce data). Subsp. *halperini* HAUSMANN, 1996 (Israel) from March to December (HAUSMANN 1997a). Larval stages unknown.

BIOLOGY: Larva of subsp. *halperini* probably living on *Gonocytisus pterocladus* and *Genista fasselata* (HAUSMANN 1997a).

HABITAT: Unknown. From 0 m up to 100 m above sea-level (limited data).

SIMILAR SPECIES: The Turkish and south-eastern European populations of *Pseudoterpna pruinata* differ in the greenish coloration, and in the much more diffuse and not dentate postmedial line. *P. rectistrigaria* WILTSHIRE, 1848 from Cyprus (allopatric) is characterized by the straight postmedial line of the forewing. *P. coronillaria axillaria* (GUENÉE, [1858]) from Lebanon and southern Turkey (allopatric) is smaller, postmedial line very fine, forewing apex rounded. The slender hindtibia is characteristic for both subspecies, however the male tarsus of *axillaria* is less shortened than in *cinerascens*.

REMARKS: In habitus and genitalia *P. coronillaria cinerascens* stands closer to *P. pruinata* than the other subspecies of *P. coronillaria* do. Occasional hybridization seems possible. One ♂ from western Turkey in coll. Lederer (MNHU) shows both intermediate wing pattern and intermediate genitalia.

179

180

Text-figs 179-180: Frons of *Pseudoterpna* species. Text-fig. 179: *P. coronillaria* (HÜBNER, [1817]). Text-fig. 180: *P. corsicaria* (RAMBUR, 1833).

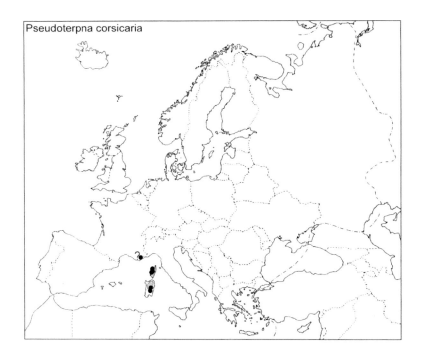

Pseudoterpna corsicaria

17. *Pseudoterpna corsicaria* (RAMBUR, 1833)

Hemithea corsicaria RAMBUR, 1833: Annls Soc. ent. Fr. 2 (1): 32, pl. 2, fig. 6 (Corsica). Lectotype ♂ (coll. HERBULOT in ZSM, examined).
Pseudoterpna corsicaria var. *ramburaria* OBERTHÜR, 1916: Lép. Comp. 12: 75, pl. 382, figs 3205, 3206 (Corsica). Syntypes ♂♀ ('coll. BELLIER') not traced. Not explicitly introduced at infrasubspecific rank.
Unavailable names (infrasubspecific): *alba*: BUBACEK (1926); *obscura*: BUBACEK (1926). - (incorrect subsequent spellings): *rambouraria*: BYTINSKI-SALZ (1934).

DIAGNOSIS: Wingspan ♂♀ 24-30 mm. Forewings comparatively slender, ground colour white or grey, sometimes with brownish tinge. Dark grey forms commoner in females, white forms commoner in males. Ante- and postmedial lines well marked, broad, dark brown, strongly dentate. Cell spot of forewing distinctly marked, sometimes divided into two dots. Frons slightly convex, ochreous, often mixed up with dark grey scales. Palpi whitish, upperside and last segment black, length 1.5 times diameter of eye. Antennae bipectinate in ♂, longest branches 3-4 times width of flagellum. Abdominal crests usually well developed, dark grey. Setal patches of ♂ sternum A3 strongly developed, almost fused into one. Length of ♂ hindtarsus ca. 2/3 length of hindtibia. Further features under generic diagnosis.

MALE GENITALIA (see also Text-figs 171-178): Harpe triangular, ventral margin strongly elongated towards tip of valva, with a few short spines only. Length of aedeagus 1.8-2.0 mm only. Distal cornutus broad, laterally with small teeth, terminally curved. Central cornutus resembling that of *Pseudoterpna pruinata*.

FEMALE GENITALIA: Antrum rectangular, without deep dorsal excision. Sclerotization of oral part of ductus bursae less extended than in *Pseudoterpna coronillaria*. Corpus bursae comparatively small, globular.

133

DISTRIBUTION: Endemic to mountains of Corsica and Sardinia, absent from coastal plains. Occurrence in southern France (1♂ in ZSM) requires confirmation.

PHENOLOGY: Univoltine. Main flight period: early June to late July, exceptionally late May or early August. Bivoltine phenology erroneously stated by PROUT (1912a). Probably over-wintering as larva.

BIOLOGY: Larva oligophagous (1) on Papilionaceae: Recorded on *Genista corsica* (MILLIÈRE 1867; PROUT 1912a; CULOT 1917) and *Calycotome villosa* (SPULER 1903 as 'Cytisus lanigerus'). ♀ ratio at light about 30%.

HABITAT: Mountainous slopes with host-plants. Mainly from 700 m up to 1,700 m above sea-level, sometimes down to 500 m.

SIMILAR SPECIES: *Pseudoterpna corsicaria* and *P. coronillaria* occur sympatrically in Corsica and Sardinia, the latter is easily distinguished by its black frons (compare LUTRAN 1984; Text-figs. 179-180). The larva bears a typical white triangular dorsal pattern (PROUT 1912a).

Tribus **Geometrini** STEPHENS, 1829

Several genera in the Palaearctic (mainly eastern part) and Indo-Pacific regions.

DIAGNOSIS: Forewing apex often falcate. Venation similar to that of Pseudoterpnini: Rs and M1 of hindwing separate. Two hindwing anal veins (A2 and A3) present. M3 and CuA1 separate in both wings. Frenulum present in ♂, absent in ♀. As in most species of Pseudoterpnini last segment of palpus slender ('naked') and long, especially in ♀. Antennae bipectinate to tip in ♂, filiform with very short cilia in ♀. Hindtibia of both sexes with four spurs, in ♂ often with pencil. Head and thorax bushy haired. Abdominal crests absent. Ansa of tympanal organ (Text-fig. 157) tapered at apex (*Geometra papilionaria*) or - in some Asiatic species - widened into hammer-headed plate. Sometimes with big globular lacinia.

MALE GENITALIA: Uncus often reduced, but present in some Asiatic genera. Gnathos tapered, strongly sclerotized. Socii greatly elongate, slender, strongly sclerotized and widely spread.

FEMALE GENITALIA: Without oblique ovipositor lobes (HOLLOWAY 1996), in contrast to other tribes of subfamily.

IMMATURE STAGES: Larval morphology quite different from that typical for many other Geometrinae (under generic diagnosis).

134

Geometra LINNAEUS, 1758

=Hipparchus LEACH, [1815]

TYPE-SPECIES: *Phalaena papilionaria* (LINNAEUS, 1758). 16 species in the Palaearctic and the northern parts of Indo-Pacific. Genus name validated by Opinion 450 and Official List of Generic Names in Zoology No. 1058.

DIAGNOSIS: Venation: A3 of hindwing very short, vestigial. Proboscis well developed. Head and thorax densely covered by green 'hairs'. Further features under diagnosis of tribe.

MALE GENITALIA: Uncus vestigial. Valva usually simple. Aedeagus without cornuti. Sternum A8 caudally with two short projections or lobes.

IMMATURE STAGES: Head of larva dorsally only slightly notched, even segment T1 lacking the diagnostic projections of most other Geometrinae. Dorsum of abdominal segments with typical warts and humps. Brownish coloured in late autumn and winter, green in springtime (crypsis).

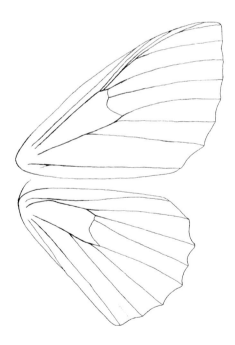

Text-fig. 181: Venation of *Geometra papilionaria* (LINNAEUS, 1758).

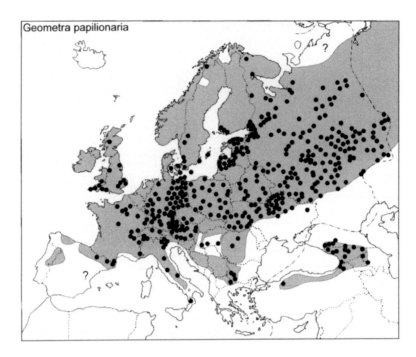

Geometra papilionaria

18. *Geometra papilionaria* (LINNAEUS, 1758)

Phalaena (Geometra) papilionaria LINNAEUS, 1758: Syst. Nat. (Ed. 10) 1: 522 (type locality not stated, [Europe]). Type(s) in LSL.
Phalaena prasinaria HUFNAGEL, 1767: Berlin. Magazin 4: 506 (Germany: Berlin region). Type(s) lost. Junior synonym.
Geometra albopunctata MATTUSCHKA, 1805: Raupen- und Schmett.-Tab. für Ins.-Sammler: 125 ([Europe]?). Type(s) not traced. Junior synonym.
Unavailable names (infrasubspecific): *alba*: GILLMER (1909); *cuneata*: BURROWS (1905); *deleta*: BURROWS (1905); *diffluata*: MARSCHNER (1932); *lutescens*: BERGMANN (1955); *obsoleta*: OSTHELDER (1929); *subcaerulescens*: BURROWS (1905); *subobsoleta*: BURROWS (1905). - (incorrect subsequent spellings): *obsoletaria*: PROUT (1935); *papilonarius*: CURTIS (1831).

DIAGNOSIS: Wingspan ♂♀ 41-50 mm. Ground colour deep green, easily fading to white or yellowish grey. Rarely albinistic (RUHLAND 1910; RICHTER 1912). Cell spot of forewing dark green, elongate, of hindwing nearly invisible. Postmedial line of all wings dark green, sharp, dentate (consisting of lunule-shaped marks), distally bordered white. Antemedial line of forewing crenulate, dark green, proximally bordered white. Terminal field with white dotted line. Terminal line lacking. Fringe concolorous, distal half white. Frons slightly convex, green. Palpi green, length 1.5-2 times diameter of eye in ♂, 2-3 times in ♀, last segment smoothly scaled. Length of ♂ antennal branches 2-3 times width of flagellum. ♂ hindtibia without pencil. Further features under generic diagnosis.

MALE GENITALIA: Socii strongly sclerotized, very long and slender. Gnathos elongate, slender and tapered. Valvae simple. Sternum A8 caudally with two slightly convex lobes.

FEMALE GENITALIA: Apophyses anteriores and posteriores of medium length. Antrum sclerotized, cup-shaped. Ductus bursae longitudinally folded. Signum bursae very small, orally with two spines.

136

DISTRIBUTION: Palaearctic. All over Europe, with isolated populations in the south. Absent from southern Spain, Albania, Greece and from all the Mediterranean islands. Record for European Turkey (SEVEN 1991) requires confirmation. South European populations endangered by loss of original woodland, such habitats meriting protection.

Outside Europe in northern Turkey, Transcaucasus, Caucasus, Siberia, eastwards to Sachalin (subsp. *herbacearia* MÉNÉTRIÈS, 1859) and Japan (subsp. *subrigua* PROUT, 1935).

PHENOLOGY: Univoltine. Main flight period: late June to mid-August, single specimens somewhat earlier or later, rarely in May or until September due to asynchronous growth of larvae (LASS 1923). In southern Italy and Bulgaria regularly until mid-September (SCALERCIO pers. comm.; NESTOROVA 1998). Larval stages July to early June, overwintering as small larva.

BIOLOGY: Larva oligophagous on various trees and shrubs, mainly on *Betula* (var. cit.). Found on *Betula pubescens, B. pendula, Alnus glutinosa, Corylus avellana, Fagus sylvatica, Sorbus aucuparia* and *Salix caprea* (Finland: SEPPÄNEN 1954; 1970; Estonia: THOMSON 1967; Poland: BUSZKO pers. comm.; Great Britain: SKINNER pers. comm.; PORTER 1997; Germany: WEGNER, LEIPNITZ pers. comm.; SCHNEIDER 1934, BERGMANN 1955; Hungary: RONKAY pers. comm.). Recorded also on *Tilia* and various species of the genera *Salix, Fagus* and *Alnus* (var. cit.). Pupation between loosely interwoven withered leaves on the ground or between leaves of host-plant (SKOU 1986). Adults feeding on flowers (BERGMANN 1955), attracted also to bait (THOMSON 1967; URBAHN 1939). Nocturnal, readily attracted to light. ♀ ratio at light about 20-30%.

PARASITOIDS: Hymenoptera, Braconidae: *Apanteles rubripes, Meteorus versicolor* (THOMPSON 1946).

HABITAT: Preferring wet woodland, mainly birch woods and hedges on humid soils. However also in gardens and dry heathlands with birches. Attacking trees of different ages. Usually from 0 m up to 600 m above sea-level, in the southern Alps and in southern Europe sometimes up to 1,500 m.

SIMILAR SPECIES: No similar species in Europe.

Tribus **Comibaenini** INOUE, 1961
=Euchlorini HERBULOT, 1963

Several genera in the Palaearctic, Indo-Pacific and Afrotropical regions. Monophyly of Comibaenini supported by homogeneous genitalic features in both sexes and behaviour of larva. Unstalked M3 and CuA1 of hindwing reminiscent of Pseudoterpnini and Geometrini. Similar behaviour of larvae perhaps indicating sister group relationships between New World tribe Synchlorini and Old World Comibaenini (FERGUSON 1985). Covering with plant debris however known also from early instars of other Geometrinae larvae, such as *Hemithea aestivaria* (HÜBNER, 1789) (PITKIN 1996). Possible relationships also to New World tribe Nemoriini (HOLLOWAY 1996).

DIAGNOSIS: Venation variable, often within species, sometimes even within populations. M3 and CuA1 of hindwing usually unstalked. Hindwing termen rounded. Usually one anal vein, exceptionally two. Ansa of tympanal organ with medial dilatation, tapered towards apex (Text-fig. 158). Abdominal crests usually absent. Paired setal patches on sternum A3 usually present, well separated from each other, laterally positioned on sternum and often with long setae. Lateral apodemes on sternum A2 usually well developed. Antennae bipectinate in ♂, with very long branches.

MALE GENITALIA: Uncus forked (exception: *Microbaena*). Socii present. Gnathos weak or absent. Saccus bisinuous, strongly concave at base. Valvae arising from central part of clasping structure. Aedeagus usually long and slender.

FEMALE GENITALIA: Apophyses anteriores long, slightly shorter than apophyses posteriores. Antrum closely connected with surface of sternum A8 (lamella antevaginalis). Ductus bursae long and very slender. Corpus bursae extremely weakly sclerotized, preserved only by careful preparation. Signum absent.

IMMATURE STAGES: Larvae without typical subfamily characters, i.e. lacking dorsal projections on head, segments T1 and A10. Abdominal segments bearing characteristic dorsal and lateral warts.

REMARKS: A behavioural synapomorphy links all the species of this tribe worldwide: The larvae cover themselves with plant debris, such as small pieces of dry leaves and other materials (flowers), attached to a silken web on the short bristles of the surface of the skin (Text-fig. 192).

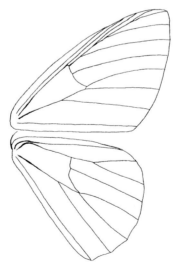

Text-fig. 182: Venation of *Comibaena bajularia* ([DENIS & SCHIFFERMÜLLER], 1775)

Comibaena HÜBNER, [1823]
=*Euchloris* auct. nec HÜBNER, [1823]; =*Phorodesma* BOISDUVAL, 1840

TYPE-SPECIES: *Geometra bajularia* [DENIS & SCHIFFERMÜLLER], 1775. Large genus with 12 species in the Palaearctic and 45 further species in the Old World tropics. Western Palaearctic *Comibaena bajularia, C. pseudoneriaria* and *C. serrulata* FLETCHER, 1963 with close relationships to each other (*Comibaena* s.str.). Chaetotaxy of caterpillars and close relationship to *Thetidia* discussed in TRÖGER (1971).

DIAGNOSIS: Bright green moths, often with red-brown markings near termen. Venation of forewing: R2-R5 and M1 stalked. M3 and CuA1 of hindwing unstalked, with exceptions in species from outside Europe. Frenulum usually present in male (plesiomorphy within Comibaenini), but rather weak, absent in ♀. Second segment of palpi bushy scaled. Hind-tibia with two proximal and two terminal spurs (plesiomorphy), pencil and one terminal projection, the latter important as differential feature to *Thetidia* and synapomorphy of *Comibaena* and *Proteuchloris*. Abdominal crests usually absent.

MALE GENITALIA: Uncus forked. Aedeagus long, slender, straight. Shape of sternum A8 diagnostic, often also that of tergum A8. Posterior margin of sternum A8 strongly sclerotized and crown-shaped in the typical group.

IMMATURE STAGES: Morphology and behaviour of larva under diagnosis of tribe. Pupa with wrinkled surface and numerous spines on tergites of last segments. The caterpillars feed on trees and shrubs (compare *Thetidia*).

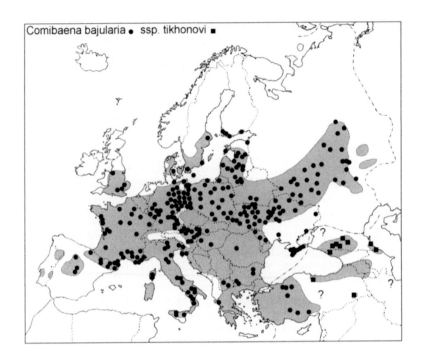

Comibaena bajularia ● ssp. tikhonovi ■

19. *Comibaena bajularia* ([DENIS & SCHIFFERMÜLLER], 1775)

Geometra bajularia [DENIS & SCHIFFERMÜLLER], 1775: Ank. syst. Werkes Schmett. Wienergegend: 97 (Austria: Vienna distr.). Type(s) lost.

‡ *Phalaena pustulata* HUFNAGEL, 1767: Berlin. Magazin 4 (5): 520, nec O.F. MÜLLER, 1764 (identity: *Hyphoraia aulica*, Arctiidae). Junior primary homonym.

Phalaena ditaria FABRICIUS, [1776]: Genera Insect.: 286 (Germany: Hamburg). Type(s) probably lost. Junior synonym.

Phalaena (Geometra) pustulataria KNOCH, 1781: Beytr. Ins. 2: 4, pl. 1, fig. 2. No separate types. Intentional change of final syllable because of bipectinate antennae. Unjustified emendation.

‡ *Phalaena glauca* FOURCROY, 1785: Entom. Paris.: 267 nec CRAMER, 1777. Junior primary homonym.

Unavailable names (infrasubspecific): *alba*: SCHNEIDER & WÖRZ (1940); *rosea*: COCKAYNE (1952); *stigmatisata*: STAUDER (1920); *tangens*: LEMPKE (1949). - (incorrect subsequent spellings): *baiularia*: BORKHAUSEN (1794); *bajuralia*: HEROLD (1844); *stigmaticata*: PROUT (1938a); *pustulsta*: NESTOROVA (1998); *pustularia*: DUPONCHEL (1829).

DIAGNOSIS: Wingspan ♂ 22-28, ♀ 24-32 mm. Forewing costa and termen convex. Ground colour bright green, readily fading. Exceptionally light rosy (PROUT 1938a; HARDONK 1954). Forewing with ante- and postmedial lines whitish, narrow and nearly parallel, lacking on hindwing. Antemedial line distally projecting below CuA1. Large ochreous spots with brown filling in fore- and hindwing tornus and in hindwing apex. Black cell spots small and sharp. Dark brown terminal line on hindwing and forewing tornus. Fringe chequered ochreous and brown. Frons flat, white. Vertex white. Palpi long, about twice diameter of eye in ♂, 2.5 times in ♀. Antennae long bipectinate to 4/5 length of flagellum in ♂, longest branches about 10 times width of flagellum. Antennae of ♀ filiform with very short cilia. Terminal projection of hindtibia (♂) covering 1/3 to 1/2 of tarsus. Abdomen green at base, caudal half white, without abdominal crests. Further features under generic diagnosis.

MALE GENITALIA: Uncus forked, sometimes slightly fused at base. Socii strongly curved. Costa of valva strongly sclerotized (to 4/5 length of valva) with lateral teeth over half length of valva. Aedeagus long, slender, straight. Posterior margin of sternum A8 crown-shaped, sometimes slightly concave. In populations from southern France, central and northern Spain membranous apex of valva longer (ca. 1/4 instead of 1/5) and posterior margin of sternum A8 strongly concave.

FEMALE GENITALIA: Sterigma characteristically folded. Corpus bursae slender, weak. Signum bursae absent.

DISTRIBUTION: European-Westasiatic (Submediterranean). Widely distributed, but often in isolated populations, in western, central and southern Europe. In the north up to southern Sweden, southern Finland and the Baltic states, in the east to the Volga plain. Absent from Ireland, Balearics, Malta, Greek islands including Crete. Also still unrecorded for Albania. Replaced by *Comibaena pseudoneriaria* WEHRLI, 1926 in southern and eastern Spain.

Outside Europe in north-western and central Turkey. Populations of north-eastern Turkey, Caucasus and Transcaucasus (subsp. *tikhonovi* HAUSMANN, 2000) transitional to the northern Iranian species *Comibaena serrulata* FLETCHER, 1963.

PHENOLOGY: Univoltine. Main flight period: early June to mid-July, single specimens somewhat earlier. In mountains and near northern border of distribution ten days later, sometimes until mid-August. Partial second generation (September) in warm years reported for the Netherlands (LEMPKE 1949; JANSSEN 1985) and Germany (URBAHN 1965). Larval stages late July to late May, overwintering as small larva.

BIOLOGY: Larva monophagous (3). Found on *Quercus robur* and *Quercus petraea* (Poland: BUSZKO pers. comm.; Germany: WEGNER pers. comm.; SCHNEIDER 1934; URBAHN 1939; BERGMANN 1955; Great Britain: SKINNER pers. comm; PORTER 1997; southern European Russia: ANIKIN et al. 2000), preferring old trees (HACKER & KOLBECK 1996). Also on sun-exposed side of young oaks in open woods (LESS 1923). Recorded also on *Fagus* (Great Britain: WATERS 1924; PROUT 1938a) and, reputedly, on *Prunus domestica* (REBEL 1910). Reared on *Prunus spinosa* (BERGMANN 1955). Young larvae feeding on oak flowers. Caterpillars covering themselves with pieces of dry foliage and oak flowers (crypsis). Pupation between folded interwoven leaves (SKOU 1986) or in a cocoon among leaf litter (PORTER 1997; KOCH 1984). Adults flying rapidly from sunset to dawn around higher parts of oak trees. Also, occasionally, attracted to light. K-strategy in central Europe.

HABITAT: In central Europe tends to be thermophilous. In sparse oak woods, oak forest fringes, mainly woods with old trees. Preferably on slightly humid soils (GELBRECHT pers. comm.; ANIKIN et al. 2000). Usually from 0 m up to 700 m above sea-level, sometimes up to 1,200 m.

SIMILAR SPECIES: *Comibaena pseudoneriaria* (allopatric) differs in the straighter transverse lines of the forewing and in the smaller spot in the forewing tornus. Ground colour darker green, densely covered with white scales. Single specimens from central Spanish populations come close to these characters. *Proteuchloris neriaria* can be distinguished by the brownish encircled double-spot of the forewing tornus, the small white abdominal spots on the tergum of abdominal segment 2, the green vertex and the regularly convex antemedial line of the forewing.

REMARKS: *Comibaena serrulata* FLETCHER, 1963, described from northern Iran, differs strikingly from the sister species *C. bajularia* in totally lacking marginal spots on the wings. It is, however, structurally closely related with just a few small differential features in the genitalia: Aedeagus somewhat shorter, costa of valva and sternum A8 less dentate. The populations from north-eastern Turkey, Caucasus and Transcaucasus are intermediate with regard to habitus and structure (subsp. *tikhonovi* HAUSMANN, 2000) and resemble both *C. pseudoneriaria* and *Proteuchloris neriaria*. The populations of the Ukraine and north-western Turkey eastwards to the Ankara region correspond well in habitus to the European populations of the nominate subspecies. The whole complex needs further study and revision.

Text-figs 183-186: Diagnostic characters (indicated) of European Comibaenini. Text-fig. 183: *Comibaena bajularia* ([DENIS & SCHIFFERMÜLLER], 1775) (Germany). Text-fig. 184: *C. bajularia tikhonovi* HAUSMANN, 2000 (north-eastern Turkey). Text-fig. 185: *C. pseudoneriaria* WEHRLI, 1926. Text-fig. 186: *Proteuchloris neriaria* (HERRICH-SCHÄFFER, 1852).

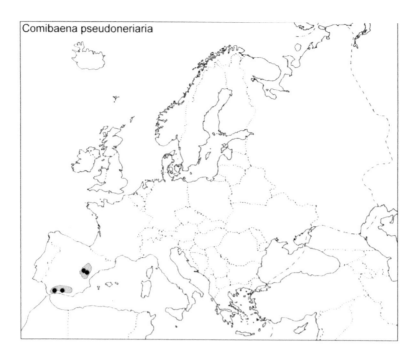

20. *Comibaena pseudoneriaria* WEHRLI, 1926

Comibaena pustulata f. (an spec.?) *pseudoneriaria* WEHRLI, 1926: Ent. Z. Frkft. 39 (40): 163 (Spain,
Andalusia: Geniltal). Holotype ♂ (ZFMK; without abdomen). Raised to species rank by PROUT (1935a).

DIAGNOSIS: Wingspan ♂ 23-24, ♀ 24-29 mm. Ground colour deep green, readily fading,
densely covered with white scales, more than in *Comibaena bajularia*. Forewing with ante-
and postmedial lines white, narrow, rather straight and nearly parallel to each other, lacking
on hindwing. Antemedial line comparatively straight below CuA1. Spot of forewing tornus
very small, only distally from postmedial line. Spot of hindwing tornus nearly absent. Tornus
spots of both wings slightly larger in eastern Spanish than in southern Spanish populations.
Cell spots small and sharp, blackish. Terminal line, fringe, abdomen, antenna and hindtibia as described for *Comibaena bajularia*. Frons flat, white. Palpi long, about 1.5 times
diameter of eye in ♂, 2.0-2.5 times in ♀. Further features under generic diagnosis.

MALE GENITALIA: Costal spinules over 1/3 length of valva only, membranous apex 1/4 length
of valva. Posterior projections of sternum A8 standing very close to each other, with deep medial notch (one specimen examined from eastern Spain, no ♂ available from southern Spain).

FEMALE GENITALIA: Without clear distinguishing characters from *Comibaena bajularia*.
Anterior part of lamella antevaginalis rather broad (specimens examined from southern
and eastern Spain).

DISTRIBUTION: Endemic to mountains of eastern and southern Spain.

PHENOLOGY: Univoltine. Main flight period: early June to mid-July, mainly in early July.
Probably overwintering as larva.

BIOLOGY: Larva still unknown, probably living on *Quercus* species.

HABITAT: Captured from shrubs of *Quercus* (WEHRLI 1926). Montane species: from 1,000 m up to 1,800 m above sea-level.

SIMILAR SPECIES: *Comibaena bajularia* (allopatric) differs in the projection of the antemedial line of the forewing below CuA1, and in the much larger tornus spots. The wings are more yellowish coloured and less covered with white scales. Subsp. *tikhonovi* from southwestern Asia resembles *Comibaena pseudoneriaria* in habitus. *Proteuchloris neriaria* (allopatric; see below) can be distinguished by the brownish encircled double spot of the forewing tornus, the brown cell spots, the green vertex and one conspicuous white spot on the abdomen.

REMARKS: *Comibaena bajularia* and *C. pseudoneriaria* are closely related allopatric vicariants. Further investigation is necessary to clear up whether the taxon would be better treated at species rank or downgraded to subspecies level. As hitherto no topotypical males have been dissected, it is actually unknown whether 'true' *C. pseudoneriaria* is structurally different from *C. bajularia*. The populations from central and northern Spain match well *C. bajularia* from the rest of Europe, but the wing pattern of single specimens is similar to that of *C. pseudoneriaria* in these populations (individual forms).

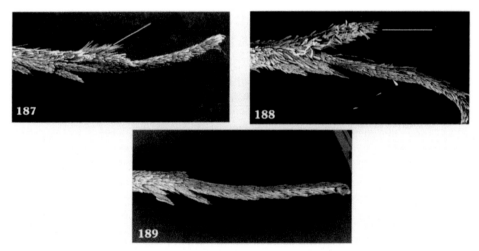

Text-figs 187-189: Hindlegs in Comibaenini (SEM). Text-fig. 187: *Comibaena bajularia* ([DENIS & SCHIFFERMÜLLER], 1775). Text-fig. 188: *Proteuchloris neriaria* (HERRICH-SCHÄFFER, 1852). Text-fig. 189: *Thetidia smaragdaria* (FABRICIUS, 1787).

Proteuchloris HAUSMANN, 1996

TYPE-SPECIES: *Geometra (Phorodesma) neriaria* (HERRICH-SCHÄFFER, 1852). One species only. External appearance, structure of ♂ hindtibia and host plant specialization suggest sister group relationship to *Comibaena*, ♂ genitalia and absence of frenulum however support closer relationships to *Thetidia*.

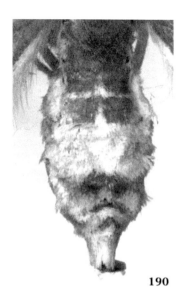

Text-figs 190-191: Abdomen of *Comibaena bajularia* ([DENIS & SCHIFFERMÜLLER], 1775) (Text-fig. 190) and *Proteuchloris neriaria* (HERRICH-SCHÄFFER, 1852) (Text-fig. 191).

DIAGNOSIS: Venation of forewing: R2-R5 and M1 unstalked. Frenulum absent in both sexes, synapomorphy of *Proteuchloris* and *Thetidia*. Proboscis developed. Hindtibia with two pairs of spurs and one terminal projection covering 1/3 to 1/2 of tarsus as in *Comibaena*.

MALE GENITALIA: Uncus forked. Valva simple, very similar to that of *Thetidia*. Aedeagus long, slender, straight. Posterior and anterior margin of sternum A8 strongly sclerotized, posteriorly with medial notch.

IMMATURE STAGES: Morphology and behaviour of larva under diagnosis of tribe.

Text-fig. 192: Larva of *Proteuchloris neriaria* (HERRICH-SCHÄFFER, 1852) (photo: LEIPNITZ).

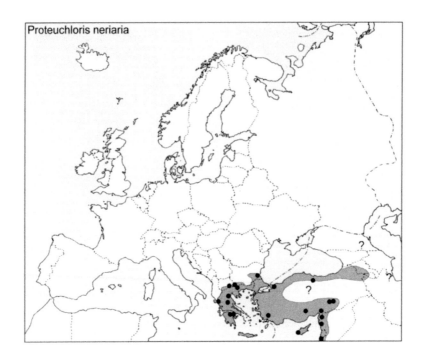

Proteuchloris neriaria

21. *Proteuchloris neriaria* (HERRICH-SCHÄFFER, 1852)

‡ *neriaria* HERRICH-SCHÄFFER, 1848: Syst. Bearb. Schmett. Eur. 3: pl. 70, fig. 429. Non binominal.
Geometra (Phorodesma) neriaria HERRICH-SCHÄFFER, 1852: Syst. Bearb. Schmett. Eur. 6 (55): 62
(Greece: Crete). Type(s) not traced. Binominal because referring to p. 9, vol. 3.

DIAGNOSIS: Wingspan ♂ 25-28, ♀ 28-34 mm, autumn generation smaller, sometimes 20 mm only. Forewing apex rounded. Ground colour deep green, colour readily fading. Spring generation more yellowish green. Forewing with ante- and postmedial lines whitish, narrow, lacking on hindwing. Postmedial line of forewing straight, strongly curved towards base near costa. Antemedial line regularly convex. Forewing tornus with double spot small, whitish and dark brown encircled. Cell spots dark brown, larger than in *Comibaena bajularia*. Terminal line brown, sharp and uninterrupted. Fringe chequered white and brown. Dorsum of abdominal segment 2 with a conspicuous white, often reddish encircled spot, which is flat, not crest-forming (Text-fig. 191). Frons flat, white with greenish tinge. Palpi long, 2.2-2.8 times diameter of eye, third segment 'naked' and longer than in *Comibaena bajularia*. Antennae long bipectinate in ♂, longest branches about 10 times width of flagellum. Antennae of ♀ dentate with very short forked 'branches'. Further features under generic diagnosis.

MALE GENITALIA: Uncus projections as long as socii. Valvae and aedeagus simple without sclerotizations or processes. Posterior margin of sternum A8 bilobed.

FEMALE GENITALIA: Antrum round to sub-rhombic, laterally slightly furrowed.

DISTRIBUTION: East-Mediterranean. Southern Balkan countries including Crete.
Outside Europe in Turkey, Armenia, Cyprus and the Levant.

146

PHENOLOGY: Bivoltine. Main flight period: mid-May to mid-July and early September to early October. In the Middle East trivoltine from mid-March to late October. Phenology of the larval stages unknown.

BIOLOGY: Larva monophagous (3) on *Quercus*. Reared on various *Quercus* species (Bulgaria: GELBRECHT, MÜLLER, LEIPNITZ pers. comm.; Cyprus: WIMMER pers. comm.). In Israel probably living exclusively on *Q. ithaburensis* (HAUSMANN 1997a). Adults nocturnal. ♀ ratio at light low, about 20%.

HABITAT: In south-eastern Europe on dry slopes with shrubs of *Quercus ilex* (LEIPNITZ pers. comm.). Probably K-strategy. ♀ flight activities on low level. From 0 m up to 800 m above sea-level.

SIMILAR SPECIES: *Comibaena bajularia* is usually easy to recognize by the strongly projecting (below CuA1) antemedial line of the forewing, larger spots of forewing tornus without brownish border, the white vertex, and by the absence of the small white abdominal spot. In north-western Turkey, Caucasus and Transcaucasus, however, the two species can easily be confused (see above).

Thetidia BOISDUVAL, 1840
=Euchloris HÜBNER, [1823] nom. praeocc.

TYPE-SPECIES: *Thetidia plusiaria* BOISDUVAL, 1840. About 20 species almost exclusively in the Palaearctic. Generic and subgeneric status controversial (RAINERI 1994; HAUSMANN 1996). Three groups *Thetidia* s.str., *Antonechloris* and *Aglossochloris* regarded as subgenera in the present study. Full generic separation not justified by venation characters, stalked forewing veins R2-R5 and M1 being typical for *Thetidia* sensu stricto, but rarely occurring also in *Aglossochloris* and *Antonechloris*.

DIAGNOSIS: Frenulum and proboscis usually vestigial. Palpi long, 2nd segment bushy scaled. Abdominal crests absent. Hindtibia of both sexes with four spurs. Proximal spurs of ♀ with tendency to reduction in some species. Hindtibia of ♂ without terminal projection, diagnostic with regard to *Comibaena* and *Proteuchloris*.

MALE GENITALIA: Uncus deeply forked. Valvae simple, without sclerotizations or processes.

FEMALE GENITALIA: Ductus bursae very weakly sclerotized. Ostium bursae region with a large sclerotized plate, often rhombic, probably homologous to the 'sterigma' (lamella ante- and postvaginalis). Closely attached to this plate a cup-shaped, strongly sclerotized structure, probably the 'antrum'.

IMMATURE STAGES: Morphology and behaviour of the larva under diagnosis of tribe. Chaetotaxy of larva described in TRÖGER (1971).

REMARKS: Larva in contrast to the preceding genera feeding on low (perennial) plants. Pupation in a flimsy cocoon between leaves of food-plant or near its base.

Subgenus *Thetidia* BOISDUVAL, 1840

Subgenus with one species only. Wing colour, proboscis and structure of uncus in ♂ genitalia indicate a closer relationships to *Aglossochloris* than to *Antonechloris*.

DIAGNOSIS: Venation of forewing: Sc and R1 fused, with exceptions. R2-R5 and M1 distinctly stalked, whilst unstalked (with exceptions) in the subgenera *Antonechloris* and *Aglossochloris*. Frenulum absent in both sexes. Proboscis very short and weak. Antennae of ♂ bipectinate to 2/3 of length.

MALE GENITALIA: Projections of uncus long and widely separated from each other. Posterior margin of sternum A8 rounded.

FEMALE GENITALIA: Length of apophyses posteriores twice length of apophyses anteriores. Antrum situated on posterior half of sterigma, with very deep posterior excavation. Sterigma very long and narrow, length about 2-3 times breadth.

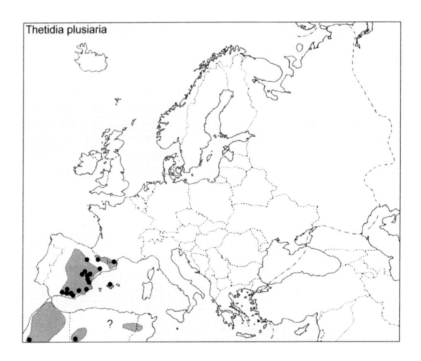

22. *Thetidia plusiaria* BOISDUVAL, 1840

Thetidia plusiaria BOISDUVAL, 1840: Genera Index Meth. eur. Lepid.: 189 (Spain, Andalusia). Type series discussed by HERBULOT (1987). 1♂ from GUENÉE collection in ZFMK, probably the BOISDUVAL type.
Thetidia plusiaria ab. *powellaria* OBERTHÜR, 1916: Lép. Comp. 12: 92; fig. 3249 (Algeria: Lambèse). Type(s) not traced. Not explicitly introduced at infrasubspecific rank. Junior synonym.
Unavailable names (infrasubspecific): *simplificata*: SCHLEPPNIK (1933). - (incorrect subsequent spellings): *plusaria*: MAZEL & PESLIER (1997).

DIAGNOSIS: Wingspan ♂ 22-27, ♀ 24-29 mm, autumn generation smaller, sometimes 20 mm only. Also smaller on Mallorca. Ground colour of forewing deep green, exceptionally orange, colour of hindwing white. Antemedial and postmedial lines of forewing white, crenulate, with white arrow spots near termen. Cell spot a white circle. White streaks on forewing vein A1 proximally and distally of antemedial line. Cell spot of hindwing small, sharp, green. A sharp green line close to termen. Terminal line green, interrupted by veins. Fringe of all wings chequered green and white. Markings clear and sharp on underside, ground colour of hindwing being green. Frons flat, light green. Length of palpi 1.5-1.8 times diameter of eye, in ♂ longer than in ♀. Antennae long bipectinate in ♂, longest branches about 7 times width of flagellum. Antennae of ♀ dentate. Further features under generic diagnosis.

MALE AND FEMALE GENITALIA: See subgeneric diagnosis.

DISTRIBUTION: West-Mediterranean. Spain, Mallorca, rarely recorded from south-western France (Pyrenees). In Spain often very abundant.
   Outside Europe in Morocco and Algeria.

PHENOLOGY: Usually bivoltine, under warm conditions trivoltine (DOMÍNGUEZ & BAIXERAS 1995). Main flight period: early June to late July; early September to late September, sometimes also in early May. Phenology of larval stages unclear. Overwintering probably as pupa (DOMÍNGUEZ & BAIXERAS 1995).

BIOLOGY: Larva oligophagous on Compositae. Found on *Artemisia herba-alba* and *Santolina chamaecyparissus* (Spain: DOMÍNGUEZ & BAIXERAS 1995; KORB: specimen in ZSM), preferring *Santolina*. Recorded on *Artemisia herba-alba* also by BOLLAND (1975). Successfully reared on *Artemisia campestris*, *Artemisia maritima* and *Achillea millefolium* (LEIPNITZ pers. comm.).

HABITAT: Mountainous habitats, open or bushy slopes and dry uncultivated areas with food-plants. In Morocco even in the Atlantic plains (RUNGS 1981). ♀ ratio at light low, about 20%. In Spain usually from 900 m up to 2,200 m above sea-level, but sometimes common even at 300 m (e.g. Penalba, northern Spain).

SIMILAR SPECIES: *Thetidia (Aglossochloris) correspondens* (allopatric) differs in the unchequered fringe, the forewing cell spot without green centre, and in the absent hindwing cell spot.

Subgenus *Aglossochloris* PROUT, 1912

TYPE-SPECIES: *Phorodesma fulminaria* LEDERER, 1870 (Iran). Seven species from the Levant to central Asia. Some authors consider *Aglossochloris* as a separate genus.

DIAGNOSIS: See also descriptions of genus and nominotypical subgenus. Venation rather variable in all species of the subgenus. Forewing: R2-R5 and M1 usually unstalked. Mainly distinguished from subgenera *Thetidia* and *Antonechloris* by structure of ♂ hindtibia with only one terminal pair of short spurs present. Some species, however, with vestigial proximal spurs. Hindtibia of ♀ with two terminal spurs tending to reduction. Proboscis absent or (rarely) vestigial. Antennae bipectinate to 4/5 of length in ♂.

MALE GENITALIA: In type-species very similar to subgenus *Antonechloris*, all other representatives with uncus projections widely separated from each other.

FEMALE GENITALIA: Semicircular chitinous clasp (antrum) situated on anterior half of sterigma.

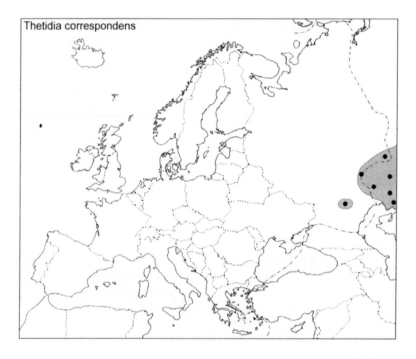

Thetidia correspondens

23. *Thetidia correspondens* (ALPHÉRAKY, 1883)

Phorodesma fulminaria var. *correspondens* ALPHÉRAKY, 1883: Horae Soc. ent. ross. 17 (3/4): 157, pl. 9, fig. 85 (western China, Kuldja distr.: Ili plain). Lectotype ♂, herewith designated to stabilize nomenclature (ZISP, examined, pl. 8, fig. 7).

DIAGNOSIS: Wingspan ♂ 22-26, ♀ 25-28 mm. Ground colour of forewing deep green, hindwing white. Antemedial and postmedial lines of forewing white, crenulate, with a row of rounded white spots close to termen. Cell spot large, white, somewhat elongate. Hindwing without cell spot. An undulate green line close to termen, getting more distinct near anal margin. Terminal line green, uninterrupted. Fringe unchequered, green at base, distal half white. Markings clear and sharp also on underside, here ground colour of hindwing green. Frons flat, light green, white near palpi. Venation: R2-R5 and M1 of forewing shortly stalked or connate. Proboscis usually absent, in one female from southern Urals present, length nearly 1 mm. Length of palpi: ♂ 1.5 times, ♀ twice diameter of eye. Antennae long bipectinate in ♂, longest branches about 5-6 times width of flagellum. Antennae of ♀ shortly bipectinate, length of branches = width of flagellum. Hindtibia of both sexes with terminal spurs only. Further features under generic diagnosis.

MALE GENITALIA: Uncus projections stout, inwardly curved, widely separated from each other. Valvae broad and rounded. Aedeagus comparatively straight and short. Posterior margin of sternum A8 flat.

FEMALE GENITALIA: Sterigma rhombic, posteriorly very broad, very shortly extending beyond antrum.

DISTRIBUTION: Sarmatian. Southern Urals, southern Volga Plain.
   Outside Europe in western Siberia, mountains of central Asia, southwards to Tadzhikistan and Uzbekistan.

PHENOLOGY: Univoltine. Main flight period: late May to early July. Larval stages unknown.

BIOLOGY: Larval host-plant unknown.

HABITAT: Open land habitats, steppes, uncultivated areas. Vertical distribution poorly known, scarce data from 0 m up to 300 m above sea-level, in central Asia up to 1,200 m.

SIMILAR SPECIES: *Thetidia plusiaria* (allopatric) differs in the chequered fringe, the green centre of the forewing cell spot, the presence of the hindwing cell spot etc. Several central Asian species of the subgenus *Aglossochloris* are similar, *T. correspondens* being the only one with the white terminal spots of the forewing evenly rounded and the postmedial line of the forewing underside broad and dentate towards the cell spot.

REMARKS: ZHURAVLEV (1910) reports *Thetidia (Aglossochloris) fulminaria* (LEDERER, 1870) from the Uralsk district. The occurrence there cannot be excluded, but no authentic material was available for examination. Possibly a misidentified *T. correspondens*.

Subgenus *Antonechloris* RAINERI, 1994

TYPE-SPECIES: *Phalaena smaragdaria* FABRICIUS, 1787. About ten species in the Palaearctic. Differential diagnosis *see* also RAINERI (1994).

DIAGNOSIS: See also descriptions of genus and nominotypical subgenus. Venation: Rather variable in all congeneric species. Forewing: R2-R5 and M1 usually unstalked. Frenulum usually absent in both sexes. Proboscis developed, but short and with strong tendency to reduction, mainly in the Middle East species. Antennae of ♂ bipectinate almost to tip. Hindtibia of both sexes with two pairs of spurs.

MALE GENITALIA: Uncus projections standing close to each other. Aedeagus curved and slender.

FEMALE GENITALIA: Semicircular chitinous clasp (antrum) situated on posterior half of sterigma.

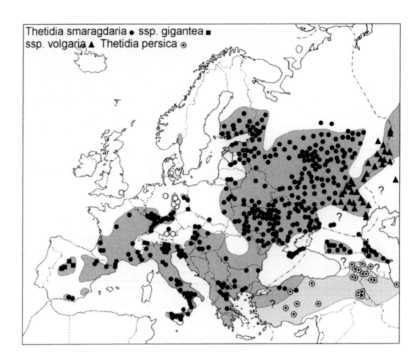

Thetidia smaragdaria ● ssp. gigantea ■
ssp. volgaria ▲ Thetidia persica ◉

24. *Thetidia smaragdaria* (FABRICIUS, 1787)

Phalaena smaragdaria FABRICIUS, 1787: Mantissa Insect. 2: 192 (Austria). Type(s) probably lost.
Euchloris smaragdaria maritima PROUT, 1935: in SEITZ Macrolep. 4, Suppl.: 17 ([southern] England).
Type(s) in BMNH. Junior synonym.
Unavailable names (infrasubspecific): *alinea*: BURROWS (1900); *caeruleoviridis*: BURROWS (1900); *obsoleta*: BURROWS (1900); *unilinea*: BURROWS (1900); *viridis*: BURROWS (1900). - (incorrect subsequent spellings): *aliena*: PROUT (1938a).

DIAGNOSIS: Wingspan ♂ 25-29, ♀ 27-34 mm, some males from southern Switzerland up to 35 mm. Second generation smaller, sometimes 20 mm only. Ground colour emerald green, hindwing brighter green, sometimes white at base and costa. Forewing costa yellowish. Antemedial line of forewing white, undulate and convex. Postmedial line white, slightly crenulate, rather straight or slightly convex, oblique towards apex. Cell spot of forewing large, whitish, round. Hindwings with an indistinct white line close to termen. Both wings often with a row of very small white dots along termen. Fringe bright green, distal half white. Underside green, cell spots distinctly marked, postmedial lines usually thin. Frons flat, green, white towards proboscis. Venation of forewing: R1 arising separately below apex of cell. Rs and M1 of hindwing usually unstalked. M3 and CuA1 usually connate. Length of palpi: ♂ 1.5 times, ♀ twice diameter of eye. Antennae of ♂ bipectinate, longest branches about 6-8 times width of flagellum. Antennae of ♀ usually filiform, exceptionally bipectinate at base with extremely short branches. Further features under generic diagnosis.

MALE GENITALIA: Projections of uncus usually long (in Sicily often short), sometimes even longer than socii, outwardly curved. Valvae comparatively slender. Posterior margin of sternum A8 slightly tapered.

FEMALE GENITALIA: Shape of sterigma variable, but usually slender and posteriorly rounded. Length of apophyses posteriores 1.5 times to twice length of apophyses anteriores.

DISTRIBUTION: Species Palaearctic. Nominate subspecies Submediterranean (Euro-Caucasian). Widely distributed nearly all over southern and eastern Europe, but isolated populations in Spain. Mosaic pattern of distribution in western and central Europe, an isolated population even in southern Norway. Absent from Portugal and all Mediterranean islands except Sicily. In central Spain and south-eastern European Russia with separate subspecies (see below). The population in Great Britain on the verge of extinction (PORTER 1997) or already extinct (REID 1998 pers. comm.). Severe protection of habitats necessary in eastern Germany (GELBRECHT 1995).

Outside Europe in Caucasus, mountains of central Asia (subsp. *anomica* PROUT, 1935) over Mongolia (subsp. *mongolica* STAUDINGER, 1897) to Far East Asia and Japan (subsp. *amurensis* PROUT, 1935). With regard to the Turkish populations of *Thetidia persica* HAUSMANN, 1996 see remarks to subsp. *volgaria*.

PHENOLOGY: Usually univoltine in western, central and northern Europe with main flight period mid-June to mid-July, exceptionally until early August. In southern Europe obligatorily bivoltine, mid-May to early July and mid-August to late September. In southern Italy emerging even in late April. Larval stages in univoltine populations late July to mid-June, but under warm conditions a very few rapidly growing larvae (late June to early August) developing to a partial 2nd brood from mid-August to mid-September. In southern Europe caterpillars in June-July and from September to early May. Overwintering as half-grown larva between leaves at base of food-plant (REID 1986; PORTER 1997).

BIOLOGY: Larva oligophagous (1) on Compositae. Found on *Achillea millefolium* (Finland: SEPPÄNEN 1954; 1970; Austria: specimen in ZSM) and *Artemisia absinthium* (Poland: BUSZKO pers. comm.). In literature recorded from open sunny localities mainly on *Achillea millefolium*, but also on *Tanacetum (Chrysanthemum) vulgare*, *Artemisia*, *Senecio* etc. (central Europe: var. cit.), mainly feeding on flowers. Caterpillar in Great Britain found exclusively on *Artemisia maritima* (PROUT 1913a; 1935a; REID 1986; PORTER 1997; SKINNER pers. comm.), reared on *Artemisia abrotanum* (ALLAN 1949; REID 1986; PORTER 1997 etc.) and other Compositae (ALLAN 1949). Adults nocturnal, but also easily flushed up in sunshine. ♀ ratio at light very low, about 5-10 %.

PARASITOIDS: Hymenoptera, Ichneumonidae: *Ichneumon quadrialbatus* (THOMPSON 1945).

HABITAT: Xerothermic open localities: In central Europe heathlands with many perennial plants, near rivers, on slopes, dams or other hot localities with rocky or stony soils. In England salt marshes, occurring only at sea-level near the coast. Survival of larva possible on meadows flooded at high tide (CARTER & HARGREAVES 1986). Less specialized in various open land habitats in southern Europe. Usually from 0 m up to 900 m, in the southern Alps up to 1,200 m, in southern Europe rarely up to 2,200 m above sea-level.

SIMILAR SPECIES: *Thetidia sardinica* (allopatric) from Sardinia differs in the distinctly marked and strongly dentate transverse lines of the forewing, and in the larger cell spots on upper- and underside. *T. persica* HAUSMANN, 1996 from Turkey, Transcaucasus, northern Iran and Turkmeniya see subsp. *volgaria*.

REMARKS: Some authors considered the southern English population as a separate subspecies ('*maritima*'). There are however no remarkable and constant differential features in habitus and structure.

Thetidia smaragdaria gigantea (MILLIÈRE, 1874)

Geometra smaragdaria var. *gigantea* MILLIÈRE, 1874: Iconogr. Chen. Descr. Lép. inédits 3 (35): 423, pl. 152, figs 16-18 ('Spain'). Holotype ♀ not traced.
Phorodesma smaragdaria var. *castiliaria* STAUDINGER, 1892: Dt. ent. Z. Iris 5: 141 (Spain, Castilia: St. Ildefonso). Syntypes 2♂2♀ (MNHU, examined). Junior synonym.

DIAGNOSIS: Larger than nominate subspecies, wingspan ♂ 28-35, ♀ 33-37 mm. Paler than nominate subspecies, postmedial line thinner, antemedial line often hardly visible. Similar differences on underside. Venation: Rs and M1 of hindwing usually stalked. Antennae bipectinate in ♂, longest branches about 6-8 times width of flagellum. Antennae of ♀ dentate or bipectinate with very short branches (0.5-1.0 times width of flagellum).

MALE AND FEMALE GENITALIA: No constant differences from nominate subspecies.

DISTRIBUTION: Subspecies endemic to the mountains of central Spain.

PHENOLOGY: Univoltine. Main flight period: early July to mid-August.

BIOLOGY: Larva recorded on *Achillea millefolium* (Spain: GUMPPENBERG 1892) (see also nominate subspecies). ♀ ratio at light very low, about 5-10 %.

HABITAT: Subspecies montane, imaginal habitat undescribed for subspecies. Usually from 1,100 m up to 2,000 m above sea-level.

SIMILAR SPECIES: No similar species in the area of distribution.

REMARKS: Well isolated from southern European *Thetidia smaragdaria* geographically and by its biological characteristics as a univoltine, montane subspecies.

Text-figs 193-196: Diagnostic characters (indicated) in Western Palaearctic *Thetidia (Antonechloris)* species. Text-fig. 193: *T. smaragdaria* (FABRICIUS, 1787); Text-fig. 194: *T. smaragdaria volgaria* (GUENÉE, [1858]); Text-fig. 195: *T. persica* (HAUSMANN, 1996) from western Turkey. Text-fig. 196: *T. sardinica* (SCHAWERDA, 1934).

Thetidia smaragdaria volgaria (GUENÉE, [1858])

Geometra volgaria GUENÉE, [1858]: in BOISDUVAL & GUENÉE, Hist. nat. Insectes (Spec. Gén. Lépid.) 9: 344 (Russia, southern Urals: Orenburg). Replacement name, no separate types.
‡ *Geometra prasinaria* EVERSMANN, 1837: Bull. Soc. imp. Nat. Moscou 10 (2): 52 (Russia, southern Urals: Orenburg; Type(s) in ZISP) nec [DENIS & SCHIFFERMÜLLER], 1775 (identity: *Ellopia fasciaria*, Ennominae). Junior primary homonym.
Unavailable names (infrasubspecific): *obsoleta*: PROUT (1935) (also praeocc.). - (incorrect subsequent spelling): *drasinaria*: OBERTHÜR (1925).

DIAGNOSIS: Smaller than *Thetidia smaragdaria* from southern Europe, wingspan ♂ 22-27, ♀ 23-29 mm. Sometimes even smaller, exceptionally 17 mm, mainly in autumn generation. Transverse lines of forewing broader than in nominate subspecies. White cell spot of forewing usually much larger. White subterminal line of hindwing more distinctly developed, sharp. Venation: R1 of forewing often arising very near cell apex. Rs and M1 of hindwing shortly stalked, sometimes connate. M3 and CuA1 separate. Length of palpi: ♂ 1.5-1.7 times, ♀ twice diameter of eye. Antennae of ♂ bipectinate, longest branches 4-7 times width of flagellum. Antennae of ♀ bipectinate with very short branches (0.5-1.0 times width of flagellum).

MALE GENITALIA: Projections of uncus nearly as long as socii, sometimes shorter. Valva shorter than in *Thetidia smaragdaria* from southern Europe, terminally broader.

FEMALE GENITALIA: Anterior margin of tergum A8 with two tapered projections. Sterigma rhombic, i.e. posteriorly, anteriorly and laterally tapered. Chitinisation of antrum with deep posterior excavation.

DISTRIBUTION: Subspecies East-European. Southern Urals, southern Volga plain. In the Volga plain contact zone (hybridization?) with nominate subspecies.

PHENOLOGY: Bivoltine. Main flight period: mid-May to late June and late July to late September. Larval stages undescribed.

BIOLOGY: Larva found on *Artemisia nutans, Artemisia austriaca* and *Achillea mille-folium* (southern European Russia: BECKER 1867; ANIKIN et al. 2000). ♀ ratio at light very low, about 10 %.

HABITAT: Steppes. In Europe usually from 0 m up to 300 m above sea-level.

SIMILAR SPECIES: *Thetidia persica* HAUSMANN, 1996 (allopatric) from Turkey, Transcaucasus, northern Iran and Turkmeniya is usually larger, deeper green, transverse lines narrower. Antennal branches of ♀ 1.5-3 times width of flagellum. Further differential features in ♂♀ genitalia (HAUSMANN 1996b).

REMARKS: In the southern Urals very few specimens are transitional in habitus to the nominate subspecies. Such intermediate forms become commoner in the southern Russian steppes and are predominant in the Ukraine. The taxon *volgaria* is retained as a subspecies of *Thetidia smaragdaria* by VIIDALEPP (1996), ANTONOVA and KOSTJUK (1999 pers. comm.), a good species in SCOBLE (1999). Relationships and status of the three taxa *smaragdaria, volgaria* and *persica* are, however, not yet definitively cleared up and need further study.

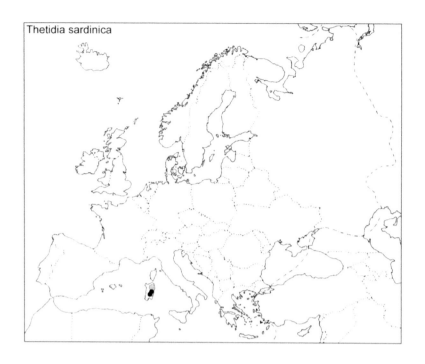

Thetidia sardinica

25. *Thetidia sardinica* (SCHAWERDA, 1934)

Euchloris sardinica SCHAWERDA, 1934: Int. Ent. Z. Guben 27 (41): 463, fig. 5 (Sardinia, Aritzo). Syntypes
  in NHMW.
Unavailable names (misidentification): *prasinaria*: sensu TURATI (1913) nec EVERSMANN, 1837 (identity:
  Thetidia smaragdaria volgaria). - (infrasubspecific): *bytinskii*: SCHAWERDA (1934); *schleppniki*:
  BYTINSKI-SALZ (1937). - (incorrect subsequent spelling): *bytinski*: PROUT (1935a).

DIAGNOSIS: Wingspan ♂♀ 25-30 mm. Ground colour of forewing deep emerald green.
Hindwing light green at base, getting darker in terminal area. Ante- and postmedial lines
of forewing more distinctly marked and much more zigzagging than in *Thetidia smaragdaria*. Cell spot of forewing larger, whitish, round. White subterminal line of hindwing
distinctly marked, but fine. Fringe light green, distal half white. Underside similar to
that of *Thetidia smaragdaria volgaria*, unicolorous green, cell spots distinctly marked,
mainly that of hindwing, postmedial lines distinct. Venation of forewing: R1 arising quite
separately from apex of cell. Hindwing: Rs and M1 stalked. M3 and CuA1 usually connate, sometimes separate. Frenulum absent. Frons flat, green, white towards proboscis.
Proboscis present, short. Length of palpi: ♂ 1.3-1.6 times, ♀ 1.8-2.0 times diameter of
eye. Antennae of ♂ bipectinate, longest branches about 6-7 times width of flagellum.
Antennae of ♀ slightly dentate. Further features under generic diagnosis.

MALE GENITALIA: Uncus projections usually shorter than socii. Basal excavation of saccus
shallow. Valva broad, rectangular at tip. Aedeagus short (1.1-1.2 mm). Posterior margin
of sternum A8 rounded.

FEMALE GENITALIA: Sterigma very broad, posteriorly rounded. Length of apophyses posteriores twice length of apophyses anteriores.

DISTRIBUTION: Endemic to the mountains of Sardinia.

PHENOLOGY: Univoltine. Main flight period: mid-June to early August. One strongly asynchronous single generation from May to September (KRÜGER 1913), partial second brood from July to August reported by BYTINSKI-SALZ (1937). Growth of larvae strongly asynchronous, occurring nearly throughout the year (KRÜGER 1913). Overwintering as larva.

BIOLOGY: Larva recorded on *Santolina* spp. (PROUT 1935; KRÜGER 1913). Reared on flowers of *Achillea millefolium* (LEIPNITZ pers. comm.), growing very slowly (KRÜGER 1913). Pupation attached to stalks of host-plant (KRÜGER 1913). ♀ ratio at light very low, about 10%.

HABITAT: Montane species. On open or bushy slopes with food-plants. From 900 m up to 1,700 m above sea-level.

SIMILAR SPECIES: *Thetidia smaragdaria* (allopatric) differs in the less marked and straighter transverse lines of the forewing. The cell spots on the upper- and underside are smaller.

Tribus **Hemistolini** INOUE, 1961

Provisional tribal concept as proposed in HAUSMANN (1996b) including 17 Palaearctic, Afrotropical and Indo-Pacific genera. Pupal morphology reveals similarities between Hemistolini and Comibaenini (PATOCKA 1995). Detailed phylogenetic studies may reveal the polyphyletic or paraphyletic nature of this tribe (HAUSMANN 1996b). HOLLOWAY (1996) combines Hemistolini, Comostolini, Jodini, Thalerini, Hemitheini and Microloxiini in 'Hemitheini, subtribe Hemitheiti', but this is disputed by PITKIN (1996). The frenulum is present in the genera *Chlorocoma* and *Prasinocyma* from Australia and Africa, but their genitalia are rather similar to those of *Hemistola*. The absent frenulum in *Hemistola* is evidently not sufficient to establish a separate tribe. The loss of the frenulum must have taken place several times independently in the evolution of the Geometrinae (FERGUSON 1985; HAUSMANN 1996b).

DIAGNOSIS: Venation: Forewing: R2-R5 and M1 typically stalked. M3 and CuA1 usually arising at same point or slightly separate (exception: *Xenochlorodes*). Hindwing: Sc+R1 and Rs fused or appressed at one point only (exception: *Xenochlorodes*). Rs and M1 stalked. M3 and CuA1 usually connate with tendency to be shortly stalked. Usually one hindwing anal vein only, exceptionally two. Frenulum often absent in both sexes. Palpi very slender and typically short in both sexes. Antennae bipectinate in ♂. Hindtibia of ♂ not dilated, number of spurs varying. Abdominal crests sometimes weakly developed (*Hemistola*) or absent. Setal patches on ♂ sternum A3 usually well developed. Ansa of tympanal organ medially dilated, tapered towards apex.

MALE GENITALIA: Uncus simple, stout, terminally forked in some extralimital species. Socii semi-membranous and broad, often of medium length, but sometimes nearly as long as uncus or vestigial. Gnathos elongate, usually strongly sclerotized, indicating relationships to Comostolini and Jodini. Saccus typically simple, but forked and concave between in *Xenochlorodes*. Sternum A8 often with two projections posteriorly. Valva often ventrally excavated.

FEMALE GENITALIA: Papillae anales and membrane between the papillae characteristically corrugated.

Text-fig. 197: Venation of *Hemistola chrysoprasaria* (ESPER, 1795).

Hemistola WARREN, 1893

TYPE-SPECIES: *Hemistola rubrimargo* WARREN, 1893 (India: Darjeeling). About 40 species in the Palaearctic and Indo-Pacific, four species in southern Africa and Madagascar. Genitalia of type-species similar to those of *Hemistola chrysoprasaria* (see below), but distinguished by coremata rather than a stout spine at base of valva.

DIAGNOSIS: Hindwing termen rounded or slightly angled at M3. Venation of forewing: Sc often anastomosing with R1. R2-R5 and M1 typically stalked. M3 and CuA1 usually connate or slightly separate. Hindwing: Rs and M1 stalked. M3 and CuA1 connate or slightly stalked (e.g. in type-species). Frenulum absent in both sexes. Antennae of both sexes usually bipectinate. Hindtibia of both sexes with four spurs. Setal patches on male sternum A3 usually well developed (exception: European species).

MALE GENITALIA: See description of tribe. Anterior margin of valva medially notched as in many Jodini. Base of valvula with one process. Coremata at base of valva developed in some species (e.g. in type-species), but modified to stoutly sclerotized spines in many others. Saccus simple.

FEMALE GENITALIA: Corpus bursae pyriform, strongly sclerotized. Lamella antevaginalis often divided into paired spinose patches (very similar in Afrotropical genus *Lophostola*).

IMMATURE STAGES: Morphology of eggs see PEKING (1953). Head capsule of larva dark (Text-fig. 154). Head and segment T1 with paired dorsal projections. Tergum of last segment (A10) with single anal process. Skin strongly granular, with small, white granules. Pupa very slender, for morphology see PATOCKA (1995).

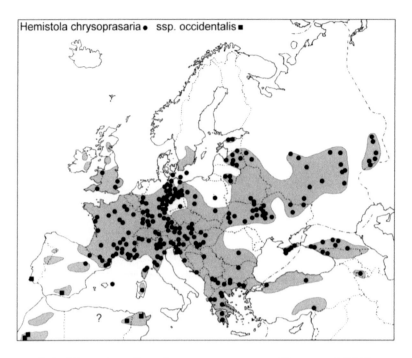

Hemistola chrysoprasaria ● ssp. occidentalis ■

26. *Hemistola chrysoprasaria* (ESPER, 1795)

Phalaena (Geometra) chrysoprasaria ESPER, 1795: Schmett. in Abbild. 5 (1): 37; ibd. [1801] pl. 5, fig. 1 (Germany, Hessen: Frankfurt a.M.). Lectotype ♂ (LMW, examined: HACKER 1999).
Phalaena nayas FOURCROY, 1785: Entomol. Paris. 2: 276 (France: Paris). Type(s) in Mus. Autun (F). (Senior) synonym according to SCOBLE (1999). Conspecificity to be confirmed by examination of type(s).
‡ *Geometra vernaria*: sensu [DENIS & SCHIFFERMÜLLER], 1775: Ank. syst. Werkes Schmett. Wienergegend: 97, nec LINNAEUS, 1761 (identity: *Jodis lactearia*). Misidentification, explicitly referring to 'L.[INNAEUS]'.
‡ *Hemistola immaculata*: sensu auct. nec THUNBERG, 1784 (identity: *Prasinocyma immaculata*, Geometrinae, from South Africa, lectotype discussed in KARSHOLT & NIELSEN 1985). Misidentification.
‡ *Hemistola biliosata*: sensu auct. nec VILLERS, 1789 (identity: *Hylaea fasciaria* f. *prasinaria*, Ennominae). Misidentification, discussed in HAUSMANN (1996a).
‡ *Phalaena (Geometra) aeruginaria*: sensu BORKHAUSEN, 1794: Eur. Schmett. 5: 43, nec [DENIS & SCHIFFERMÜLLER], 1775 (identity: *Jodis lactearia*). Misidentification.
‡ *Phalaena lucidata* DONOVAN, 1794: The natural history of British Insects 3: 67, pl. 97, nec FABRICIUS, 1781. Junior primary homonym.
‡ *Geometra volutaria*: sensu HAWORTH, 1809: Lep. Brit. 2: 298. Misidentification of *Phalaena volutata* FABRICIUS, 1775 (identity: *Chlorissa viridata*) and unjustified emendation (change of final syllable because of antennal structure).
Hemistola chrysoprasaria ab. *dentigera* PROUT, 1913: in SEITZ, Macrolep. 4: 30 (type locality not stated). Type(s) in BMNH. Junior synonym.
Unavailable names (infrasubspecific): *albifusa*: COCKAYNE (1950); *contracta*: CHALMERS-HUNT (1961). - (incorrect subsequent spellings): *inmaculata*: GOMEZ DE AIZPURUA (1974); *verbaria*: HUEMER & TARMANN (1993); *wernaria*: GARTNER (1866).

DIAGNOSIS: Wingspan ♂ 25-33, ♀ 30-39 mm, in second generation sometimes 23 mm only. Forewing costa, termen and apex rounded, hindwing slightly angled at vein M3. Ground colour light green with turquoise green tinge. Transverse lines white. Antemedial line of forewing strongly convex, strongly inwardly projecting near inner margin. Postmedial line of both wings slightly convex. Breadth of medial field rather variable. Cell spots lacking on all the wings. Fringe green at base, distal half shining white. Underside green with postmedial lines slightly marked. Frons flat, pale brown with reddish tinge. Vertex white. Palpi ochreous, slender, length about diameter of eye or slightly longer, especially in ♀. Proboscis well developed, thick. Antennae of ♂ bipectinate nearly to tip, longest branches 3.5-4 times width of flagellum. Antennae of ♀ shortly bipectinate, longest branches slightly exceeding width of flagellum. Further features under generic diagnosis.

MALE GENITALIA: Uncus long, stout. Socii broad, as long as uncus. Gnathos tapered, long. Centre of valva ('valvula') with short, broad, semi-membranous process ('harpe'). In southern Turkish specimens process of valvula narrower than in European populations. Base of valva with single, long, stout, tapered spine, connected by sclerotized ridge with ventral margin of valva. Medial notch of anterior margin of valva shallow. Aedeagus with two fields of cornuti, distal smaller than central. Shape of paired projection of sternum A8 quite variable.

FEMALE GENITALIA: Paired spinose patch on lamella antevaginalis large, with long spines. Semi-membranous plate near ostium bursae (lamella postvaginalis) rounded in most European populations, tapered to an angle of 90° in specimens from Macedonia and southern Turkey.

DISTRIBUTION: Palaearctic. Widely distributed in western and central Europe, in the north to southern Sweden and Estonia, in the south to central Spain, northern Portugal, Menorca (DANTART et al. 1993), Corsica, Sardinia, Toscana and Greece, in the east to southern Russia and southern Urals. The records for Denmark may refer to rare migrants from Germany. Absent from the Greek islands. Often in isolated local populations. In southern Spain and southern Portugal as the subsp. *occidentalis* (see below). In Sicily, southern and central Italy replaced by the allopatric sister species *Hemistola siciliana* (see below). In eastern Germany expanding northwards since the 1950s (KOCH 1984; MÜLLER & GELBRECHT 1992).

Outside Europe in north-western Africa (subsp. *occidentalis*), Turkey, Caucasus, Georgia, in the east over the Siberian mountains (subsp. *intermedia* DJAKONOV, 1926), the mountains of central Asia (subsp.? *lissas* PROUT, 1912) to East Asia (nominate subspecies according to VIIDALEPP 1996).

PHENOLOGY: Univoltine in central Europe. Main flight period: mid-June to late July, exceptionally somewhat earlier or later, very rarely with partial second generation (Germany: URBAHN 1965). In southern Europe bivoltine: late May to mid-July and early August to early September, sometimes until mid-October. In the southern European Mountains (Pyrenees, Balkans) univoltine as in central Europe. In univoltine populations larval stages from late July to early June. In bivoltine populations quickly growing larvae from June to July. Overwintering as larva in the ground (LASS 1923).

BIOLOGY: Oligophagous (1) on some Ranunculaceae: Eggs and larvae found on *Clematis vitalba* (Germany: SCHNEIDER 1934; PEKING 1953; BERGMANN 1955; WEGNER 1996; LEIPNITZ pers. comm.; Poland: BUSZKO pers. comm.; Great Britain: SKINNER pers. comm.; Austria: SCHWINGENSCHUSS 1953), *C. alpina and C. integrifolia* (Germany: LEIPNITZ pers. comm.) and *Pulsatilla pratensis* (Germany: GELBRECHT pers. comm.; Estonia: VIIDALEPP pers. comm.). Recorded also on ornamental species such as *Clematis recta, C. viticella* (var. cit.) and *Pulsatilla vulgaris* (FIBIGER & SVENDSEN 1981; SKOU 1986; SVENSSON

1993). Eggs flat, positioned in piles of about 5-20 eggs on twigs of host-plant. Pupation between loosely interwoven leaves on the host-plant (SKOU 1986; JØRGENSEN & SKOU 1982). Adults nocturnal, attracted to light. ♀ ratio at light high, about 50%.

HABITAT: In central Europe thermophilous. Inhabiting hedges, gardens, forest fringes, river valleys, glades, and other sun-exposed localities with the host-plant. Often on limestone. From 0 m up to 1,500 m above sea-level.

PARASITOIDS: Hymenoptera, Braconidae: *Microgaster reconditus* (GUMPPENBERG 1892). - Diptera, Tachinidae: *Zenillia bisetosa* (THOMPSON 1946), *Exorista aemula* (GUMPPENBERG 1892). Larvae often parasitized (BERGMANN 1955).

SIMILAR SPECIES: Exceptionally small specimens can resemble large females of *Xenochlorodes olympiaria*, but there is no real overlap in size, in *X. olympiaria* the hindwing termen is rounded, the transversal lines are less convex, the ♀ antennae nearly filiform and both sexes have only two hindspurs. *Chlorissa asphaleia* WILTSHIRE, 1966 from northern Turkey and Armenia also feeds on *Clematis* but differs clearly in the ciliate ♂ antennae, the more tailed hindwing and in the antemedial line of the forewing that meets the inner termen at right angles. The genitalia are totally different (HAUSMANN 1996b).

REMARKS: The taxon *lissas* PROUT, 1912 from central Asia may be a distinct species (HAUSMANN 1996).

Hemistola chrysoprasaria occidentalis WEHRLI, 1929

Hemistola chrysoprasaria occidentalis WEHRLI, 1929: Ezheg. gosud. Muz. N.M. Mart'yanova 6 (1): 9 (Tunisia: Ain-Draham nr. Tunis). Lectotype ♂, herewith designated (to stabilize nomenclature) from Tunis, Ain-Draham, gen.prp. 4037 (ZFMK, examined, pl. 8, fig. 8) with paralectotypes from Spain, Andalusia: Sierra Nevada, Ronda (ZFMK).

DIAGNOSIS: Wingspan ♂ 26-30, ♀ 32-35 mm. Antemedial line nearly invisible. Postmedial line of forewing usually less concave than in nominate subspecies, slightly dentate in lectotype from Tunisia, straight in Spanish and Moroccan specimens. Veins somewhat lighter than ground colour, especially in terminal area. External morphology as in nominate subspecies, but antennae of both sexes with shorter pectinations, longest branches about three times width of flagellum in ♂, 0.8-1.0 times in ♀.

MALE GENITALIA: As in nominate subspecies, but semi-membranous process of valvula slender. Spine at base of valva slightly shorter and narrower, curved. Medial notch of anterior margin of valva rather deep. Aedeagus with two patches of cornuti, distal usually larger than central. Paired projection of sternum A8 tapered, outwardly curved. Lectotype from Tunisia corresponding well genitalically to examined males from southern Spain and Morocco.

FEMALE GENITALIA: Paired spinose patches on lamella antevaginalis smaller, spines shorter than in nominate subspecies. Both patches widely separated from each other. Semi-membranous plate near ostium bursae (lamella postvaginalis) flat, rounded.

DISTRIBUTION: Subspecies West-Mediterranean. Mountains of southern Spain, southern Portugal.

Outside Europe in Morocco (Middle and Great Atlas), Algeria, Tunisia.

PHENOLOGY: Univoltine (scarce data). Main flight period: late June to late July. Larval stages unknown.

BIOLOGY: Larva probably feeding on *Clematis*, but at present without authentic host-plant records for this subspecies.

HABITAT: Undescribed. Montane: Mainly from 1,200 m up to 1,600 m above sea-level.

SIMILAR SPECIES: See nominate subspecies.

REMARKS: All the populations from central Spain and Portugal need further investigation, to clarify, whether they are attributable to this subspecies or not.

198   199

Text-figs 198-199: Diagnostic characters (indicated) of European *Hemistola* species. Text-fig. 198: *H. chrysoprasaria* (ESPER, 1795). Text-fig. 199: *H. siciliana* PROUT, 1935.

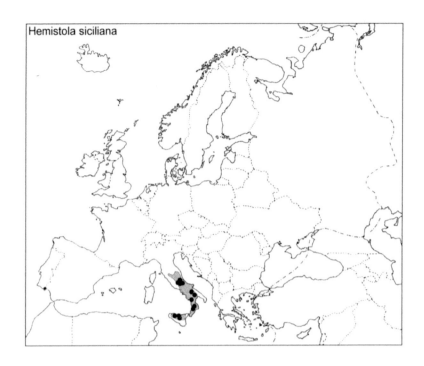

27. *Hemistola siciliana* PROUT, 1935

Hemistola chrysoprasaria siciliana PROUT, 1935: in Seitz, Macrolep. 4, Suppl.: 19 (Italy, Sicily: Taormina, Ficuzza, Busambra). 8 syntypes (BMNH, 1♀ examined).

DIAGNOSIS: Wingspan ♂ 27-30, ♀ 30-38 mm. Wing colour and wing pattern similar to that of *Hemistola chrysoprasaria*, but ground colour more yellowish green, lacking the bluish turquoise tinge. Postmedial line of forewing dentate on veins. Antemedial line weakly marked, with strong distal projection below CuA1. Veins lighter, mainly in distal half of wings. Underside with postmedial line very weakly marked. External morphology as in *H. chrysoprasaria*.

MALE GENITALIA: Basal spine of valva usually not connected by sclerotized ridge with anterior margin of valva. Socii and membranous process of valvula narrower than in *H. chrysoprasaria*. Medial notch of anterior margin of valva rather deep. Aedeagus without cornuti, in central Italian specimens exceptionally present (hybrids?). Sternum A8 with two tapered projections widely separated from each other.

FEMALE GENITALIA: Paired spinose patches of lamella antevaginalis close to each other, with many long spines. Semi-membranous plate near ostium bursae (lamella postvaginalis) subrectangular.

DISTRIBUTION: Adriato-Mediterranean. Sicily, central and southern Italy.

PHENOLOGY: Bivoltine in the lowlands (scarce data): late May to late July; late August to late September. Perhaps univoltine in the mountains. Larval stages unknown.

BIOLOGY: Larvae probably feeding on *Clematis*, but as yet no authentic host-plant records for this species.

HABITAT: Forest fringes and hedges with *Clematis*. From 0 m up to 1,300 m above sea-level.

SIMILAR SPECIES: Large females of *Xenochlorodes olympiaria* from Sicily somewhat resemble small *Hemistola siciliana*, but the former can easily be distinguished by the white dorsal line on the green abdomen and the absence of proximal spurs in both sexes. *H. chrysoprasaria* (allopatric) differs in the more bluish ground colour, the straight post-medial line of the forewing, and in the presence of cornuti in the ♂ genitalia.

REMARKS: Raised to species rank in HAUSMANN (1996).

Xenochlorodes WARREN, 1897

TYPE-SPECIES: *Xenochlorodes pallida* WARREN, 1897 [=*Xenochlorodes olympiaria cremonaria* (STAUDINGER, 1897)]. Subgenus *Xenochlorodes* with three species in the Western Palaearctic and one in South Africa. Subgenus *Hissarica* VIIDALEPP, 1988 with two or three species in central Asia (HAUSMANN 1996b). Placement in Hemistolini tentative. Some features, such as forewing venation and absence of ♂ setal patches on sternum A3, reminiscent of Microloxiini.

DIAGNOSIS: Hindwing termen rounded. Venation (subgenus *Xenochlorodes*): Both R1 and R2 of forewing usually fused with Sc, to form areole, distally not reaching costa. R1 and common stalk of R2-R5-M1 usually arising from cell at the same point. M3 and CuA1 shortly stalked. Hindwing: Sc+R1 and Rs fused over 2/3-3/4 length of cell. Rs and M1 distinctly stalked. M3 and CuA1 stalked. Discocellular veins strongly angled. Frenulum absent in both sexes. Proboscis usually developed, length ca. 2 mm, absent in subgenus *Hissarica* from central Asia and in African species. Palpi slender, short. Eyes quite distant from each other. Frons flat, reddish brown. Vertex white. Antennae of ♂ bipectinate, in ♀ with very short branches or filiform. Hindtibia (♂ ♀) not dilated, with two long terminal spurs of unequal length. Abdominal crests absent. Paired setal patch on ♂ sternum A3 absent. Tergum A2 of ♂ interiorly with pair of long, spatulate apodemes of unknown function.

MALE GENITALIA: Uncus and gnathos strongly sclerotized, pincer-shaped. Gnathos strongly tapered. Socii membranous, very weakly developed. Saccus concave, but convex in *Hissarica*. Valva as in many other Hemistolini, ventrally excavated, without spinose processes. Basal coremata present. Posterior margin of sternum A8 bilobed, with narrow emargination.

FEMALE GENITALIA: Simple: Lamella antevaginalis weakly sclerotized. Ductus bursae strongly sclerotized. Signum bursae absent.

IMMATURE STAGES: Larva slender. Head cleft. Segment T1 with typical paired dorsal processes. Anal process present, tapered.

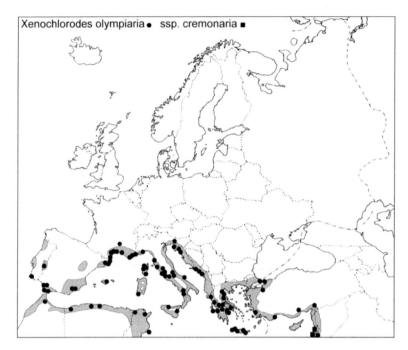

Xenochlorodes olympiaria ● ssp. cremonaria ■

28. *Xenochlorodes olympiaria* (HERRICH-SCHÄFFER, 1852)

‡ *olympiaria* HERRICH-SCHÄFFER, 1851: Syst. Bearb. Schmett. Eur. 3: pl. 87, fig. 539. Non binominal.
Geometra (Eucrostes) olympiaria HERRICH-SCHÄFFER, 1852: Syst. Bearb. Schmett. Eur. 6 (55): 63 (north-western Turkey: Olymp near Brussa). Type(s) not traced. Binominal because referring to p. 9, vol. 3.
Geometra beryllaria MANN, 1853: Verh. zool.-bot. Ver. Wien 3: 76 (Croatia, Dalmatia: Spalato). Type(s) not traced (NHMW?). Junior synonym (HAUSMANN 1996b).
Nemoria aureliaria MILLIÈRE, 1864: Iconogr. Chen. Descr. Lép. inédits 2 (12): 37, pl. 55, figs 1, 2 (southern Italy). Type(s) not traced. Junior synonym.
Unavailable names (infrasubspecific): *minor*: SCHWINGENSCHUSS (1926). - (incorrect subsequent spelling): *eureliaria*: PROUT (1913).

DIAGNOSIS: Wingspan ♂ 18-21, ♀ 20-25 mm, second generation 15-19 mm. Forewing costa strongly convex, termen nearly straight. Ground colour bright green with bluish tinge. Fresh specimens exceptionally pale green (mainly in south-eastern Europe) and reminiscent of subsp. *cremonaria* (STAUDINGER, 1897) from the Levant. Forewing costa white, often with yellowish or reddish tinge. Antemedial and postmedial lines of forewing white, indistinct. Antemedial line convex. Postmedial line usually rather straight, oblique. Postmedial line of hindwing slightly convex. Cell spots and terminal line absent. Fringe green at base, distal half white. Abdomen in fresh specimens green with white dorsal line. Palpi reddish brown fading to ochreous, short in both sexes (0.8-1.0 times diameter of eye). Antennae of ♂ bipectinate nearly to tip, length of longest branches about 4-5 times width of flagellum. Antennae of ♀ very shortly bipectinate or almost filiform, branches not exceeding width of flagellum. Hindtarsus of ♂ not shortened. Further features under generic diagnosis.

MALE GENITALIA: Aedeagus long and slender, with 2-5 external spines. Lobes at posterior margin of sternum A8 rounded, in subsp. *cremonaria* from the Levant sub-rectangular.

166

FEMALE GENITALIA: Ductus bursae strongly sclerotized, slightly widened near ostium bursae and in the oral half, near corpus bursae narrow. Antrum cup-shaped. Corpus bursae membranous, pyriform or globular.

DISTRIBUTION: Mediterranean. Widely distributed along Mediterranean coasts from Spain to Greece including all islands except Malta.

Outside Europe from Morocco to Libya, Turkey, western Syria. Subsp. *cremonaria* in Cyprus, Lebanon, northern and central Israel.

PHENOLOGY: Bivoltine. Main flight period: Early May to late June; late July to late September; in Crete until mid-October. Larval stages early June to mid-July; mid-August, overwintering, to mid-April.

BIOLOGY: Larva oligophagous (2) on Oleaceae, reputedly also on Rhamnaceae. Found on *Phillyrea angustifolia* (Spain: BODI, LEIPNITZ pers. comm.), *P. latifolia* (Israel: HALPERIN & SAUTER 1992; HAUSMANN 1997a), *Olea europaea* (MILLIÈRE 1874) and *Ligustrum lucidum* (Crete: LEIPNITZ pers. comm.). Recorded on *Phillyrea latifolia* (Europe: var. cit.) and *Rhamnus* spp., probably *R. lycioides* (Israel: HALPERIN & SAUTER 1992; HAUSMANN 1997a). Reared by the author on *Ligustrum vulgare*. Adults nocturnal. ♀ ratio at light 25-40%.

HABITAT: Xerothermophilous. *Olea-Ficus* plantations near the coasts, Mediterranean macchia, rocky, sun-exposed slopes. From 0 m up to 500 m above sea-level, in Israel up to 800 m.

SIMILAR SPECIES: Large females from Sicily somewhat resemble small *Hemistola siciliana*. The latter can easily be distinguished by the presence of the proximal spurs and the absence of the white dorsal line on the green abdomen. *Hierochthonia pulverata* (WARREN, 1901) from southern Turkey and Lebanon, and *H. semitata* (PÜNGELER, 1901) from Israel are very similar, but differ in the length of the ♀ antennal branches (twice width of flagellum), the slightly sharper and oblique postmedial lines of the forewing, and in the absence of the proboscis. The genitalia of both sexes are strikingly different.

Tribus **Comostolini** INOUE, 1961

At least five genera with many species in the Old World. One species, *Eucrostes dominicaria* GUENÉE, [1858], in the Neotropical region. Some features, such as ♂ genitalia, reminiscent of Hemistolini. Structural details of antennae, hindtarsus etc. similar to equivalents in Jodini. Genus *Eucrostes* matching diagnosis of Comostolini well (HAUSMANN 1996b). Alternative tribal concept in HOLLOWAY (1996) see notes to Hemistolini.

DIAGNOSIS: Venation: M3 and CuA1 stalked in both wings. Rs and M1 of hindwing stalked. Hindwing termen rounded or slightly angled at M3. Frenulum absent in both sexes. Proboscis present. Palpi very slender. Third segment of ♀ palpi often elongate. Antennae bipectinate in ♂, last 1/4-1/2 filiform, ciliate. Antennae of ♀ usually filiform, or with very short pectination (*Eucrostes*). Hindtibia usually with two pairs of spurs in both sexes, but one in *Eucrostes*. Hindtarsus usually short. Ansa of tympanal organ very long and narrow. Abdominal crests absent. Setal patches on ♂ sternum A3 weakly developed or absent.

MALE GENITALIA: Uncus long, simple or forked at apex. Gnathos present, tapered, often long. Basal coremata of valvae usually absent. Base of valva with spinulose lobes. Aedeagus long and slender. Sternum A8 and tergum A8 usually simple.

FEMALE GENITALIA: Signum often present.

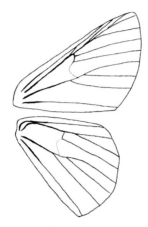

Text-fig. 200: Venation of *Eucrostes indigenata* (VILLERS, 1789).

Eucrostes HÜBNER, [1823]

TYPE-SPECIES: *Geometra fimbriolaria* HÜBNER, [1817] [=*Eucrostes indigenata* (VILLERS, 1789)]. Genus with 10 species nearly all over the world, mainly in tropical and subtropical zones. Frequent misspellings in subsequent literature: *Eucrostis, Euchrostis, Euchrostes*.

DIAGNOSIS: Small species with robust body. Hindwing rounded. Yellowish green, fringe reddish. Venation of forewing: Sc and R1 fused. M1 arising from below cell apex, diagnostic against other Comostolini-genera. R2-R5 and M1 usually separate. Hindwing: Sc+R1 and Rs touching, but not fused, at one point. Rs and M1 usually stalked. Discocellulars often weak between M2 and M3 on all wings. Frons flat. Proboscis present, but weak. Second and third segment of palpi elongate in ♀. Antennae short, in ♂ bipectinate with long branches, in ♀ with very short branches or in some species, nearly filiform. Setal patches on sternum A3 absent. Hindtibia in both sexes with two spurs, not dilated, tarsus not shortened. Further features under diagnosis of tribe.

MALE GENITALIA: Uncus terminally forked. Socii semi-membranous, long. Gnathos well sclerotized, slender, tapered. Valva with chitinous crest instead of basal coremata. Aedeagus long, slender, vesica spinulose in central part. Sternum and tergum A8 simple.

FEMALE GENITALIA: Apophyses anteriores very short, apophyses posteriores of medium length. Lamella antevaginalis sclerotized to resemble longitudinally folded ribbon. Ostium bursae funnel-shaped. Corpus bursae membranous and tender, more or less pyriform. Signum absent.

IMMATURE STAGES: Larva (pl. 8, fig. 1) with small dark head capsule, dorsally without notch. Segment T1 with paired red dorsal processes, red dorsal processes also on segments A1-A5 and A8. Pupa slender, with long cremaster.

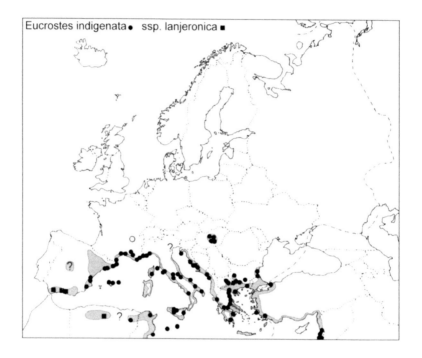

29. *Eucrostes indigenata* (VILLERS, 1789)

Phalaena (Geometra) indigenata VILLERS, 1789: Caroli Linn. Ent. 2: 383, pl. 6, fig. 19 (France: 'old Rhone valley', [near Lyon]). Type(s) not traced.
Geometra fimbriolaria HÜBNER, [1817]: Samml. Eur. Schmett. 5: pl. 91, fig. 468 ('Europe'). Type(s) lost. Junior synonym.
Fidonia indigenaria TREITSCHKE, 1827: Schmett. Eur. 6 (1): 269. No separate types. Unjustified emendation: Intentional change of final syllable because of bipectinate antennae.
Eucrostis nudilimbaria MABILLE, 1880: Bull. Soc. Ent. Fr. (1879) p. CLV [155] (Corsica, Saint Florent). Holotype ♂ (coll. HERBULOT in ZSM, examined). Junior synonym.
‡ *Phalaena verginalis* COSTA, [1841]: Cat. Lep. Regno Nap., in Dizion. Univ. Agric.: [13]. Incorrect (multiple) original spelling of *Phalaena virginalis*.
‡ *Phalaena virginalis* COSTA, [1841]: Cat. Lep. Regno Nap., in Dizion. Univ. Agric.: [15], nec GEOFFROY in FOURCOY, 1785 (identity: *Scopula virginalis*, senior synonym of *Scopula caricaria* REUTTI, 1853). Junior primary homonym.
Eucrostes indigenata lanjeronica HAUSMANN, 1996: Nota lepid. 19 (1/2): 43, fig. 148 (southern Spain, Andalusia: Lanjeron). Holotype ♀ (ZSM, examined). Valid at subspecific rank.
Unavailable names (infrasubspecific): *pulchra*: MARIANI (1937). - (incorrect subsequent spelling): *indignata*: GOMEZ DE AIZPURUA (1974).

DIAGNOSIS: Wingspan ♂ 14-16, ♀ 18-20 mm, in subsequent generations much smaller: ♂ usually 10-13, ♀ 14-18 mm. Forewing apex and termen rounded. Ground colour bright green with yellow tinge, especially towards terminal area. Forewing costa yellow. Post-medial line white, fairly indistinct, slightly convex, in southern Spanish specimens (subsp. *lanjeronica* HAUSMANN, 1996) more distal than in nominate subspecies. Antemedial line lacking. Cell spots red-brown, sharp and very small. Fringe and terminal line red-brown. Terminal line red-brown, in subsp. *lanjeronica* slightly narrower than in nominate subspecies. Terminal line proximally bordered yellow, exceptionally not ('f. *nudilimbaria*'). Underside of forewing with conspicuous orange streak at costa. Venation of hindwing: Rs and M1 in western Mediterranean populations shortly stalked or sometimes connate, distinctly stalked in populations from Tunisia, Sicily, southern Italy and eastern Mediterranean, transitional in central Italy. Abdomen unicolorous green. Frons reddish brown or orange. Vertex white with some yellow or green scales. Proboscis weak, short (1-2 mm). Palpi pale ochreous, length in ♂ 1.0 times in ♀ about 1.5 times diameter of eye, third segment 'naked' with appressed scales. Antennae bipectinate over 2/3 length of flagellum, length of longest branches in ♂ 4-5 times width of flagellum, in ♀ slightly exceeding width of flagellum (nominate subspecies) or 2-3 times width of flagellum (subsp. *lanjeronica*). Further features under generic diagnosis.

MALE GENITALIA: Uncus comparatively long and slender, terminally forked. Shorter and broader in subsp. *lanjeronica*. Basal crest of valva with very small spines.

FEMALE GENITALIA: See generic diagnosis.

DISTRIBUTION: Mediterranean, constituting a Mediterranean-African species-complex with its sister species *Eucrostes disparata* (WALKER, 1861). Widely distributed along coasts of southern Europe from north-western Spain, Mallorca and south-eastern France to Greece and Bulgaria, even in some inland localities such as Hungary and Macedonia. Present on nearly all the Mediterranean islands, even on Crete, but not yet recorded from the other Greek islands. Probably extinct from type locality near Lyon. Subsp. *lanjeronica* HAUSMANN, 1996 distributed in southern Spain.
   Outside Europe in Algeria (subsp. *lanjeronica*), Tunisia (nominate subspecies), Libya, Turkey, and the Levant down to Israel.

PHENOLOGY: Bivoltine or, often, trivoltine (PROUT 1913a; CULOT 1919). From late April to mid-October, main flight period from mid-May to late June and mid-August to mid-September. In Malta until early November. Larval stages usually June to July and September, overwintering, to early May.

BIOLOGY: Larva monophagous (2): Recorded on *Euphorbia* spp. (var. cit.), mainly on *E. spinosa* (GUMPPENBERG 1892; PROUT 1913a). In Malta on *E. pinea* (VALLETTA 1973). Reared on *E. cyparissias, E. virgata* and *E. platyphyllos*, preferring flowers and green seeds (LEIPNITZ, GELBRECHT pers. comm.). Larvae recorded sucking 'juice' of *Euphorbia* plants (PROUT 1913a), hence plant chemistry perhaps important for defence strategy of *Eucrostes indigenata* by making it unpalatable. Low flight activity in ♀. K-strategy. ♀ ratio at light usually about 30%.

170

HABITAT: Xerothermophilous. Coastal salt marshes and other open habitats. Mainly in hot coastal lowlands from 0 m up to 700 m, exceptionally up to 1,200 m above sea-level.

SIMILAR SPECIES: No similar species in the area of distribution. This species is very closely related to its Afrotropical sister species *Eucrostes disparata* (WALKER, 1861). The latter differs in ♂ genitalia in the shorter and thicker uncus, the basal crest of the valva bears longer spines.

REMARKS: Females from north-eastern Spain have short antennal branches like those of the nominate subspecies. Examined females from central Algeria with long branches, more than twice width of flagellum, in eastern Algeria 1.5 times only.

Tribus **Jodini** INOUE, 1961

Introduced as 'Jodiini', an incorrect original spelling. Provisional tribal concept (HAUSMANN 1996b) with four Palaearctic and Indo-Pacific genera. Many features reminiscent of Comostolini. Alternative tribal concept in HOLLOWAY (1996) see introductory notes to Hemistolini. Wider concept of Hemitheini sensu VIIDALEPP (1996) with Jodini included. Differential characters of Jodini against Thalerini and Hemitheini see below, the last two being closely related to each other.

DIAGNOSIS: Venation of forewing: M3 and CuA1 usually unstalked. Hindwing: Termen tailed at M3, sometimes also angled at M1. Sc+R1 and Rs touching (not fused) at one point. Rs and M1 stalked. M3 and CuA1 usually stalked. Frenulum usually absent, but present in genus *Gelasma*. Palpi long in both sexes. Proboscis present. Antennae bipectinate in ♂, last 1/4-1/2 filiform, ciliate. Pectination of the appressed, untidy type (HOLLOWAY 1996; Text-fig. 82), diagnostic against Thalerini and Hemitheini. Antennae of ♀ usually filiform. Hindtibia (♂ ♀) with four spurs, ♂ tarsus short. Abdominal crests absent. Setae of sternum A3 usually in a circular central patch, diagnostic against Thalerini and Hemitheini. Lateral apodemes on ♂ sternum A2 well developed. Ansa of tympanal organ very long and narrow.

MALE GENITALIA: Socii present, but often very weak and/or short. Gnathos present, usually slightly sclerotized only. Valva slender, typically with ventral excavation and 'oblique groove running from the transtilla to the centre of the ventral margin' (HOLLOWAY 1996). Basal coremata of valvae usually present, but absent in European *Jodis* species. Posterior margin of sternum A8 strongly sclerotized, notched medially (with exceptions).

FEMALE GENITALIA: Ovipositor lobes with setae on papillate projections situated in two narrow oblique rows. Apophyses anteriores very short, apophyses posteriores of medium length. Ductus bursae sclerotized and corrugated. Corpus bursae usually with bicornute signum, connected with ductus bursae at posterior termen.

Jodis HÜBNER, [1823]

TYPE-SPECIES: *Geometra aeruginaria* [DENIS & SCHIFFERMÜLLER], 1775 [=*Jodis lactearia* (LINNAEUS, 1758)]. Two species in the western Palaearctic, over 30 Indo-Pacific and eastern Palaearctic species. Subsequent misspelling: *Iodis*.

DIAGNOSIS: Wing colour light green or whitish green, often shining or slightly transparent, not 'chequered' as in most other Jodini. Colour very quickly fading. Cell spots darker, close to antemedial line. Hindwing termen tailed at M3, not angled at M1. M3 and CuA1 of forewing connate in European species. M3 and CuA1 of hindwing stalked in European species. Frenulum absent. Third segment of ♀ palpi slightly elongate. Hindtibia (♂ ♀) with four spurs, in ♂ with pencil, covering ca. 1/4 of tarsus.

MALE GENITALIA: Aedeagus slender, without cornuti in European species, with basal stalk long and slender. Basal coremata of valvae absent, however present in some tropical species (HOLLOWAY 1996). Posterior margin of sternum A8 strongly sclerotized with medial notch. Further features under diagnosis of tribe.

IMMATURE STAGES: Larva very slender. Head deeply notched dorsally (half length of head capsule). Paired projections on segment T1 rather short, tergum of last segment A10 extended to long, single anal process. Skin comparatively smooth. Pupa very slender, for morphology see PATOCKA (1995).

Text-fig. 201: Venation of *Jodis lactearia* (LINNAEUS, 1758).

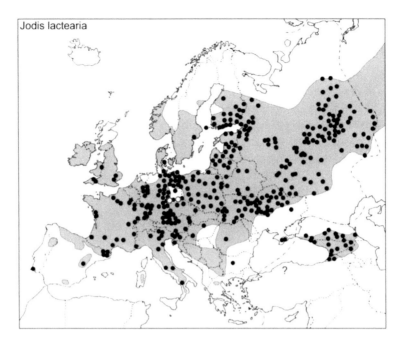

Jodis lactearia

30. *Jodis lactearia* (LINNAEUS, 1758)

Phalaena (Geometra) lactearia LINNAEUS, 1758: Syst. Nat. (Ed. 10) 1: 519 ('Europe', [Sweden?]). Type(s) in LSL.

Phalaena (Geometra) vernaria LINNAEUS, 1761: Fauna Suecica (Ed. 2): 323 ([Sweden]: 'Dalekarlia' [Dalarna]). Syntype(s) in NHRS. Junior synonym.

Geometra aeruginaria [DENIS & SCHIFFERMÜLLER], 1775: Ank. syst. Werkes Schmett. Wienergegend: 314 (Austria: Vienna distr.). Type(s) lost. Junior synonym.

Phalaena semipectinata RETZIUS, 1783: Gen. Spec. Insect.: 43. ([Europe]). Type(s) not traced. Synonym of *J. lactearia* (WOLF 1988; LERAUT 1997) or of *J. putata* (PROUT 1912b; SCOBLE 1999).

Phalaena pomona FOURCROY, 1785: Entomol. Paris. 2: 264 (France: Paris). Type(s) in Mus. Autun (F). Junior synonym of *Jodis lactearia* according to WOLF (1988). However possibly conspecific with *Hemistola chrysoprasaria* (ESPER, 1795) according to PROUT (1912). Examination of types necessary.

Phalaena lactea FOURCROY, 1785: Entomol. Paris. 2: 273 (France: Paris). Type(s) in Mus. Autun (F). Junior synonym.

Phalaena syrene FOURCROY, 1785: Entomol. Paris. 2: 286 (France: Paris). Type(s) in Mus. Autun (F). Identity of type to be examined. Junior synonym of *J. lactearia* (LERAUT 1997) or of *Chlorissa viridata* (SCOBLE 1999).

Phalaena (Geometra) decolorata VILLERS, 1789: Linn. Ent. 2: 385 ('Europe, France'). Type(s) not traced. Junior synonym.

Jodis norbertaria RÖSSLER, 1877: Stettin. ent. Zeit. 38 (7-9): 365 (Spain, Basque Country: Bilbao). Holotype ♂ not traced. Junior synonym.

Jodis alliata HÖFNER, 1880: Jahresb. naturh. Landesmus. Kärnthen 14: 266. Type(s) not traced. Junior synonym. Reputedly synonym of *J. putata* (WOLF 1988; SCOBLE 1999). 'Straight transverse lines' however clearly referring to *J. lactearia*. Fresh specimens with dark colour occurring also in *J. lactearia*.

Unavailable names (misidentifications): *putata*: sensu STEPHENS (1850) nec LINNAEUS, 1758; *putataria*: sensu ESPER (1795) nec LINNAEUS, 1767 (identity: *Jodis putata*). - (infrasubspecific): *approximata*: LEMPKE (1949); *mediofasciata*: SCHWINGENSCHUSS (1953). - (incorrect subsequent spellings): *lactoaria*: CURÒ (1878); *sirene*: VILLERS (1789); *syrena*: FOURCROY (1785).

DIAGNOSIS: Wingspan ♂♀ 18-23 mm. Ground colour in fresh specimens pale, bluish green, almost appearing transparent. In collection specimens, but also in nature, quickly fading to white. Transverse lines white. Antemedial line often developed on both wings, regularly convex. Postmedial line of hindwings convex, of forewing straight, near costa inwardly projecting. Transverse lines invisible in faded specimens. Cell spots elongate, darker, but hardly visible. Fringe concolorous. Frons flat, pale brown or whitish. Palpi whitish, very slender, length ca. 1.5-2 times diameter of eye. Antennae bipectinate in ♂, last 1/4-1/2 filiform, ciliate, longest branches 10 times width of flagellum. Antennae of ♀ filiform. Hindtarsus short in ♂, length 1/3 tibia. Further features under generic diagnosis.

MALE GENITALIA: Uncus broad. Saccus long. Valva with oblique groove large. Juxta posteriorly tapered. Posterior margin of sternum A8 with paired, tapered, strongly spinulose projections. Setae of sternum A3 long, arising from large circular central patch.

FEMALE GENITALIA: Very similar to those of *Jodis putata*, but usually larger. Ductus bursae often longer, signum slightly larger. Identification on the base of ♀ genitalia often problematic.

DISTRIBUTION: Palaearctic. Europe except northern Scandinavia. Occurrence in central and southern Spain (RIBBE 1912) requiring confirmation. Absent from most parts of southern and eastern Balkans, and from nearly all Mediterranean islands except Sardinia. In southern Europe often endangered due to isolation of populations.

Outside Europe in the Caucasus and Transcaucasus, in the east to Japan.

PHENOLOGY: In central and northern Europe univoltine. Main flight period: mid-May to late June, exceptionally somewhat earlier or later. Under warm conditions partial second brood in late August to mid-September (Germany: BERGMANN 1955; KOCH 1984; KRAUS 1993). In Hungary and southern Europe usually bivoltine, main flight periods mid-April to early June and early July to mid-August, sometimes until mid-September. Generations not sharply separated (FORSTER & WOHLFAHRT 1981). Univoltine in the southern European mountains. Larval stages in univoltine populations mid-July to late September, in bivoltine populations June to July and early August to early October. Overwintering as pupa.

BIOLOGY: Larva polyphagous on various trees, shrubs and herbaceous plants, in central and northern Europe slightly preferring *Betula* (var. cit.). Found on *Betula* (various species), *Quercus robur, Malus domestica, Sorbus aucuparia, Crataegus, Corylus avellana, Tilia cordata, Salix caprea, Prunus padus, Fraxinus excelsior, Vaccinium uliginosum*, and *V. myrtillus* (Finland: SEPPÄNEN 1954; 1970; Poland: BUSZKO pers. comm.; Great Britain: SKINNER pers. comm.; PORTER 1997; BURROWS 1940a; Germany: WEGNER pers. comm.; BERGMANN 1955; URBAHN 1939), and *Rhododendron* (Alps: OSTHELDER 1929). Recorded also on *Alnus, Carpinus betulus, Fagus sylvatica, Rubus, Populus tremula* (var. cit.) and *Genista* (REBEL 1910; ALLAN 1949; FORSTER & WOHLFAHRT 1981). Pupation in loose cocoon between leaves and twigs of host-plant. Adults flying for a short period at dusk. Attraction to light almost exclusively confined to males, rarely approaching close to light source.

PARASITOIDS: Hymenoptera, Braconidae: *Apanteles caberae*. - Ichneumonidae: *Meso chorus pictilis* (THOMPSON 1945).

HABITAT: In central Europe mainly in forest-rides, glades and forest-fringes in deciduous wetland woods. Preferring birch forests and birch marshes with many undergrowing shrubs (BERGMANN 1955; SKOU 1986; WEGNER pers. comm.). Mesophilous in eastern Germany and Estonia, mainly in deciduous woods of 'broad-leaved type', i.e. *Quercus, Tilia, Fraxinus, Corylus, Acer* etc. (VIIDALEPP, GELBRECHT pers. comm.). In France reported as associated with Quercetalia pubescenti (CHAPELON 1992), Often on soils with high ground water. Mainly from 0 m up to 700 m above sea-level, in the southern Alps, southern Italy and Bulgaria up to 1,500 m.

SIMILAR SPECIES: *Jodis putata* differs in the more yellowish green, less transparent, less shining ground colour, and in the strongly jagged transverse lines.

REMARKS: '*J. norbertaria* RÖSSLER, 1870' was described from northern Spain, from specimens taken in August, as being characterized by a darker ground colour and dark medial area. Some examined northern Spanish specimens, from the spring generation, match the nominate subspecies well in structure and habitus and do not deserve a separate name.

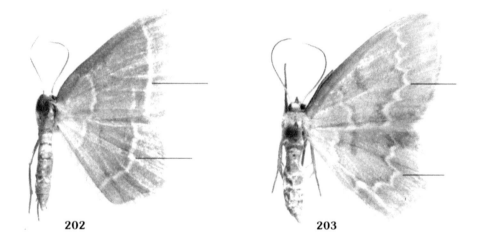

202   203

Text-figs 202-203: Diagnostic characters (indicated) of European *Jodis* species. Text-fig. 202: *J. lactearia* (LINNAEUS, 1758). Text-fig. 203: *J. putata* (LINNAEUS, 1758).

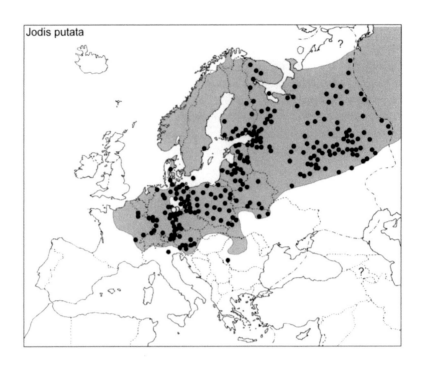

Jodis putata

31. *Jodis putata* (LINNAEUS, 1758)

Phalaena (Geometra) putata LINNAEUS, 1758: Syst. Nat. (Ed. 10) 1: 523 (Europe). Type(s) in LSL.
Phalaena (Geometra) putatoria LINNAEUS, 1761: Fauna Suecica (Ed. 2): 323. Unjustified emendation: Intentional change of final syllable because of bipectinate ♂ antennae.
Phalaena (Geometra) putataria LINNAEUS, 1767: Syst. Nat. (Ed. 12) 1 (2): 859. Unjustified emendation: Correction of print error 'putatoria'.
‡ *Phalaena vernaria*: sensu HUFNAGEL, 1767: Berlin. Magazin 4: 506, nec LINNAEUS, 1761 (identity: *Jodis lactearia*). Misidentification.
Phalaena undulosa FOURCROY, 1785: Entomol. Paris. 2: 286 (France: Paris). Type(s) in Mus. Autun (F). Junior synonym of *J. putata* according to LERAUT (1997).
Phalaena (Geometra) micantaria ESPER, 1795: Schmett. in Abbild. 5 (1): p. 28; pl. 2, figs. 7-8 (Germany, Saxonia: Leipzig). Lectotype ♂ (LMW; examined: HACKER 1999). Junior synonym.
Geometra putaria FISCHER VON RÖSLERSTAMM, 1838: in FREYER: Neu. Beitr. Schmett. 2: 31. No separate types. Unjustified emendation: Intentional change of final syllable because of bipectinate ♂ antennae.

DIAGNOSIS: Wingspan ♂ ♀ 17-21 mm. Ground colour whitish green, in fresh specimens with faint bluish tinge and appearing slightly cloudy. Colour in nature and collection specimens quickly fading to a very pale bony yellow. Transverse lines white, often darkly bordered. Postmedial lines strongly jagged. Transverse lines slightly visible even in most faded specimens. Cell spots elongate, darker. Fringe concolorous. Frons flat, brownish green. Palpi whitish, very slender, length 1.5-2.2 times diameter of eye. Antennae bipectinate in ♂, last 1/4-1/2 filiform, ciliate, longest branches 8 times width of flagellum. Antennae of ♀ filiform. Hind-tarsus of ♂ short, length 1/3 tibia. Further features under generic diagnosis.

MALE GENITALIA: Uncus comparatively slender. Saccus shorter. Valva with oblique groove usually smaller than in *Jodis lactearia*, slender. Juxta not tapered posteriorly, often bilobed. Posterior margin of sternum A8 with rounded, slightly spinulose projections. Setose patch of sternum A3 smaller than in *J. lactearia*, with shorter setae.

176

FEMALE GENITALIA: Very similar to those of *J. lactearia*, but usually smaller. Ductus bursae often shorter, signum slightly smaller.

DISTRIBUTION: Palaearctic. Central and northern Europe except Great Britain, Ireland and Hungary. In the south from central France and northern Italy to Slowenja and Romania. Outside Europe recorded from Armenia (to be confirmed), in the east across Siberia to the Far East, Korea and Japan (subsp. *orientalis* WEHRLI, 1923).

PHENOLOGY: Univoltine. Main flight period: mid-May to late June, sometimes even in early May or until mid-July.Larval stages early July to early September, exceptionally late June. Overwintering as pupa.

BIOLOGY: Larva oligophagous. Found on *Vaccinium myrtillus* (Finland: SEPPÄNEN 1954; 1970; Poland: BUSZKO pers. comm.; Russia: ANIKIN et al. 2000; Germany: WEGNER pers. comm.; KRAUS 1993; BERGMANN 1955; OSTHELDER 1929), *V. uliginosum* (Finland: SEPPÄNEN 1954; 1970) and *Ledum palustre* (Germany: GELBRECHT pers. comm.) with clear preference for *V. myrtillus* (var. cit.). Recorded also on *Alnus* spp. (Finland: VALLE 1946), requiring confirmation. Pupation in loose cocoon between leaves and twigs of host-plant. Adults flying at dusk, also attracted to light and easily flushed during daytime. Often visiting flowers of *Frangula alnus* ('*Rhamnus frangula*'; SCHNEIDER 1934).

HABITAT: Woods on dry or slightly humid soils, but also in bogs with *Ledum palustre* (GELBRECHT pers. comm.). Often common in old oak or pine woods with extensive occurrence of *Vaccinium*. Higher tolerance of dryness than *Jodis lactearia*. In Estonia characteristically inhabiting the boreal type of deciduous forests, i.e. *Betula, Populus, Sorbus* etc. (VIIDALEPP pers. comm.). K-strategy. From 0 m up to 1,300 m above sea-level, in the southern Alps up to 1,600 m (FORSTER & WOHLFAHRT 1981).

SIMILAR SPECIES: *Jodis lactearia* differs in the more transparent wings, the shining, whitish green ground colour with turquoise tinge, and in the straight, not jagged transverse lines.

Tribus **Thalerini** HERBULOT, 1963
=Chlorochromini DUPONCHEL, [1845]

Chlorochromini postulated as valid name for both Thalerini and Hemitheini (SEVEN 1991), but validation of Thalerini preferable (HAUSMANN 1993, HOLLOWAY 1996a). Actual tribal concept sensu stricto including four mainly Palaearctic genera with Indo-Pacific representatives (HAUSMANN 1996b). Alternative tribal concept (HOLLOWAY 1996) see introductory notes to Hemistolini. Some features of genus *Bustilloxia* indicating relationships to Thalerini (HAUSMANN 1995a), but sclerotization of transtilla and shape of sternum A8 strongly differing from equivalents in typical Thalerini and suggesting isolated systematic position.

The tribes Hemistolini, Comostolini, Jodini, Thalerini, Hemitheini and Microloxiini are probably monophyletic within the subfamily Geometrinae. There are some, mostly tropical, genera showing mixed up feature combinations that are typical for different tribes.

Within this group of tribes the Hemitheini and the Thalerini should be regarded as 'sister tribes'. European taxa of both tribes differ constantly in the structure of the ♂ antenna and the apodemes of sternum A2. On the other hand, the genus *Culpinia* unites typical Hemitheini features, such as the number of ♀ hindtibial spurs and the presence of ♂ frenulum, and typical Thalerini features, such as certain details in the ♂ genitalia, the absence of setae on ♂ sternum A3, wing colour and hindwing margin. Some species of the Hemitheini genera *Chlorissa* and *Diplodesma* are linked with the Thalerini by ♂ genitalic features, such as the costal process of the valva and the narrow transtilla.

DIAGNOSIS: Venation: Forewing: R2-R5 and M1 usually stalked. Hindwing: Rs and M1 stalked. M3 and CuA1 usually unstalked. Margin of hindwing strongly concave between M1 and M3. Frenulum absent in both sexes or present in ♂ of genera *Culpinia* and *Bustilloxia*. Frons red, orange or reddish brown. Palpi short in both sexes, very slender. Antennae of both sexes usually bipectinate to tip, 'normal' type, diagnostic against Jodini. Hindtibia not dilated, usually with pair of terminal spurs only (both sexes). Abdominal crests absent. Setal patches of sternum A3 absent, diagnostic with regard to Jodini and Hemitheini. Lateral apodemes on sternum A2 absent, whilst present in Hemitheini.

MALE GENITALIA: Socii present, long, sometimes strongly sclerotized. Gnathos weak or absent. Transtilla very slender, elongate and U-shaped (exception: *Bustilloxia*). Valvae long, slender, usually with basal coremata. Costa of valva medially with strongly sclerotized process. Second process near base of valva usually present, but not in *Bustilloxia* and *Kuchleria*. Sternum A8 and tergum A8 usually simple (exception: *Bustilloxia*).

FEMALE GENITALIA: Sterigma circular or semi-circular. Ductus bursae short, corrugated. Corpus bursae subterminally connected with ductus bursae.

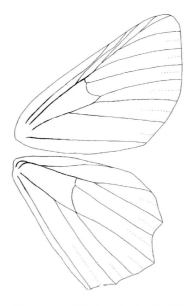

Text-fig. 204: Venation of *Thalera fimbrialis* (SCOPOLI, 1763).

Thalera HÜBNER, [1823]

TYPE-SPECIES: *Phalaena thymiaria* LINNAEUS, 1767 [=*Thalera fimbrialis* (SCOPOLI, 1763)].
Seven species in the Palaearctic and the 'temperate', northern parts of the Indo-Pacific.

DIAGNOSIS: Venation of forewing: R1 often fused with R2 over short distance. R2-R5 and
M1 usually stalked. M3 and CuA1 distinctly separate. Hindwing: Sc+R1 and Rs fused over
1/4 length of cell, whilst not fused in other Thalerini. M3 and CuA1 unstalked. Fringe
chequered. Proboscis well developed. Antennae (♂♀) bipectinate. Hindtibia (♂♀) with
pair of terminal spurs only. Further features under diagnosis of tribe.

MALE GENITALIA: Saccus elongate. Basal process of valva present, length diagnostic
between species. Aedeagus long, without cornuti. Sternum A8 and tergum A8 simple.

FEMALE GENITALIA: Sterigma stoutly sclerotized, irregularly cup-shaped, with large ventral
lobe. Ductus bursae short, leading to subapical part of corpus bursae. The latter very long,
slender, weakly sclerotized, upper part corrugated. Signum bursae absent.

IMMATURE STAGES: Larva (Text-fig. 25) slender. Head and segment T1 with long, paired,
dorsal processes. Last segment (A10) with long, pointed anal process (HAGGETT 1954).

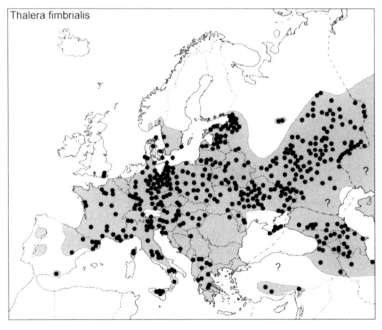

Thalera fimbrialis

32. Thalera fimbrialis (SCOPOLI, 1763)

Phalaena fimbrialis SCOPOLI, 1763: Ent. Carniolica: 216 (Slovenia: Carnia). Type(s) lost.
Phalaena (Geometra) thymiaria LINNAEUS, 1767: Syst. Nat. (Ed. 12) 2 (1): 859 ([Germany?]). Type(s)
  in LSL. Junior synonym.
‡ *Phalaena fimbriata* HUFNAGEL, 1767: Berlin. Mag. 4 (6): 604, nec SCHREBER, 1759 (identity: *Noctua
  fimbriata*, Noctuidae). Junior primary homonym.

Geometra bupleuraria [DENIS & SCHIFFERMÜLLER], 1775: Ank. syst. Werkes Schmett. Wienergegend: 97 (Austria: Vienna distr.). Type(s) lost.

Phalaena fimbriaria ROTTEMBURG, 1777: Naturf. XI: Junior synonym according to WOLF (1988). Unjustified emendation: Intentional change of final syllable because of bipectinate antennae.

‡ *Phalaena ditaria*: sensu VILLERS, 1789: Linn. Ent. 2: 302, nec FABRICIUS, 1777 (identity: *Comibaena bajularia*). Misidentification.

‡ *Phalaena sesquistriata*: sensu FABRICIUS, 1794: Ent. Syst. 3 (2): 130, nec KNOCH, 1793 (identity: *Campaea margaritata*, Ennominae). Misidentification.

Phalaena (Geometra) albaria ESPER, 1806: Schmett. in Abbild. 5 (10): 268, pl. 47, figs 3, 4 (Germany, Thuringia: Erfurt). Lectotype ♂ (LMW, examined: HACKER 1999). Junior synonym.

Thalera fimbrialis var. *moskovita* GUMPPENBERG, 1892: Nova Acta Acad. Leop. 64: 508 (type locality not stated). Type(s) not traced. Not explicitly infrasubspecific. Junior synonym.

Unavailable names (misidentification): *viridata*: sensu GRONOVIUS (1781) nec LINNAEUS, 1758 (identity: *Chlorissa viridata*). - (infrasubspecific): *approximata*: LEMPKE (1949); *fasciata*: HARTWIEG (1951); *obsoleta*: SKALA; *ochracea*: KOLOSSOW (1936); *tangens*: LEMPKE (1949). - (incorrect subsequent spellings): *bipleuraria*: PRINS (1998); *bupleuvraria*: OBERTHÜR (1925); *buplevraria*: ESPER (1794); *fimbialis*: BUSZKO et al. (1996); *moscovita*: PROUT (1912b).

DIAGNOSIS: Wingspan variable, ♂♀ 24-34 mm, in reared specimens sometimes 22 mm only. In general smaller towards the north and bigger towards the south in Europe. In central Asia larger, 27-35 mm. Hindwing angled at M1 and M3, concave between. Ground colour deep green with slight yellow tinge. Colour readily fading to yellowish green. Costa of forewing yellowish. Postmedial and antemedial lines white, distinctly developed, the latter strongly convex on forewing, lacking on hindwing. Cell spots dark green, lunulate, often indistinct. Fringe chequered white and reddish brown, very rarely unchequered (f. *moskovita*). Postmedial line and cell spots well marked on underside. Frons flat, dark red brown. Palpi (♂♀) slender, somewhat shorter than diameter of eye. Antennae of ♂ bipectinate to tip, longest branches 5-7 times width of flagellum, in specimens from Italy and Balkan countries 7-9 times. Antennae of ♀ shortly bipectinate, longest branches twice, in southern Europe 2-3 times width of flagellum. Further features under generic diagnosis.

MALE GENITALIA: Uncus long, slender, tapered. Socii broad at base, tapered towards apex. Costal process of valva large, stoutly sclerotized, laterally dentate. Basal process of valva present, flat, rounded. In southern European specimens membranous apex of valva usually longer and more slender than in specimens from elsewhere, distinctly exceeding costal process.

FEMALE GENITALIA: Characteristic longitudinal foldings between base of papillae anales. Apophyses anteriores comparatively long, half length of apophyses posteriores, in Asian specimens slightly longer.

DISTRIBUTION: Eurasiatic. Distributed nearly all over Europe. Recorded in southern Finland, but perhaps not breeding at site. Absent from central and northern parts of Scandinavia, Ireland, Sardinia, Malta, Crete and the other Greek islands. Often with great fluctuations of abundance in local populations. Endangered in Great Britain due to its isolated occurrence (southern Kent), specially protected by law.

Outside Europe in Turkey, Lebanon, Caucasus, Transcaucasus, northern Iran, southern Siberia, in the east to Dahuria (VIIDALEPP 1996). In central Asia subsp. *magnata* FUCHS, 1903. Records from far East Asia (eastern Mongolia, Korea, Amur, Primorje) referring to sister species *Thalera chlorosaria* (GRAESER, 1890) (PROUT 1935a; VIIDALEPP 1996).

PHENOLOGY: Usually univoltine. Main flight period: mid-June to early August, single specimens until mid-August. In the south-eastern Alps sometimes emerging from late May with partial second generation late August to early September. In southern Italy exceptionally until mid-October (PARENZAN 1994). Larval stages August to early June, overwintering as larva on the ground (LASS 1923).

BIOLOGY: Larva polyphagous, preferring perennial plants. Frequently found on *Achillea millefolium*, but also on *Calluna vulgaris, Daucus carota, Rumex acetosella, Genista tinctoria, Senecio jacobaea, S. erucifolius, Artemisia campestris* (Great Britain: SKINNER pers. comm.; HAGGETT 1954; Germany: WEGNER pers. comm.; SCHNEIDER 1934; BOLDT 1929). Recorded also on *Bupleurum falcatum, Thymus serpyllum, Solidago virgaurea, Galium, Anemone, Hypericum perforatum* (var. cit.), *Euphorbia cyparissias* (GUMPPENBERG 1892), sometimes also on trees and shrubs such as *Prunus spinosa, Crataegus, Betula, Rubus* (var. cit.). Reared on *Campanula medium* and *Hypericum* (LEIPNITZ pers. comm., BERGMANN 1955). Reputedly refusing *Thymus* in captive rearings (HAGGETT 1954). Larvae often resting on grass (KETTLEWELL 1953). Adults attracted to light and bait, but also diurnal, easily flushed during daytime.

HABITAT: Xerothermophilous. Dry open habitats on stony, rocky or sandy soils, heathland, steppes, often on dams. In Estonia abundant on dry parts of bogs with *Calluna* (VIIDALEPP pers. comm.). Usually from 0 m up to 800 m above sea-level, in the southern Alps, Apennine mountains and Balkan countries up to 1,400 m, Sierra Nevada (southern Spain) at about 1.600 m, in northern Iran up to 3,100 m.

SIMILAR SPECIES: *Hemithea aestivaria* can easily be distinguished by the shape of the hindwing, with an angle at M3 only, the cell spot touches the antemedial line, the underside is silky shining without markings, and the ♂ antennae are not bipectinate (see Text-figs 210-213).

REMARKS: *Thalera fimbrialis magnata* FUCHS, 1903 from the mountains of central Asia is slightly larger and the transverse lines on both upper- and underside are less distinctly marked. Structurally it matches the nominate subspecies from Europe well. The populations from south-western Asia would be better assigned to the nominate subspecies. Extensive descriptions of the rearing and the early stages in HAGGETT (1954), KETTLEWELL (1953), COCKAYNE (1953) and HAWKINS (1953). Larva see Text-fig. 25.

Dyschloropsis WARREN, 1895

TYPE-SPECIES: *Jodis impararia* GUENÉE, [1858]. One species only. Erroneously synonymized with *Holoterpna* (Pseudoterpnini) by VOJNITS (1976). Re-established as separate genus in Thalerini by HAUSMANN (1996b).

DIAGNOSIS: Forewing slender, costa straight, apex pointed. Hindwing termen rounded, very slightly angled at M1 and M3. Ground colour of forewing green, hindwing white. Fringe unchequered. Venation: R2-R5 and M1 of forewing on short common stalk.

Hindwing: Rs and M1 stalked. M3 and CuA1 unstalked. Frenulum absent in both sexes. Proboscis present, rather short. Palpi very slender, short. Antennae (♂♀) bipectinate. Hindtibia (♂♀) with pair of terminal spurs only. Further features under diagnosis of tribe.

MALE GENITALIA: Uncus stout, but particularly narrow near tip. Socii slender, strongly sclerotized. Gnathos weak. Transtilla very slender. Saccus short. Valvae very slender, basal coremata vestigial. Costal process of valva comparatively small, rounded. Basal process of valva weak, short, tapered. Aedeagus long, without cornuti, with deep, subterminal excavation. Sternum A8 and tergum A8 simple.

FEMALE GENITALIA: Apophyses anteriores shorter than in *Thalera*, 1/5 length of apophyses posteriores. Ductus bursae short, corrugated, leading to subterminal part of broad, pyriform corpus bursae. Signum bursae absent.

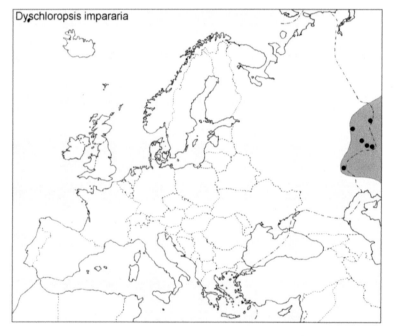

33. *Dyschloropsis impararia* (GUENÉE, [1858])

Jodis impararia GUENÉE, [1858]: in BOISDUVAL & GUENÉE, Hist. nat. Insectes (Spec. Gén. Lépid.) 9: 354 (Russia: Ural mts.). Described from LEDERER collection (holotype ♂), material from Orenburg present in MNHU.
‡ *Eucrostis imparata* HERRICH-SCHÄFFER, [1861]: Neu. Schmett. Eur. 3: 27, fig. 136. Incorrect subsequent spelling.

DIAGNOSIS: Wingspan ♂♀ 30-31 mm, in populations from Dahuria sometimes 25-27 mm only. Ground colour of forewings light green with slight yellow tinge. Postmedial line oblique, straight and narrow. No further wing pattern. Hindwing white with green tinge towards termen. Cell spots lacking. Fringe concolorous, unchequered. Underside whitish green without pattern. Frons flat, red brown. Palpi (♂♀) red brown, slender and short, not exceeding diameter of *eye*. Antennae of ♂♀ bipectinate almost to tip, longest branches in ♂ 4 times width of flagellum, in ♀ 1-1.5 times. Fore- and mid-legs red brown. Further features under generic diagnosis.

MALE AND FEMALE GENITALIA: See generic diagnosis.

DISTRIBUTION: Mongolian. South-eastern European Russia, southern Urals.

Outside Europe in the steppes of Kazakhstan, mountains of central Asia and Mongolia, in the east to Dahuria.

PHENOLOGY: Univoltine. Main flight period: early June to early July, in Dahuria until mid-July. Larval stages undescribed.

BIOLOGY: Larva found on *Spiraea* and *Prunus (Amygdalus)* (European Russia: ANIKIN et al. 2000). ♀ ratio at light high, usually over 50%.

HABITAT: Steppes, in Dahuria also in pine forests. From 0 m up to 500 m above sea-level, in central Asia up to 1,500 m.

SIMILAR SPECIES: *Bustilloxia saturata iberica* (allopatric) differs in the green colour of the hindwings, which can, however, exceptionally be white also in this species. *Holoterpna diagrapharia* PÜNGELER, 1900 from Transcaspia, Turkmeniya and northern Iran differs in the rectangular hindwing tornus, the distinctly green terminal area of the hindwing with a white postmedial line, the green frons and in the short pectination of the ♂ antennae.

Bustilloxia EXPOSITO, 1979

TYPE-SPECIES: *Eucrostes saturata* BANG-HAAS, 1906. One species only. Revision of generic concept in HAUSMANN (1995a). Genus phylogenetically isolated from other Thalerini, with some features reminiscent of Thalerini, such as red frons, shape of hindwing margin, certain genitalic features. Other characters not matching the tribal diagnosis well, such as frenulum, transtilla, sternum A8. Shape of uncus very similar to that of *Dyschloropsis*.

DIAGNOSIS: Forewing costa straight, apex somewhat pointed. Hindwing termen slightly concave between M1 and M3. Venation: Sc, R1 and R2 of forewing free. R2-R5 and M1 on short common stalk. M3 and CuA1 usually separate. Hindwing: Sc+R1 and Rs appressed at one point. Rs and M1 shortly stalked. M3 and CuA1 shortly stalked. Frenulum strong in ♂, absent in ♀. Frons red. Palpi (♂♀) very short. Proboscis vestigial, very short. Antennae of both sexes bipectinate to tip, flagellum dorsally with many red brown scales. Hindtibia (♂♀) with pair of terminal spurs only. Abdominal crests absent.

MALE GENITALIA: Uncus stout, but particularly narrow near tip. Gnathos weak. Transtilla extended, not U-shaped as in the other Thalerini. Costa of valva at base with inwards projecting sclerites (compare Pseudoterpnini). Valvae long and slender. Costal process of valva strongly sclerotized. Aedeagus long and slender, without internal cornuti, but subterminally with small external spine. Sternum A8 with two long, tapered, strongly sclerotized processes.

FEMALE GENITALIA: Base of papillae anales longitudinally corrugated, transversly between. Apophyses anteriores about 1/2 length of apophyses posteriores. Sterigma (lamella antevaginalis) strongly sclerotized, cordate. Corpus bursae cylindrical, broad even near ostium bursae, corrugated there. Ductus bursae very short, 'hidden' behind vaginal plate and leading to subapical part of corpus bursae.

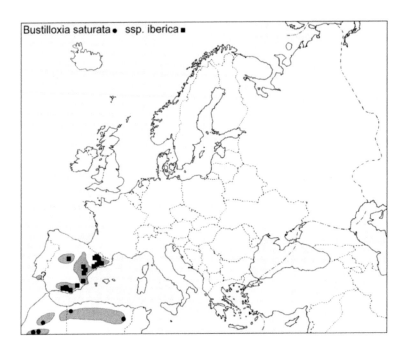

34. *Bustilloxia saturata* (BANG-HAAS, 1906)

Eucrostes saturata BANG-HAAS, 1906: Dt. ent. Z. Iris 19: 137, pl. 5, fig. 18 (Algeria: southern Oran). Syntypes 2♂, not traced.
Microloxia saturata crassilineata ZERNY, 1936: Mém. Soc. Sci. nat. Phys. Maroc 42: 71, pl. 1, figs 28, 29 (Morocco, Great Atlas: Tachdirt). Lectotype ♂ (NHMW, examined). Junior synonym (HAUSMANN 1995a).
Bustilloxia saturata iberica HAUSMANN, 1995: Atalanta 25 (3/4): 593 (Spain, prov. Granada: Diezma). Holotype ♂ (ZSM, examined). Valid at subspecific rank.
Unavailable names (incorrect subsequent spelling): *suturata*: PROUT (1912b).

DIAGNOSIS: Wingspan ♂ 24-29, ♀ 28-32 mm, in the second generation often 20-23 mm only. Ground colour deep green, exceptionally with slight reddish tinge, rarely light green or white on hind- or even on forewing. Forewing costa whitish ochreous. Antemedial lines lacking. White postmedial lines rather distant from termen, on forewing slightly curved mainly near costa, on hindwing convex, rather narrow. Postmedial line of forewing distinctly marked and broad in nominate subspecies, fine in subsp. *iberica* HAUSMANN, 1995. Postmedial line of hindwing obsolete. Fringe concolorous at base, distal half white. Underside with postmedial line well marked in ♂ of nominate subspecies, less or lacking in ♀ and in subsp. *iberica*. Frons slightly convex, deep red brown. Vertex whitish. Palpi slender, white, last segment red. Length of palpi not exceeding diameter of eye. Antennae bipectinate almost to tip, longest branches in ♂ 3-4 times width of flagellum, 2.5-3.5 times in subsp. *iberica*, 2-2.5 times in ♀. Further features under generic diagnosis.

MALE GENITALIA: See generic diagnosis. Variable in shape of valva and sternum A8.

FEMALE GENITALIA: See generic diagnosis. Sterigma heart-shaped, with posterior lobes angled laterally in nominate subspecies, rounded in subsp. *iberica*.

DISTRIBUTION: West-Mediterranean. Nominate subspecies in the mountains of Morocco and Algeria. Subsp. *iberica* in the southern, central and eastern parts of the Iberian peninsula.

PHENOLOGY: Bivoltine. Main flight period: early June to late July; August to September. Larval stages unknown.

BIOLOGY: Larva reputedly on *Bupleurum spinosum* (Morocco: ZERNY 1936; PROUT 1938a). ♀ ratio at light low, 10-30%. Adults readily attracted to light. Males diurnal, flying in the afternoon (ZERNY 1936).

HABITAT: Mountainous habitats, open to partially wooded slopes with many perennial plants. In Europe usually from 1,000 m up to 2,800 m above sea-level, exceptionally down to 200 m, in North Africa from 1,600 m up to 2,700 m.

SIMILAR SPECIES: *Kuchleria insignata* differs in the absent or dotted postmedial line, the green frons, the white dorsal scales of the antennae, and in the long ♀ palpi.

205   206

Text-figs 205-206: Diagnostic characters (indicated) of European Thalerini. Text-fig. 205: *Bustilloxia saturata* (BANG-HAAS, 1906). Text-fig. 206: *Kuchleria insignata* HAUSMANN, 1995.

Kuchleria HAUSMANN, 1995

TYPE-SPECIES: *Kuchleria insignata* HAUSMANN, 1995. Tentatively placed between *Bustilloxia* and Hemitheini, mainly due to external similarity. Phylogenetically isolated and not closely related to Hemitheini (compare structure of antennae, of sternum A2, etc.). Revisions: PROUT (1935a), HAUSMANN (1995a).

DIAGNOSIS: Venation characteristic and fairly constant: Sc, R1 and R2 of forewing without fusions. R2-R5 and M1 usually shortly stalked. Sc+R1 and Rs of hindwing appressed at one point. Rs and M1 shortly stalked. M3 and CuA1 shortly stalked. Frenulum strongly developed in ♂, completely lacking in ♀. Frons green. Palpi of ♀ elongated. Proboscis vestigial. Antennae of both sexes bipectinate almost to tip. Basal, dorsal scales of flagellum white, covering inner row of pectinations. Hindtibia of both sexes with pair of terminal spurs only. Abdominal crests lacking. Setal patches of sternum A3 lacking. Sternum A2 without lateral apodemes, diagnostic difference from Hemitheini.

MALE GENITALIA: Socii long. Gnathos present. Valvae without processes. Costa of valva sclerotized in proximal half. Transtilla expanded, with two anterior projections. Juxta small. Aedeagus with narrow basal stalk and characteristic patches of small cornuti. Posterior margin of sternum A8 slightly concave.

FEMALE GENITALIA: Lamella postvaginalis without setal patches. Ductus bursae corrugated, comparatively broad.

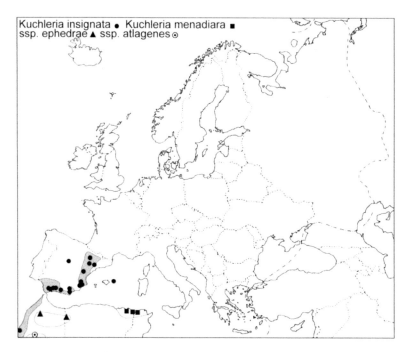

Kuchleria insignata ● Kuchleria menadiara ■
ssp. ephedrae ▲ ssp. atlagenes ⊙

35. *Kuchleria insignata* HAUSMANN, 1995

Kuchleria insignata HAUSMANN, 1995: Atalanta 25 (3/4): 584, figs 5,15,41,47 (Spain, prov. Alicante: Cabezon de Oro-Busot). Holotype ♂ (ZSM, examined).
‡ *Microloxia menadiara*: sensu PROUT, 1912 (partim): Cat. Lep. 14: 135 ('?Gibraltar') nec THIERRY-MIEG, 1893. Misidentification, name referring to North African sister species.
Unavailable names (incorrect subsequent spellings): *menadara*: PROUT (1935a); *menadaria*: STAUDINGER (1901) (referring to sister species); *menadiaria*: PROUT (1913).

DIAGNOSIS: Wingspan ♂ 22-26 mm, ♀ 26-29 mm. Males of subsequent generations smaller, sometimes only 17 mm. Hindwing termen rounded. Ground colour deep green, comparatively dark, but readily fading to yellowish green. Forewing costa usually orange or red brown. Postmedial lines whitish, very narrow, usually reduced to a series of dots, sometimes completely lacking. Antemedial lines and cell spots absent. Fringe concolorous at base, distal half white. Underside green with forewing costa orange. Forelegs red, femur green. Frons flat, green. Palpi slender, white, third segment reddish. Length of palpi in ♂ 1.1-1.3 times diameter of eye, in ♀ twice. Antennae of both sexes bipectinate almost to tip, longest branches in ♂ 3-4 times width of flagellum, in ♀ about 2.5-3 times. Further features under generic diagnosis.

MALE GENITALIA: Aedeagus without distal patch of microcornuti, exceptionally a very few small cornuti there. In North African sister species *Kuchleria menadiara* (THIERRY-MIEG, 1893) distal patch of microcornuti large. Central cornuti in the latter longer than in *K. insignata*.

FEMALE GENITALIA: Apophyses anteriores half length of apophyses posteriores. Corpus bursae elongate, pyriform. Sterigma rounded. In *K. menadiara* apophyses anteriores shorter and sterigma triangular.

187

DISTRIBUTION: West-Mediterranean. Eastern and southern Spain, local populations also in the province Madrid (central Spain). Recently recorded from Menorca, Balearic Islands (HONEY pers. comm.).

Outside Europe along western coast of Morocco. In eastern Algeria and northern Tunisia replaced by the allopatric sister species *K. menadiara*, with two subspecies occurring in Morocco, Great Atlas (*atlagenes* PROUT, 1935), and Middle Atlas to western Algeria (*ephedrae* PROUT, 1935).

PHENOLOGY: Trivoltine. Main flight period: mid-March to mid-April; mid-June to early August; early September to mid-November.

BIOLOGY: Larva and host-plant relationships unknown. Larva of *K. menadiara* found on *Ephedra major* (Morocco: PROUT 1935a: '*E. nebrodensis*'). ♀ ratio at light low, about 10%.

HABITAT: Hot, dry open habitats. In lowlands from 0 m up to 700 m above sea-level. Closely related *K. menadiara atlagenes* occurring in the Great Atlas between 1,200 and 2,100 m.

SIMILAR SPECIES: The allopatric sister species *Kuchleria menadiara* differs in the genitalia of both sexes, the postmedial line is usually more distinct. *K. therapaena* (PROUT, 1924) from eastern Algeria and Tunisia differs in the white forewing costa, and in the more pointed apex. *Bustilloxia saturata* can easily be distinguished by the red frons and foreleg femur, the distinctly developed postmedial line (but exceptionally weak or lacking), the red brown dorsal scales on the antennae and by the much shorter ♀ palpi. Large females of *Microloxia herbaria* may be similar too, but their antennae have very short pectination.

REMARKS: *Kuchleria menadiara* has been repeatedly reported from Sardinia (BYTINSKI-SALZ 1934; RAINERI & ZANGHERI 1995; MÜLLER 1996). All Sardinian specimens that were determined as *K. menadiara* in various examined collections proved to be females of *Microloxia herbaria*. The collection of BYTINSKI-SALZ (Hamburg) does not contain any *K. menadiara*, but 15 specimens of *M. herbaria* from Sardinia (WEGNER pers. comm.). Thus the species has to be removed from the fauna list of Europe (confirmed by RAINERI, MÜLLER pers. comm.).

Tribus **Hemitheini** BRUAND, 1846

Seven Palaearctic genera, with many additional (tropical) genera matching the diagnosis well. Nearly 50 Nearctic and Neotropical Hemitheini species listed and characterized by FERGUSON (1985) and PITKIN (1996). Some species of genera *Chlorissa* and *Diplodesma* with genitalic features reminiscent of Thalerini, see taxonomic notes to Thalerini. Hemitheini and Thalerini linked by 'sister-tribe-relationship', united by some authors (e.g. SEVEN 1991). For an alternative tribal concept (HOLLOWAY 1996) see introductory notes to Hemistolini.

DIAGNOSIS: Venation rather variable. Hindwing: Sc+R1 and Rs fused or appressed at one point. Rs and M1 stalked. M3 and CuA1 usually stalked. Hindwing termen angled at M3 or rounded. Only one anal vein (A2) present. Frenulum present in ♂, 'weak' (PROUT

1912a), in ♀ completely lacking. Frons flat, in European species red or brown, in tropical species often green. Palpi of ♀ usually very long. Antennae of both sexes filiform, ciliate, but bipectinate in males of some extralimital genera. Hindtibia of ♂ with one pair of terminal spurs, in ♀ proximal spurs sometimes present. Abdominal crests usually absent. Setal patches of ♂ sternum A3 usually well developed. Typical group of genera (*Hemithea, Idiochlora, Chlorissa, Diplodesma, Phaiogramma* and others) with strongly sclerotized, slender, lateral apodemes on sternum A2.

MALE GENITALIA: Uncus slender, comparatively long. Socii membranous or slightly sclerotized, usually similar to uncus in shape and size, often closely appressed to it. Gnathos weak or absent. Transtilla often narrow and U-shaped as in Thalerini, sometimes extended as flat sclerite. Valvae with large basal coremata. Costal and basal processes of valva well developed as in Thalerini. Aedeagus with slender basal stalk, distal half broad, usually with cornuti or sclerites. Sternum A8 in ♂ sometimes with tapered posterior projection(s).

FEMALE GENITALIA: Lamella postvaginalis usually with setose zone, or paired setose projections. Ductus bursae strongly corrugated. Corpus bursae often pyriform. Signum absent.

IMMATURE STAGES: Egg flat, disc-shaped, somewhat elongate. Larva (see Text-fig. 24) slender. Skin granulose, very shortly setose. Paired projections of medium length on head and segment T1. Tergum of last segment (A10) extended into long, single anal process. Pupa fairly slender.

Text-fig. 207: Venation of *Phaiogramma etruscaria* (ZELLER, 1849).

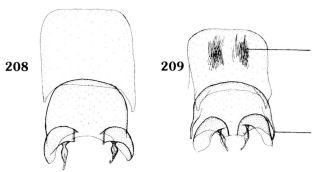

208 **209**

Text-figs 208-209: Differential features (indicated) of Thalerini and Hemitheini in ♂ abdomen, sternum A1-A3. Text-fig. 208: *Thalera fimbrialis* (SCOPOLI, 1763). Text-fig. 209: *Hemithea aestivaria* (HÜBNER, 1789).

Hemithea DUPONCHEL, 1829

TYPE-SPECIES: *Geometra aestivaria* HÜBNER, 1789. More than 30 Palaearctic, Indo-Pacific and Australian species, type-species also occurring in North America.

DIAGNOSIS: Ground colour usually dirty green, with dark grey tinge. Venation of forewing without secondary fusions of radial veins, R1 free, separate from R2-R5. R2-R5 and M1 usually connate. M3 and CuA1 usually connate. Hindwing: Sc+R1 and Rs appressed at one point. Rs and M1 shortly stalked. M3 and CuA1 stalked. Hindwing termen strongly tailed or, rarely, angled at M3. Postmedial lines usually dentate. Fringe often distinctly chequered. Proboscis well developed. Frons red brown, with a few exceptions. Palpi long, third segment elongate in ♀. Antennae (♂♀) filiform, ciliate, but bipectinate in ♂ of some Indo-Pacific species. Hindtibia of ♂ with two, of ♀ usually with four spurs. Hindleg of ♂ with tibial pencil and shortened tarsus. Dark abdominal crests on tergum A3 and A4 usually present. Setal patches well developed on ♂ sternum A3. Further features under diagnosis of tribe.

MALE GENITALIA: Gnathos present, but weak. Transtilla extended as flat sclerite, but more strongly sclerotized at lateral and caudal margins in some species, resembling the U-shaped transtilla of Thalerini and *Chlorissa*. Saccus cruciform. Ornamentation of valva on medial ridge over basal half of valva, with basal, inwards directed projection. Shape and length of this projection diagnostic. Valvae without sclerotized projections at costa and base (compare Thalerini, genus *Chlorissa*). Aedeagus long, without cornuti, but with characteristic sclerites in central and terminal part of vesica. Sternum A8 and tergum A8 usually simple, exceptionally bearing specific characters (HOLLOWAY 1996).

FEMALE GENITALIA: Lamella postvaginalis with setose zone. Ductus bursae and corpus bursae often hardly distinguishable, anterior part ('corpus bursae') membranous, posterior part ('ductus bursae') strongly sclerotized, corrugated. Signum bursae absent.

IMMATURE STAGES: For egg and larval morphology under diagnosis of tribe. For morphology of pupa see PATOCKA (1995)

REMARKS: The first larval instars show similar behaviour to the Comibaenini (PROUT 1913a): They cover themselves with silken threads and consequently become covered with plant debris. There is a strong tendency towards flower-feeding by the caterpillars (HOLLOWAY 1996).

210

211

212

213

Text-figs 210-213: Diagnostic characters (indicated) of *Thalera fimbrialis* (SCOPOLI, 1763) (Text-fig. 210 upperside of wings; 211 underside) and *Hemithea aestivaria* (HÜBNER, 1789) (Text-fig. 212 upperside; 213 underside).

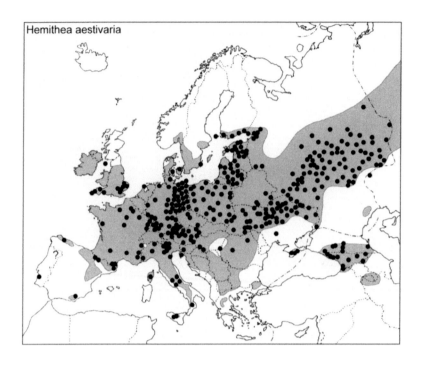

Hemithea aestivaria

36. *Hemithea aestivaria* (HÜBNER, 1789)

Phalaena (Geometra) aestivaria HÜBNER, 1789: Beitr. Schmett. 1 (4): 22, pl. 3, f. R. ('Europe': 'Niemierow'?). Type(s) lost.

‡ *Phalaena (Geometra) strigata* MÜLLER, 1764: Fauna Insect. Fridr.: 51, nec SCOPOLI, 1763 (identity: *Cabera pusaria*, Ennominae). Junior primary homonym.

‡ *Geometra thymiaria*: sensu [DENIS & SCHIFFERMÜLLER], 1775: Ank. syst. Werkes Schmett. Wienergegend: 97, nec LINNAEUS, 1767 (identity: *Thalera fimbralis*). Misidentification.

‡ *Phalaena (Geometra) fimbriata*: sensu GLEDITSCH, 1775: Forstwissensch. 1: 519, nec HUFNAGEL, 1767 (identity: *Thalera fimbralis*). Misidentification.

‡ *Phalaena (Geometra) viridata*: sensu KLEEMANN, 1777: Mader's Raupenkal.: 37, nec LINNAEUS, 1758 (identity: *Chlorissa viridata*). Misidentification.

‡ *Geometra bajularia*: sensu LANG, 1789: Verz. Schmett. (ed. 2): 174, nec [DENIS & SCHIFFERMÜLLER], 1775 (identity: *Comibaena bajularia*). Misidentification.

‡ *Phalaena (Geometra) vernaria*: sensu ESPER, 1795: 5: pl. 1, fig. 9 nec LINNAEUS, 1761 (identity: *Jodis lactearia*). Misidentification.

‡ *Macaria thymiata* CURTIS, 1826: Brit. Ent. 3: 132. Unjustified emendation of misidentified *thymiaria* sensu [DENIS & SCHIFFERMÜLLER], 1775 (see above). Not available for this species.

Nemoria alboundulata HEDEMANN, 1879: Horae Soc. ent. ross. 14: 511, pl. 3, fig. 8 (Russia, Amur: Blagoweschtschensk). Type(s) not traced. Junior synonym.

Unavailable names (infrasubspecific): *krajniki*: KOMAREK (1950). - (incorrect subsequent spellings): *aestiviaria*: BUSZKO et al. (1996); *strigiata*: BACHMETJEV (1902); *striata*: BURESCH (1910).

DIAGNOSIS: Wingspan ♂ 22-28, ♀ 25-31 mm. Hindwing strongly angled at M3. Ground colour dirty greyish green. Costa of forewing yellow, chequered with brown. Antemedial and postmedial lines well developed, undulate, fine and white, bordered dark green towards medial area. Antemedial line of hindwing absent. Cell spots indistinct, dark green, lunate, touching antemedial line. Terminal line very narrow, dark brown or dark green, uninterrupted. Fringe chequered yellowish white and dark brown. Underside silky shining, without markings. Frons flat, slightly convex towards proboscis, red brown. Vertex white. Palpi long, red brown, basal segment bushy scaled and ochreous, length in ♂ 1.5 times, in ♀ 2-2.5 times diameter of eye. Antennae of ♂ dentate, densely ciliate, length of cilia = width of flagellum. Antennae of ♀ filiform, with very short cilia. Hindtarsus of ♂ short, 0.4 times length of tibia. Hindtibia of ♀ with four spurs. Dark abdominal crests on tergum A3 and A4, sometimes lost, mainly in worn specimens. Paired setal patch of ♂ sternum A3 with long setae. Further features under generic diagnosis.

MALE GENITALIA: Uncus long, tip very slender. Socii narrow. Transtilla extended as flat sclerite, strongly sclerotized at lateral and caudal margins, notched anteriorly. Shape of central ridge of valva bisinuous, caudal margin and basal end spinulose. Sternum A8 small, concave on both anterior and posterior margins.

FEMALE GENITALIA: Lamella postvaginalis with pair of setose projections, well separated from each other. Antrum divided into two lateral sclerites.

DISTRIBUTION: Holarctic. Widely distributed in Europe. Absent from northern Scandinavia, hot lowlands of southern Europe and from all Mediterranean islands, except Corsica. In the east over Crimea and central Russia to central and southern Urals.

Outside Europe in the Caucasus, Transcaucasus, northern Iran. In the east from western Siberia to Japan. Also recorded from North America.

PHENOLOGY: Univoltine. Main flight period: early June to mid-July, in the north of the distribution area usually one month later. Late females exceptionally until mid-August. In southern Europe sometimes emerging from late May, exceptionally with partial second generation early August to early September. Partial second brood reported also for Germany and northern Europe (KOCH 1984). In Japan regularly bivoltine (INOUE 1971). Larval stages early August to early June, overwintering as young larva.

BIOLOGY: Larva polyphagous on various deciduous trees and shrubs (var. cit.). Found on *Prunus padus*, *P. spinosa*, *Rosa*, *Clematis vitalba*, *Lonicera xylosteum*, *Corylus avellana*, *Crataegus*, *Carpinus betulus*, *Salix caprea*, *Alnus incana*, *Betula pendula*, *B. pubescens*, *Rhamnus cathartica*, *Frangula alnus*, *Quercus robur* (Poland: BUSZKO pers. comm.; URBAHN 1939; Great Britain: SKINNER pers. comm.; PORTER 1997; Germany: WEGNER, LEIPNITZ, GELBRECHT pers. comm.; OSTHELDER 1929; SCHNEIDER 1934; BERGMANN 1955; KRAUS 1993; Austria: SCHWINGENSCHUSS 1953), *Ribes rubrum* and *Sorbus aucuparia* (Finland: SEPPÄNEN 1954; 1970). Recorded also on *Rubus caesius*, *Tilia*, *Ligustrum*, *Castanea*, *Viburnum* and *Vaccinium myrtillus* (var. cit.). Wide spectrum of host-plants reported by DOGANLAR & BEIRNE (1979) and NAKAJIMA & SATO (1979), including unusual plant genera such as *Larix*, *Aralia* and *Rhododendron*. Young autumnal larvae reputedly feed on *Artemisia vulgaris*, *Rumex*, *Potentilla* and other low herbs (CARTER & HARGREAVES 1986; ALLAN 1949). Pupation in a loose cocoon between leaves on the ground. Adults feed at flowers and flowering trees, e.g. *Tilia*, readily attracted to bait. Main activity at dusk, but also nocturnal and attracted to light. ♀ ratio at light low, 10-20%.

PARASITOIDS: Hymenoptera, Ichneumonidae: *Casinaria ischnogaster*, *Labrorychus clandestinus*, *Phobocampe obscurella*, *Stenichneumon rufinus* (THOMPSON 1946).

HABITAT: Various habitats with deciduous trees, mainly woods on soils with high ground water, shrub-rich forest fringes, hedges, gardens, parks. From 0 m up to 700 m above sea-level, in the southern Alps and the Apennines up to about 1,500 m. In southern Spain up to 1,800 m (RIBBE 1912).

SIMILAR SPECIES: *Thalera fimbrialis* can easily be distinguished by the concave excision in the hindwing termen, the position of the cell spot, the green underside with marked transverse lines, the short palpi, and by the bipectinate antennae (Text-figs 210-213).

REMARKS: Monography of this species: BOLTE & MUNROE (1979). Larva see Text-fig. 24.

Chlorissa STEPHENS, 1831
=*Nemoria* auct. nec HÜBNER, 1818

TYPE-SPECIES: *Phalaena viridata* LINNAEUS, 1758. About 25 Palaearctic and Indo-Pacific species, one in the Neotropical region. 16 Afrotropical species currently placed in this genus, but probably none of them congeneric (HAUSMANN 1996c). Genitalic features of *Chlorissa* reminiscent of Thalerini, more than in any other Hemitheini genus. Clear differential features in shape of valva and harpe, and in structure of ♀ genitalia do not allow downgrading of *Chlorissa* to subgenus of *Hemithea*, as proposed by PROUT (1913a: 24; 1933b: 116). Some genitalic features, such as narrow socius, contents of aedeagus, however, suggest sister-group relationship between *Chlorissa* and *Hemithea* against *Phaiogramma* as outgroup. Genus *Diplodesma (Idiochlora)* differing in stalked R1-R5 of forewing with R1 fused to Sc (PROUT 1934b: 117). Important literature: URBAHN (1964), DANTART (1990), HAUSMANN (1996b); REZBANYAI-RESER (1999).

DIAGNOSIS: Ground colour usually unicolorous, deep green. Fringe white on distal half in European species. Venation of forewing: Sc and R1 often fused, exceptionally R1 free. R1 arising separately from R2-R5. R2-R5 and M1 usually shortly stalked. M3 and CuA1 often stalked. Hindwing: M3 and CuA1 distinctly stalked. Termen slightly angled at M3, with exceptions. Frons green. Palpi of medium length, slightly longer in ♀ than in ♂. Proboscis well developed. Antennae of both sexes filiform, ciliate. Hindtibia of ♂ with pencil and pair of terminal spurs, of ♀ with two pairs of tibial spurs. Hindtarsus of ♂ usually shortened. Abdominal crests absent or very small.

MALE GENITALIA: Socii very narrow. Transtilla U-shaped, narrow. Costal process of valva prominent. Aedeagus with fairly long basal stalk. Vesica bearing some sclerites. Sternum A8 with strongly sclerotized medial tooth.

FEMALE GENITALIA: Signum bursae absent. Lamella antevaginalis developed as large single sclerite ('sinus vaginalis') in European species, lamella postvaginalis with paired setose projections. Corpus bursae connected with ductus bursae subterminally. Corpus bursae elongate, reminiscent of Thalerini. ♀ genitalia of North American genus *Hethemia* matching this diagnosis well.

IMMATURE STAGES: For egg and larval morphology under diagnosis of tribe.

194

REMARKS: *Chlorissa pretiosaria* (STAUDINGER, 1877) reputedly occurs in southern European Russia (ANIKIN et al. 2000), but the identification needs verification before adding this species to the European fauna list.

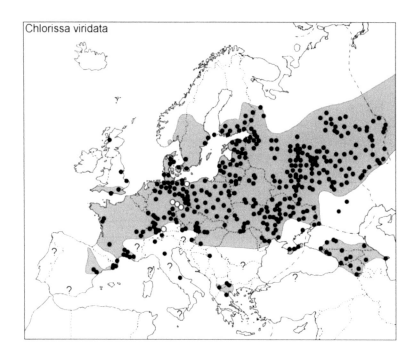

Chlorissa viridata

37. *Chlorissa viridata* (LINNAEUS, 1758)

Phalaena (Geometra) viridata LINNAEUS, 1758: Syst. Nat (Ed. 10) 1: 523 ([Europe]). Type(s) in LSL.
Phalaena volutata FABRICIUS, 1775: Syst. Ent.: 635 (Germany). Type(s) probably lost (not traceable in ZMUC). Junior synonym.
Phalaena herbacea FOURCROY, 1785: Entomol. Paris. 2: 282 (France: Paris). Type(s) in Museum Autun (F), requiring examination. Junior synonym of *C. viridata* (WOLF 1988; LERAUT 1997) or *Hemithea aestivaria* (SCOBLE 1999).
‡ *Geometra viridaria* HÜBNER, [1799]: Samml. Eur. Schmett. 5: pl. 2, fig. 11. Incorrect subsequent spelling.
‡ *Geometra vernaria*: sensu HAWORTH, (1809): Lep. Brit. 2: 300, nec LINNAEUS, 1761 (identity: *Jodis lactearia*). Misidentification. Moreover junior primary homonym nec [DENIS & SCHIFFERMÜLLER], 1775 (identity: *Hemistola chrysoprasaria*).
‡ *melinaria* HERRICH-SCHÄFFER, 1848: Syst. Bearb. Schmett. Eur. 3: pl. 67, fig. 413. Non binominal.
Geometra (Hemithea) melinaria HERRICH-SCHÄFFER, 1852: Syst. Bearb. Schmett. Eur. 6 (55): 63 (Russia: Urals). Binominal because referring to p. 10, vol. 3. Type(s) not traced. Junior synonym.
‡ *Geometra chloroticaria*: sensu TENGSTRÖM, 1859: Bidr. Finl. Naturk. 3: 147, nec *cloraria* HERRICH-SCHÄFFER, 1848 (fig. 362; identity: *Phaiogramma etruscaria*). Misidentification and emendation.
Geometra prasinata WERNEBURG, 1864: Beitr. Schmett. 1: 225. Unnecessary replacement name for '*viridata* SCOP. et TR. nec L.'. WERNEBURG's '*viridata* L.' misidentified (identity: *Hemithea aestivaria*).
Unavailable names (misidentifications): *cloraria*: sensu STEPHENS (1831) nec HÜBNER, [1813] (identity: *Chlorissa cloraria*); *herbaria*: sensu BOISDUVAL (1840) nec HÜBNER, [1813] (identity: *Microloxia herbaria*). - (infrasubspecific): *approximata*: LEMPKE (1949); *caerulescens*: BURROWS (1908); *concavilinea*: BURROWS (1908); *mathewi*: BANKES (1907); *olivaceomarginata*: BURROWS (1908); *rosearia*: CULOT (1910); *rufotincta*: BURROWS (1908); *subobsoleta*: LEMPKE (1949). - (incorrect subsequent spellings): *chloraria*: PROUT (1913C); *coerulescens*: REBEL (1910); *insignata*: REBEL (1910).

DIAGNOSIS: Wingspan ♂♀ usually 18-22 mm, rarely 17 mm only. Ground colour bright green, sometimes with slight turquoise tinge. Very rarely orange or rosy-coloured ('f. *mathewi*' and 'f. *rosearia*'). Forewing costa usually unicolorous sand-coloured. Postmedial line white, slightly dentate, rather straight on all wings (less in ♀). Antemedial line hardly visible on forewing. Transverse lines slightly broader in populations from the Urals ('*melinaria*'). Cell spots nearly invisible, dark green, elongate, appressed to antemedial line. Frons slightly convex, reddish brown with copper tinge. Palpi with red brown scales, length in ♂ 1-1.2 times diameter of eye, in ♀ up to 1.5 times. In specimens from the Urals sometimes slightly longer. Length of antennal cilia = width of flagellum in ♂, slightly shorter in ♀. Hindtarsus of ♂ shortened. Further features under generic diagnosis.

MALE GENITALIA (see also Text-figs 218-223): Costal process of valva narrow, long, posteriorly slightly convex, strongly tapered towards end. Aedeagus with thumb-shaped chitinous plate. Terminal sclerotization of vesica tapered, usually well separated from patch of microcornuti by hyaline area of vesica. Posterior margin of sternum A8 with stout thorn of variable shape. Some features such as width of socii, shape of thorn of sternum A8 and length of saccus variable, thus not valuable as differential features separating *C. viridata* and *C. cloraria*, although mentioned in literature as important for species identification.

FEMALE GENITALIA: Lamella antevaginalis rectangular, comparatively slender, anteriorly rounded. Those features hardly visible in permanent slides, but can be seen much better in fresh preparations. Pattern and strength of sclerotization of lamella antevaginalis ('sinus vaginalis') not valuable as differential features separating *C. viridata* and *C. cloraria*, although mentioned in literature as important for species identification (URBAHN 1964; DANTART 1990). Corpus bursae elongate, slender.

DISTRIBUTION: Eurosiberian. Widely distributed in central Europe (except large areas in the Alps), the most southern parts of Scandinavia and the temperate parts of eastern Europe including Urals. In Great Britain resident only in the extreme south with vagrant specimens also in the east and north (SKINNER pers. comm.). Isolated populations in northern Spain, France, Italian peninsula, Balkan countries (mainly Macedonia). Further, unconfirmed records for Portugal, Corsica, Sicily and Malta (MÜLLER 1996). In southern Europe probably more widely distributed than marked on the map, each record however requiring accurate, critical verification. In central Europe decreasing, regionally endangered through habitat loss.

Outside Europe eastwards to mountains of central Asia (subsp. *insignata* STAUDINGER, 1901) and Dahuria (VIIDALEPP 1996). In eastern Asia and Japan replaced by the sister species *Chlorissa obliterata* (WALKER, 1862). Populations from Turkey, Caucasus and Transcaucasus (KOSTJUK, ANTONOVA pers. comm.; VIIDALEPP 1996) require taxonomical revision.

PHENOLOGY: Usually univoltine. Main flight period: mid-May to early July, rarely until mid-July, exceptionally single specimens emerging in August (partial second generation). In Spain bivoltine: May to early July; August to early September (DANTART 1990). Larval stages from late June to late August, in northern Europe one month later. Exceptionally in September and October (CULOT 1919; FORSTER & WOHLFAHRT 1981). In Spain from June to July and August to October (DANTART 1990). Overwintering as pupa.

BIOLOGY: Larva polyphagous on numerous low plants and shrubs, sometimes also on trees. Found on *Calluna vulgaris* (Finland: SEPPÄNEN 1954; 1970; Poland: BUSZKO pers. comm.; Great Britain: PORTER 1997; Germany: WEGNER pers. comm.), *Empetrum nigrum, Myrica gale* (Finland: SEPPÄNEN 1954; 1970), *Vaccinium uliginosum* (Germany: DRECHSEL, GELBRECHT pers. comm.), *Quercus* spp. (Germany: URBAHN 1964), *Betula, Salix repens* (Great Britain: SKINNER pers. comm.; PORTER 1997). Recorded also on *Potentilla, Galium, Hieracium, Artemisia, Ononis, Clematis, Prunus, Crataegus, Corylus, Rubus, Erica, Ulex, Genista* (var. cit.). Reared on *Lotus corniculatus, Crataegus, Ledum palustre* and many other plants (DRECHSEL, GELBRECHT pers. comm.; SCHNEIDER 1934). Clearly preferring *Calluna vulgaris*, at least in the north (WEGNER pers. comm.; SEPPÄNEN 1970; SKOU 1986; URBAHN 1939). Adults often found dayflying (var. cit.), and at dusk, but also nocturnal and attracted to light. ♀ ratio at light about 30%.

PARASITOIDS: Hymenoptera, Ichneumonidae: *Casinaria pallipes, Labrorychus clandestinus* (THOMPSON 1944).

HABITAT: Thermophilous. Lowland heathlands, sparse grassland on low nutrient, sandy soils. Also on slightly wet heathy clearings in woods, fringes of moors, dry places on moors covered by *Calluna* (WEGNER pers. comm.). From 0 m up to about 400 m above sea-level. Even in southern Europe usually in lowlands near coast, in Spain up to 550 m. According to FORSTER & WOHLFAHRT (1981) and NESTOROVA (1998) up to 1,400 m, but those specimens presumably misidentified.

SIMILAR SPECIES: *Chlorissa cloraria* differs mainly in the dark scales on the forewing costa. Further, less reliable, differential features are the greyish brown frons, darker ground colour and the more convex postmedial lines. In the ♂ genitalia the costal process of the valva is less tapered. Females have to be determined mainly on the base of habitus features (Text-figs 214-217). *Phaiogramma etruscaria* and *P. faustinata* can easily be distinguished by the rounded hindwings, the ground colour, being powdered with many white scales, the concolorous fringe, the long ♀ palpi and by the clearly different genitalia.

REMARKS: Larva of *C. viridata* light green (FORSTER & WOHLFAHRT 1981). Pupal wing-shaft unicolorous, dark, dorsal streak usually down to 7th abdominal segment (PATOCKA 1995). The correct identification of *C. cloraria* and *C. viridata* is often difficult and some specimens may remain doubtful, especially females. The variability of the costal process of the valva in the ♂ genitalia sometimes leads to confusion (see REZBANYAI-RESER 1999:71). All the collections must be revised, as almost everywhere there are misidentifications. Many citations in literature must be treated with caution.

Sometimes there are intermediate feature combinations: For example the Bavarian specimen figured in FORSTER & WOHLFAHRT (1981, pl. 1, fig. 21) as *C. viridata* has a greyish-brown frons. In southern Germany (near Munich) there is a population with typical ♂ genitalia of *C. cloraria*, but its habitus features (wing pattern, forewing costa, frons) are exactly those of *C. viridata*. Similar observations from a central German population are mentioned in URBAHN (1964). In some regions *C. cloraria* and *C. viridata* may hybridize in nature (HAUSMANN 1996b).

The populations of the Urals are considered as a separate subspecies ('*melinaria*') by some authors (e.g. VIIDALEPP 1996). The material studied by the author does not support this.

Text-figs 214-215: Diagnostic characters (indicated) of European *Chlorissa* species. Text-fig. 214: *C. viridata* (LINNAEUS, 1758). Text-fig. 215: *C. cloraria* (HÜBNER, [1813]).

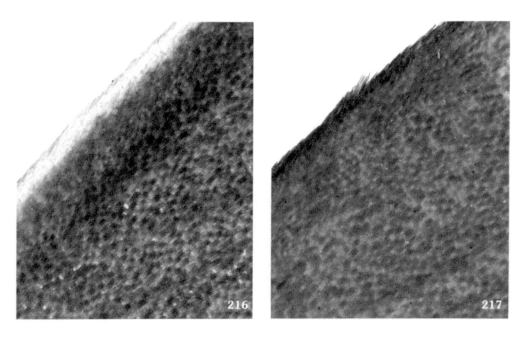

Text-figs 216-217: Forewing costa of *Chlorissa* species. Text-fig. 216: *C. viridata* (LINNAEUS, 1758). Text-fig. 217: *C. cloraria* (HÜBNER, [1813]).

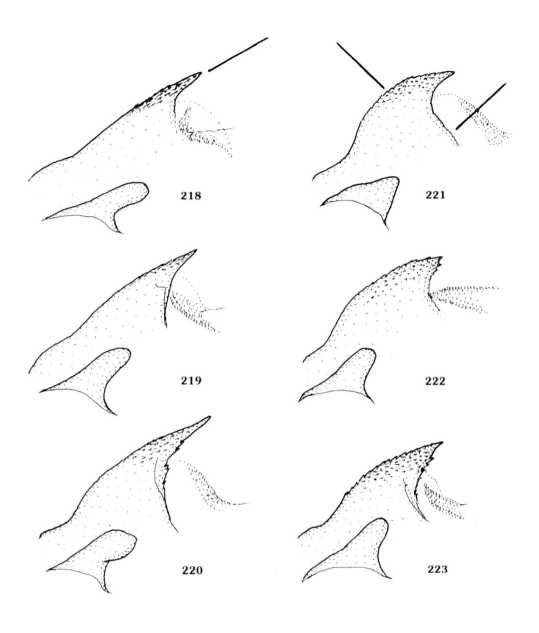

Text-figs 218-223: ♂ genitalia, costa of valva in *Chlorissa* species (Regensburg, southern Germany). Text-figs 218-220: *C. viridata* (LINNAEUS, 1758). Text-figs 221-223: *C. cloraria* (HÜBNER, [1813]).

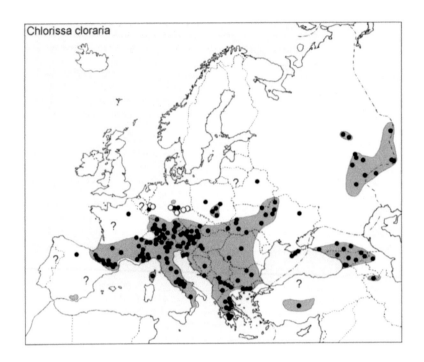

Chlorissa cloraria

38. *Chlorissa cloraria* (HÜBNER, [1813])

Geometra cloraria HÜBNER, [1813]: Samml. Eur. Schmett. 5: pl. 68, fig. 352 ('Europe'). Type(s) lost.
  HÜBNER's figure clearly shows important differential features and proves identity of type (PROUT 1913a).
Geometra chlorata TREITSCHKE, 1825: Schmett. Eur. 5 (2): 430. No separate types. Unjustified emen-
  dation: Intentional change of final syllable because of filiform antennae.
Geometra porrinata ZELLER, 1848: Stettin. ent. Zeit. 9 (9): 273 ([Germany, Thüringen: Jena]).
  Syntype(s) (BMNH). Junior synonym.
‡ *Geometra porrinaria* HEYDENREICH, 1851: Lepid. Eur. cat. meth. (ed. 3): 50. Subsequent misspelling.
  Emendation according to SCOBLE (1999).
‡ *Nemoria prasinata* WALKER, 1862: List. Lep. Ins. Brit. Mus. 26: 1556. Nomen nudum, without
  description. Identity: *C. cloraria* (PROUT 1913c).
Unavailable names (misidentifications): *viridata*: sensu TREITSCHKE (1825) nec LINNAEUS, 1758;
  viridaria: sensu LA HARPE (1853) nec HÜBNER, [1799]; *etruscaria*: sensu ZELLER, 1849 (partim:
  ♀) nec ZELLER, 1849 (♂). - (infrasubspecific): *rosea*: GUMPPENBERG (1895). - (incorrect subsequent spellings): *chloraria*: HANNEMANN (1917); *porrlnata*: HANNEMANN (1917).

DIAGNOSIS: Wingspan ♂ ♀ 17-20 mm, in Greece sometimes to 22 mm. Ground colour deep
green, usually with 'warmer' tinge and slightly darker than *Chlorissa viridata*, very rarely
pink ('f. *rosea*'). Forewing costa with many dark brown or blackish scales. Postmedial line
white, usually convex on all wings, slightly dentate. Antemedial line hardly visible on forewing. Cell spots nearly invisible, dark green, elongate, appressed to antemedial line. Frons
slightly convex, greyish brown. Palpi with grey or brown scales, length in ♂ 1-1.2 times
diameter of eye, in ♀ up to 1.5 times. Length of antennal cilia = width of flagellum in ♂,
slightly shorter in ♀. Hindtarsus of ♂ shortened. Further features under generic diagnosis.

MALE GENITALIA (see also text figs 218-223): Shape of costal process of valva variable,
but shorter and broader than in *C. viridata*, posteriorly convex, right-angled near tip.
Thumb-shaped sclerite of aedeagus on average slightly narrower and straighter than in

C. viridata. Terminal sclerotization of vesica tapered, usually closely attached to patch of microcornuti. Posterior margin of sternum A8 with stout thorn of variable shape. See also remarks to ♂ genitalia of *C. viridata.*

FEMALE GENITALIA: Lamella antevaginalis often sub-triangular, comparatively broad, anteriorly rounded. Such features hardly visible in permanent slides, but can be seen much better in fresh preparations. Corpus bursae elongate and slender. See also remarks on ♀ genitalia of *C. viridata.*

DISTRIBUTION: Submediterranean. Widely distributed in southern Europe except Portugal and Greek islands, present on Corsica. In the north up to central Germany and Poland. In the east across Ukraine and Byelorussia to southern Urals. Doubtful records for Lithuania (VIIDALEPP 1996) and Belgium (RICHARD 1950). For some regions in Germany (Niedersachsen, Nordrhein-Westfalen, Sachsen) no records since 1980 (GELBRECHT 1999).

Outside Europe in Turkey, Caucasus, Georgia, Armenia and Transcaucasus (see remarks). In the east not extending beyond Europe.

PHENOLOGY: Bivoltine. Main flight period: mid-May to late June; mid-July to mid-August. In southern Europe, the southern Alps, and in south-eastern Austria first generation one month earlier, second generation until early September. In northern Spain, however, first generation from May to early July (DANTART 1990). In mountains over 700 m often univoltine mid-June to early August. Larval stages in bivoltine populations June to July and August to September, overwintering as pupa.

BIOLOGY: Larva reputedly polyphagous on numerous low plants, shrubs and trees (var. cit.), such as *Calluna vulgaris, Corylus avellana, Crataegus oxyacantha, Rubus fruticosus, Betula* and *Alnus* (BERGMANN 1955; URBAHN 1964). Identification of species probably correct in these cases, but requires verification by dissection. Rearing very difficult on *Lotus corniculatus, Crataegus* and many other plants (DRECHSEL pers. comm.; SCHNEIDER 1934), perhaps indicating oligophagous specialization on a few, hitherto unknown, food-plants. Adults usually dayflying, mainly in the morning, also attracted to light (DANTART 1990). K-strategy. ♀ ratio at light 30%.

HABITAT: Xerothermophilous in central Europe, preferring heathlands on sandy soils (BERGMANN 1955) and limestone, in Hungary also on salt-steppes (GELBRECHT pers. comm.). In southern Europe in open habitats such as dry, sparse grassland and rocky slopes, tending to be a montane species here. From 0 m up to 1,900 m above sea-level, mainly from 400-1,200 m.

SIMILAR SPECIES: *Chlorissa viridata* differs mainly in the unicolorous bright forewing costa. Further, less reliable, differential features are the red brown frons, the lighter ground colour and the straight postmedial lines. In the ♂ genitalia the costal process of valva is strongly tapered. Females have to be determined mainly on the base of habitus features (Text-figs 214-217). *Phaiogramma etruscaria* and *P. faustinata* can easily be distinguished by the rounded hindwings, the ground colour, which is powdered with many white scales, the concolorous fringes, the long ♀ palpi and by the clearly different genitalia.

REMARKS: Larva of *C. cloraria* reddish (FORSTER & WOHLFAHRT 1981). Pupal wing shaft with pale veins, dorsal streak usually down to 9th abdominal segment (PATOCKA 1995). With regard to the problem of correct identification and the occurrence of intermediate feature combinations see remarks to *Chlorissa viridata*. In some regions *C. cloraria* and *C. viridata* may hybridize in nature. Transcaucasian populations require taxonomic revision with extensive material (HAUSMANN 1996b).

Phaiogramma GUMPPENBERG, 1887

TYPE-SPECIES: *Nemoria faustinata* MILLIÈRE, 1868. Small genus with a few species and subspecies in the Palaearctic, Afrotropical and Indo-Pacific regions. Closely related to the mainly African genus *Neromia* STAUDINGER, 1898. Important literature: LAEVER (1968); BOLLAND (1968; 1977); DUFAY (1972); DANTART (1990); HAUSMANN (1996b); REZBANYAI-RESER (1999).

DIAGNOSIS: Ground colour of wings green, powdered with many white scales. Fringe concolorous. Venation variable, similar to that of *Chlorissa*, but R1 usually free on forewing. Hindwing termen more rounded, less angled at M3. Frons flat. Proboscis well developed. Palpi of ♀ extremely long. Antennae of ♂ dentate with cilia grouped in tufts. Antennae of ♀ scarcely ciliate. Hindtibia of ♂ with one, ♀ with two pairs of spurs. Hindleg of ♂ with tibial pencil and short tarsus. Abdominal crests absent. Setal patches of ♂ sternum A3 of medium size, paired and well separate from each other.

MALE GENITALIA: Socii much dilated subapically. Transtilla expanded. Process of costa of valva absent or small. Aedeagus without sclerites on vesica, but with patches of cornuti. Posterior margin of sternum A8 slightly concave.

FEMALE GENITALIA: Apophyses short and weak. Lamella antevaginalis divided into two sclerites, surface wrinkled.

IMMATURE STAGES: For egg and larval morphology under diagnosis of tribe. Caterpillars more slender than those of *Chlorissa* and *Hemithea*. Paired projections on segment T1 and single anal process of the last segment (A10) comparatively short.

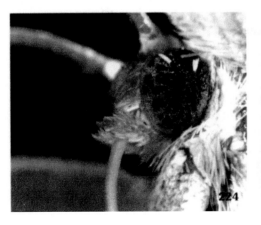 

202

Text-figs 224-226: Head of *Chlorissa-* and *Phaiogramma*-species (♀). Text-fig. 224: *C. cloraria* (HÜBNER, [1813]). Text-fig. 225: *P. etruscaria* (ZELLER, 1849). Text-fig. 226: *P. faustinata* (MILLIÈRE, 1868).

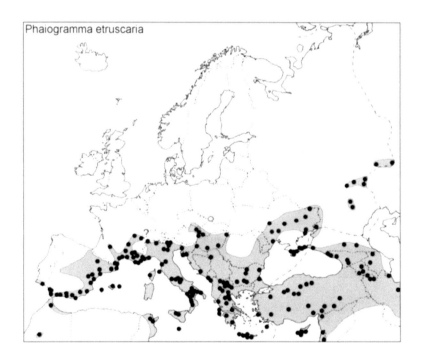

Phaiogramma etruscaria

39. *Phaiogramma etruscaria* (ZELLER, 1849)

Geometra etruscaria ZELLER, 1849: Stettin. ent. Zeit. 10 (7): 203 and index, binominal in index (Italy, Tuscany: Livorno). Syntypes 1♂3♀ in BMNH.
Nemoria pulmentaria GUENÉE, [1858]: in BOISDUVAL & GUENÉE: Hist. nat. Insectes (Spec. Gén. Lépid.) 9: 349 (southern France, Dalmatia, Italy). Syntypes, not traceable in ZFMK. Junior synonym.
Nemoria pulmentata BELLIER, 1861: Ann. Soc. Ent. Fr. (3) 8 (4): 701. No separate types. Unjustified emendation: Intentional change of final syllable because of filiform antennae.
Neromla (sic!) *palaestinensis* FUCHS, 1903: Societ. Entom. 18 (7): 51 (Israel: Jerusalem). Syntypes 11♂13♀ (LMW). Junior synonym.

Unavailable names (misidentification): *cloraria*: sensu DUPONCHEL (1830) and HERRICH-SCHÄFFER (1848) nec HÜBNER, [1813] (identity: *Chlorissa cloraria*). - (infrasubspecific): *benderi*: KOUTSAFTIKIS (1973). - (incorrect subsequent spellings): *chloraria*: DUPONCHEL (1843; moreover misidentified); *etrtuscaria*: PRINS (1998).

DIAGNOSIS: Wingspan ♂ 19-23, ♀ 20-25 mm, males of subsequent generations sometimes 17 mm only. Ground colour green, somewhat brighter than in European *Chlorissa* species, readily fading. Antemedial line of forewing often hardly visible. White postmedial line of forewing straight, on hindwing slightly crenulate, angled at CuA1. Cell spots dark green, but nearly invisible. Frons, upperside of palpi and forelegs red brown, sometimes faded to ochreous. Length of palpi 1.5-2 times diameter of eye in ♂, 2.0-2.6 times in ♀. Antennae with cilia 1.5-1.7 times width of flagellum in ♂, 0.3-0.4 times only in ♀. Hind-tarsus of ♂ about half length of tibia. Further features under generic diagnosis.

MALE GENITALIA: Socii subterminally very broad. Costa of valva sclerotized, straight or very slightly convex. Central ridge (valvula) with ventral brush and basal spine. Juxta round. Aedeagus with strongly, irregularly wrinkled surface, terminally with some small, fine cornuti. Basal stalk of aedeagus rather short. Posterior margin of sternum A8 slightly concave.

FEMALE GENITALIA: Lamella antevaginalis divided into two small, slightly wrinkled, semicircular sclerites. Corpus bursae pyriform, longitudinally folded on posterior half, irregularly wrinkled in centre, anteriorly smooth.

DISTRIBUTION: Mediterranean-Turanian. Widely distributed in southern Europe from the Iberian peninsula over southern France, southern Switzerland, Italy, Balkan countries, Ukraine to southern Urals. In the north up to eastern Austria, southern Czech Republic, southern Slowakia, Hungary. Present on all Mediterranean islands. Records from southern Finland (SOTAVALTA 1995) probably misidentified. Absent from central and western Pyrenees. Recorded for Belgium (LAEVER 1966), but probably transported there by man (PRINS 1998).

Outside Europe locally in Morocco and Tunisia, Turkey, Levant, Caucasus, Transcaucasus, over northern Iraq and Iran to Afghanistan and the mountains of central Asia.

PHENOLOGY: Bivoltine in the north and on mountains of southern Europe. Main flight period: mid-May to late June; late July to late August. Trivoltine in the lowlands of the south: late April to early June; mid-June to late July; mid-August to late September. Under favourable conditions up to five generations, late March to mid-October (e.g. in the Levant). In southern Italy and on Malta until mid-October. Larval stages in the north July and September, in the south from late May to late October. Overwintering as pupa.

BIOLOGY: Polyphagous, with slight preference for Umbelliferae. Found on *Ferula communis* (Umbelliferae) (Croatia: STADI pers. comm.), *Paliurus* (Rhamnaceae) (Dalmatia: SCHWINGENSCHUSS & WAGNER 1926) and *Rubus caesius* (Rosaceae) (Croatia: LEIPNITZ pers. comm.). Recorded also on various Umbelliferae such as '*Peucedanum, Bupleurum, Foeniculum, Seseli, Anthriscus*' (var. cit.), '*Althaea* (Malvaceae) and other herbs' (Iraq: WILTSHIRE 1957), *Lotus subbiflorus* (Papilionaceae) (GUMPPENBERG 1892), *Clematis vitalba, Quercus ilex, Rosmarinus officinalis* (Spain: DANTART 1990), and reputedly on *Quercus* (REBEL 1910; FORSTER & WOHLFAHRT 1981). Reared on *Taraxacum officinale* and *Daucus carota* (LEIPNITZ pers. comm.). Larva quickly growing, often found flower-feeding (var. cit.). Adults exclusively nocturnal. ♀ ratio at light about 30%. R-strategy. Sometimes artificially distributed by man, especially with vegetables from southern Europe.

HABITAT: Xerothermophilous, on rocky steppes and slopes, open dry grassland with many perennial plants. Associated with Festuco-Brometea, rich in Umbelliferae (CHAPELON 1992). Usually from 0 m up to 800 m above sea-level, in southern European mountains exceptionally up to 1,300 m. In eastern Turkey and central Asia xeromontane distribution mainly from 500 m up to 1,800 m.

SIMILAR SPECIES: *Phaiogramma faustinata* (MILLIÈRE, 1868) can be distinguished by the dark postmedial lines and by the well marked cell spots. Both European *Chlorissa* species have an angled hindwing termen, white distal half of fringe, shorter palpi and a darker green ground colour without white scales.

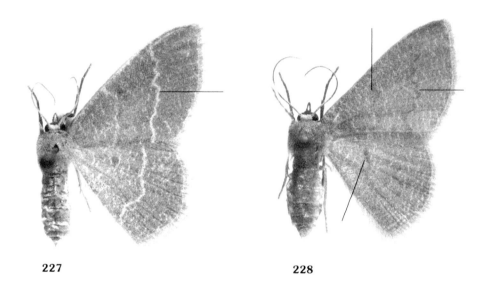

227   **228**

Text-figs 227-228: Diagnostic characters (indicated) of *Phaiogramma* species. Text-fig. 227: *P. etruscaria* (ZELLER, 1849). Text-fig. 228: *P. faustinata* (MILLIÈRE, 1868).

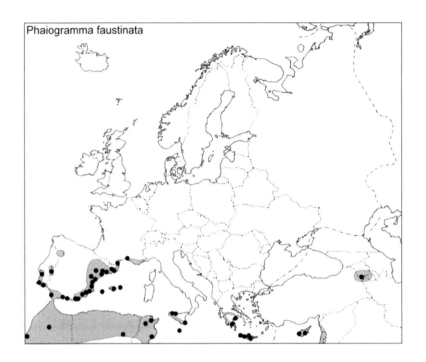

Phaiogramma faustinata

40. Phaiogramma faustinata (MILLIÈRE, 1868)

Nemoria faustinata MILLIÈRE, 1868: Ann. Soc. Linn. Lyon (N.S.) 17: 26 (Iconogr. Chen. Descr. Lép.
   inédits 2 (21): 436; pl. 96, figs 2-8) (Spain, Catalonia: Barcelona). Type(s) not traced.
Eucrostes rhoisaria CHRÉTIEN, 1909: Le Naturaliste 31: 30 (Algeria: Biskra). Lectotype ♀ (coll.
   HERBULOT in ZSM, examined). Junior synonym.
Unavailable names (misidentification): *palaestinensis*: sensu PÜNGELER in PROUT (1913a) nec FUCHS,
   1903 (identity: *Phaiogramma etruscaria*).

DIAGNOSIS: Wingspan ♂ 16-20, ♀ 18-24 mm. Ground colour green, often with slight
turquoise tinge, very rarely orange (CULOT 1919; faded pigment?). Strongly powdered
with white scales, more than in *Phaiogramma etruscaria*. Postmedial lines dentate or
crenulate, dark green, distally usually lightly bordered. Cell spots dark green, somewhat
elongated, usually clearly visible. Frons, upperside of palpi and forelegs red. Palpi long,
length in ♂ about 1.5 times diameter of eye, in ♀ 2.3-2.8 times. Antennae with cilia 1.3-
1.7 times width of flagellum in ♂, 0.3-0.4 times only in ♀. Hindtarsus of ♂ about half
length of tibia. Further features under generic diagnosis.

MALE GENITALIA: Socii usually less dilated than in *P. etruscaria*. Costa of valva sclerotized,
with short and stout dorsal spine. Central ridge (valvula) with one large ventrally directed
spine, accompanied by 3-7 smaller ones. Juxta very large, posteriorly round, anterior
margin usually with two small projections. Aedeagus with longitudinally folded surface and
two patches of large, stout cornuti. Size of patches varying. One examined ♂ from Crete
with three patches as in Levantine populations (HAUSMANN 1996b). Basal stalk of aedeagus usually long. Posterior margin of sternum A8 concave. The Indo-Pacific sister species *P. discessa* with rounded costa of valva, longest spine of valvula in distal part, aedeagus with one patch of long spines, juxta smaller (examined specimens from Nepal).

FEMALE GENITALIA: Lamella antevaginalis strongly sclerotized, widely extended, medially folded into two nearly triangular 'pouches'. Ductus bursae posteriorly widened and sclerotized. Corpus bursae sometimes rather slender. Longitudinally folded in posterior half.

DISTRIBUTION: Mediterranean, or Mediterranean-African according to taxonomical interpretation of Afrotropical populations (see remarks). Patchy distribution in the most southern parts of Europe, also in southern France and on some islands (Balearic Islands, southern Sicily, Malta and Crete). Most of the European populations endangered, their habitats meriting special protection.

Outside Europe in North Africa, including Canary Islands, Egypt, northern Sudan and Ethiopia in the east across Cyprus, Armenia, Levant, Arabian peninsula to Iraq and Iran. In Pakistan, Nepal, India and Sri Lanka replaced by the sister species *P. discessa* (WALKER, 1861). Some Afrotropical taxa closely related, but doubtfully conspecific (see remarks).

PHENOLOGY: Plurivoltine. Main flight period: late April to early November, most frequently from early September to mid-October. In Crete until late November. Outside Europe (e.g. Levant) often occurring throughout the year, mainly mid-October to mid-December. Larval stages almost throughout the year.

BIOLOGY: Larva polyphagous. Found on *Foeniculum vulgare* (Umbelliferae), *Linum grandiflorum* (Linaceae) and *Schinus terebinthifolia* (Anacardiaceae) (Morocco: RUNGS 1981) and on flowers of *Teline (Genista) monspessulana*, with caterpillars yellow instead of green (Crete: LEIPNITZ pers. comm.). Recorded also on *Rhus tripartita* (Anacardiaceae) (North Africa: PROUT 1915a, as '*Microloxia rhoisaria*'), *Acacia nilotica* (Nubia and Egypt: FLETCHER 1963; ANDRES & SEITZ 1924), *Acacia* spp. (Egypt and Israel: WILTSHIRE 1949; 1990; HALPERIN & SAUTER 1992), *Crotalaria* (Papilionaceae), *Inula* (Compositae), *Prosopis* (Mimosaceae) (Israel: HALPERIN & SAUTER 1992) and on *Rosmarinus officinalis* (Spain: var. cit.), but in captivity not accepting *Rosmarinus*, whilst growing well on *Daucus carota* and *Foeniculum vulgare* (LEIPNITZ pers. comm.). Afrotropical sister taxon '*stibolepida*' reported as pest on cotton (Nigeria: ZHANG 1994) and on other plants (SCOBLE 1999: 139). Adults nocturnal. ♀ ratio at light about 30%.

HABITAT: Hot coastal habitats. In Israel restricted to localities with annual temperature mean of more than 19°. Usually not above 200 m above sea-level, rarely up to 400 m. In Israel up to 600 m, in Yemen and Ethiopia to 1,700 m and absent from the lowlands.

SIMILAR SPECIES: *Phaiogramma etruscaria* differs in the white postmedial lines and in the less marked cell spots. Both European *Chlorissa* species have an angled hindwing termen, white distal half of fringe, shorter palpi, and a darker green ground colour without white scales.

REMARKS: *P. discessa* (WALKER, 1861) from the subtropical parts of the Indo-Pacific region is closely related, but not conspecific to *P. faustinata* (see remarks on ♂ genitalia). Wing pattern, ♂ genitalia and venation of the Afrotropical taxa *vermiculata* WARREN, 1897, *albistrigulata* WARREN, 1897, and *stibolepida* BUTLER, 1879, the last from Madagascar, seem to justify their separation from *faustinata* as one or two additional species. This complex of Afrotropical taxa will be revised separately.

Tribus **Microloxiini** HAUSMANN, 1996

About 10 genera mainly in Africa, but also in the Palaearctic and Indo-Pacific regions. Within the subfamily probably most closely related to Hemitheini-Thalerini-complex. For alternative tribal concept (HOLLOWAY 1996) see introductory notes to Hemistolini.

DIAGNOSIS: Usually small, robust-bodied. Ground colour green, often with straight post-medial lines only. Hindwing termen usually rounded. Venation: Rs and M1 of hindwing usually distinctly stalked. Very long fusion of Sc+R1 and Rs, except in type genus. Frenulum absent, or sometimes present but weak in ♂. Palpi in both sexes short, slender. Frons flat. Proboscis short or absent, rarely well developed. Antennae usually bipectinate in both sexes, in ♀ often with very short branches. Abdominal crests absent. Setal patches of sternum A3 absent. Lateral apodemes of sternum A2 absent. Hindtibia often dilated in ♂, in both sexes with pair of terminal spurs only.

MALE GENITALIA: Distal end of uncus often split dorsoventrally. Socii present, but often broad and shortened. Gnathos typically weak or absent. Valvae usually without basal coremata. Harpe arising from ventral part of valva.

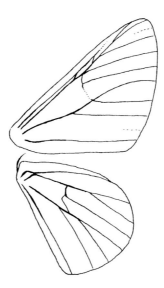

Text-fig. 229: Venation of *Microloxia herbaria* (HÜBNER, [1813])

Microloxia WARREN, 1893
=*Eucrostes* auct. nec HÜBNER, [1823]

TYPE-SPECIES: *Geometra herbaria* HÜBNER, [1813]. Six Palaearctic, Afrotropical and Indo-Pacific species. Revision: HAUSMANN (1995a).

DIAGNOSIS: Hindwing termen rounded. Venation of forewing: R1 and R2 anastomosing with Rs to form areole. Hindwing: Sc+R1 and Rs fused or appressed at one point. Rs and M1 distinctly stalked. Frenulum present in ♂, absent in ♀. Antennae bipectinate in ♂, dentate in ♀. Palpi very long in ♀. Hindtibia of ♂ not dilated. Further features under diagnosis of tribe.

MALE GENITALIA: Uncus slightly split dorsoventrally. Socii membranous, about as long as uncus. Gnathos weak. Valvae without basal coremata. Harpe fairly elongate, arising from ventral margin of valva. Aedeagus with narrow stalk, vesica longitudinally folded.

FEMALE GENITALIA: Lamella antevaginalis without special sclerotizations. Ductus bursae short, joining corpus bursae subterminally. Ductus bursae posteriorly broadened and sclerotized forming a small antrum.

IMMATURE STAGES: Larva with small, dorsally cleft head.

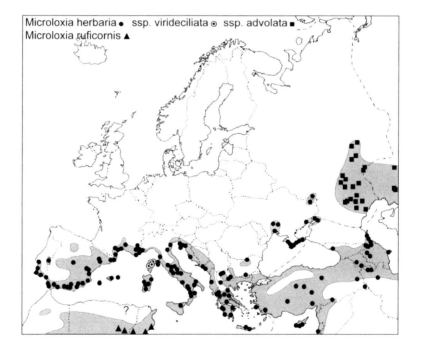

41. Microloxia herbaria (HÜBNER, [1813])

Geometra herbaria HÜBNER, [1813]: Samml. Eur. Schmett. 5: pl. 79, fig. 407 ([south-eastern Europe] (HAUSMANN, 1995a)). Type(s) lost.
Geometra graminaria ZELLER, 1849: Stettin. ent. Zeit. 10 (7): 204 and index, binominal in index (Italy, Tuscany: Ardenza). Type(s) not traced. Junior synonym.

Nemoria bruandaria MILLIÈRE, 1860: Iconogr. Chen. Descr. Lép. inédits 1 (3): 163, pl. 8 (18), figs 10-12 (France, Ardèche: Celles-les-Bains). Syntypes 2♀ not traced. Junior synonym.
Unavailable names (misidentification): *halimaria*: sensu PROUT (1912, partim: 'Gibraltar') nec CHRÉTIEN, 1909 (identity *Microloxia ruficornis*). - (infrasubspecific): *monotona*: REISSER (1926). - (incorrect subsequent spellings): *herbariaria*: PATOCKA (1995); *herbariata*: VIIDALEPP (1974).

DIAGNOSIS: Wingspan ♂ 14-17, ♀ 15-22 mm, males of subsequent generations sometimes only 12 mm. Forewing triangular, costa and termen straight, apex pointed, in ♀ termen more rounded. Ground colour deep green, readily fading. Transverse lines whitish, sometimes with yellow tinge. Postmedial line of forewing straight, close to forewing termen and parallel to it. Postmedial line of hindwing usually convex, parallel to termen. Antemedial line of forewing convex, sometimes lacking. In south-eastern Europe, Turkey to Pakistan sometimes with broader transverse lines, reminiscent of subsp. *advolata*. Cell spots absent. Fringe concolorous at base, distal half white. Underside green with indistinct postmedial line showing through from upperside. Frons white to yellow, red laterally and towards proboscis. Palpi white or ochreous, bushy scaled, last segment slender, length in ♂ 1.5 times diameter of eye, in ♀ twice. Proboscis well developed. Antennae of ♂ short, bipectinate, last fifth filiform, longest branches about 5-6 times width of flagellum. Antennae of ♀ with very short pectinations, almost dentate, densely ciliate, cilia about half width of flagellum. Hindtarsus of ♂ not shortened. Further features under generic diagnosis.

MALE GENITALIA: Socii somewhat shorter than uncus. Harpe long and slender. Saccus of medium size. Sister species *M. ruficornis* with shorter saccus and narrower aedeagus.

FEMALE GENITALIA: Corpus bursae very long and slender, with two small spinulose signa. Signa usually rounded with short spines in Europe. Spinules extremely short in specimens from Italy and Corsica. Females from Balkan countries and Turkey with numerous long spinules on elongate signa. *M. ruficornis* with corpus bursae much shorter and usually broader.

DISTRIBUTION: Mediterranean-Turanian. Coastal regions and lower valleys of Mediterranean from southern Portugal across Spain, southern France, Italy, Balkan countries across southern Ukraine to Volga Plain (subsp. *advolata* EVERSMANN, 1837). Locally distributed in central Spain. Present on nearly all Mediterranean islands, on Corsica as different subspecies (subsp. *virideciliata* BUBACEK, 1926). Populations from Elba and Tuscany (central Italy) transitional in habitus to subsp. *virideciliata*. Still unrecorded from Malta. One very doubtful record for Latvia in PROUT (1938a:218).
   Outside Europe in Turkey, Cyprus, Levant, Caucasus, Transcaucasus, northern Iran eastwards to the mountains of central Asia (partim subsp. *advolata*), Afghanistan and northern Pakistan. Allopatric sister species *M. ruficornis* WARREN, 1897 (= *halimaria* CHRÉTIEN, 1909; = *stenopteraria* TURATI, 1930) recorded from North Africa except north-western Morocco, widely distributed in the Sahara, from savannahs all over Africa, including Madagascar and South Africa, and in the east across the Arabian peninsula to southern Iraq and southern Iran. In India replaced by the allopatric 'sister species' *M. indecretata* (WALKER, [1863]), perhaps not more than a subspecies of *M. ruficornis*.

PHENOLOGY: Usually trivoltine. Main flight period: early May to mid-June; late June to early August; late August to late September, sometimes until early October. First and second generations not clearly separated. In the mountains of southern Spain most data from mid-June to mid-July (univoltine?), in the lowlands April to late October, one record from

January (southern Portugal: CORLEY pers. comm.). In the Levant mid-March to early November. Larval stages occurring nearly throughout the year.

BIOLOGY: Larva polyphagous. Found on flowers of *Helichrysum stoechas* (Compositae) (Istria: LEIPNITZ pers. comm.) and on *Teucrium polium capitatum* (Lamiaceae) (Turkey: STAUDINGER 1879; France: MILLIÈRE 1861). Recorded also on *Vernonia centaureoides* (Compositae) (SCOBLE 1999). Reared on *Artemisia* spp. (Compositae) (GELBRECHT pers. comm.). Larva of sister species *M. ruficornis* found on *Atriplex halimus* (Chenopodiaceae) (PROUT 1913a). ♀ ratio at light low, about 20%.

HABITAT: Xerothermophilous, on open grassland of hot coastal plains, rocky steppes. Usually from 0 m up to 400 m above sea-level, exceptionally up to 1,200 m. In southern Spain, Transcaucasus and in central Asia xeromontane and common also between 1,000 and 2,300 m.

SIMILAR SPECIES: Large females with indistinct wing pattern can easily be distinguished from *Kuchleria insignata* by the bipectinate antennae of the latter. The (allopatric) sister species *Microloxia ruficornis* (WARREN, 1897) is on average smaller, the transverse lines are often yellowish, and there are some differential features in ♂ ♀ genitalia (see above).

Microloxia herbaria virideciliata (BUBACEK, 1926)

Eucrostes herbaria var. *virideciliata* BUBACEK, 1926: Z. öst. EntVer. 11 (4): 35 (France, Corsica: Calacuccia). Type(s) not traced.

DIAGNOSIS: Slightly larger than nominate subspecies: Wingspan ♂ 14-18, ♀ 19-22 mm. Forewing apex more rounded and wing colour somewhat darker than in nominate subspecies. Wing pattern variable, transverse lines usually very fine, sometimes hardly visible or totally absent. Fringe often entirely green. External structure as in nominate subspecies.

MALE GENITALIA: See nominate subspecies. Saccus comparatively long.

FEMALE GENITALIA: See nominate subspecies. Spinules on signa very small, sometimes absent. Populations from Italian peninsula, however, also with very small spinules on signa.

DISTRIBUTION: Subspecies endemic to Corsica, also recorded as 'form' for Sardinia (BYTINSKI-SALZ 1934). Populations from Elba and Tuscany (central Italy) transitional in habitus between subsp. *virideciliata* and nominate subspecies.

PHENOLOGY: In the lowlands as in nominate subspecies, in the mountains probably bivoltine: early June to late July; late August to mid-September. Larval stages unknown.

BIOLOGY: Larva and host-plant relationships unknown for subspecies. ♀ ratio at light low, about 20%.

HABITAT: Mainly on hot, sun-exposed rocky slopes, but also on open grassland of coastal plains. From 0 m up to 1,000 m above sea-level.

SIMILAR SPECIES: Large females have repeatedly been misidentified as 'Kuchleria menadiara', see nominate subspecies and remarks to *K. insignata*.

REMARKS: Specimens with typical habitus are found mainly in the mountains. The validity of the subspecific rank is controversial and may be refuted.

Microloxia herbaria advolata (EVERSMANN, 1837)

Ellopia advolata EVERSMANN, 1837: Bull. Soc. imp. Nat. Moscou 10 (2): 51 (southern European Russia: Kamüschin). Lectotype ♀, herewith designated to stabilize nomenclature (ZISP, examined, pl. 8, fig. 9).
Chlorochroma advolataria DUPONCHEL, 1845: Cat. Méth. Lép. Eur.: 224 (no separate types). Unjustified emendation: Intentional change of final syllable because of bipectinate antennae.
‡ *advolaria* HERRICH-SCHÄFFER, 1848: Syst. Bearb. Schmett. Eur. 3: pl. 67, fig. 414. Non binominal.
Geometra (Hemithea) advolaria HERRICH-SCHÄFFER, 1852: Syst. Bearb. Schmett. Eur. 6: 64 (no separate types). Binominal because referring to p. 11, vol. 3. Unjustified emendation: Intentional change of final syllable because of bipectinate antennae.

DIAGNOSIS: Slightly smaller than nominate subspecies: Wingspan ♂ 14-16, ♀ 17-19 mm, males of subsequent generations sometimes 12 mm only. Habitus as in nominate subspecies, transverse lines straight, broad, clearly visible. Postmedial line distinctly marked on underside. External structure as in nominate subspecies.

MALE GENITALIA: See nominate subspecies.

FEMALE GENITALIA: See nominate subspecies. Signa elongate, with spinules of medium size, longer in specimens from Transcaucasus and eastern Turkey.

DISTRIBUTION: Subspecies Sarmatian. Volga plain and Ural river plain. Populations from south-eastern Ukraine transitional in habitus to nominate subspecies.

Outside Europe in the steppes of Kazakhstan and most northern parts of central Asian mountains, i.e. Altai, eastern Kazakhstan and northern Kirgisia. Populations from Caucasus, Transcaucasus, Turkey, northern Iran, south-western Afghanistan, Turkmeniya and Uzbekistan often intermediate in habitus between subsp. *advolata* and nominate subspecies, but closer to the latter.

PHENOLOGY: In the lowlands as in nominate subspecies, in the mountains bivoltine: late May to mid-July; late August to late September. Larval stages unknown.

BIOLOGY: Larva and host-plant relationships unknown for subspecies. ♀ ratio at light about 30 %.

HABITAT: Open grassland (steppe), rocky slopes. Typically from 0 m up to 200 m above sea-level in the steppes of southern European Russia, in the mountains of central Asia xeromontane and common between 1,000-2,300 m.

SIMILAR SPECIES: No similar species in the area of distribution.

REMARKS: The validity of the subspecific rank is controversial and may be refuted.

Colour plates

PLATE 1

(all specimens natural size)

1a. **Archiearis parthenias L.,** ♂, Praha, Czechia (ZSM)
1b. *A. parthenias* L., ♂, Coye La Forêt, N. France (ZSM/HERBULOT)
1c. *A. parthenias* L., ♂, Cibinsgebirge, Romania (ZSM)
1d. *A. parthenias* L., ♂, Vnukovo, C. Russia (ZSM)
1e. *A. parthenias* L., ♀, Kötzting, S. Germany (ZSM)
1f. *A. parthenias* L., ♀, Diessenltn., N. Austria (ZSM)
1g. *A. parthenias* L., ♀, Linz, N. Austria (ZSM)
1h. *A. parthenias* L., ♀, Cibinsgebirge, Romania (ZSM)
2a. **Archiearis notha Hbn.,** ♂, Reval, Estonia (ZSM)
2b. *A. notha* Hbn., ♂, Regensburg, S. Germany (ZSM)
2c. *A. notha* Hbn., ♂, Salzburg, C. Austria (ZSM)
2d. *A. notha* Hbn., ♂, Waidbruck, N. Italy (ZSM)
2e. *A. notha* Hbn., ♀, Wiesbaden, W. Germany (ZSM)
2f. *A. notha* Hbn., ♀, Regensburg, S. Germany (ZSM)
2g. *A. notha* Hbn., ♀, Jönschwalde, E. Germany (ZSM)
3A. **Archiearis touranginii Berce**, ♂, Autun, E. France (ZSM/HERBULOT)
3b. **Archiearis puella Esp.,** ♂, Wien, E. Austria (ZSM)
3c. *A. puella* Esp., ♂, Wien, E. Austria (ZSM)
3d. *A. puella* Esp., ♂, Groß-Enzersdorf, E. Austria (ZSM)
3e. *A. puella* Esp., ♂, Moravia (ZSM)
3f. *A. puella* Esp., ♀, Stadlau, E. Austria (ZSM)
3g. *A. puella* Esp., ♀, Wien, E. Austria (ZSM)
3h. *A. puella mediterranea* Ganev, ♂, Paratype, Rupite, SW. Bulgaria (coll. GELBRECHT)
3i. *A. puella mediterranea* Ganev, ♀, Rupite, SW. Bulgaria (coll. GELBRECHT)
4a. **Orthostixis cribraria Hbn.,** ♂, Szentkut, Hungary (ZSM)
4b. *O. cribraria* Hbn., ♀, Madonie, Sicily (ZSM)
4c. *O. cribraria* Hbn., ♂, Lithochoron, N. Greece (ZSM)
4d. *O. cribraria* Hbn., ♂, Tembi, Greece (coll. OSWALD)
5a. **Gypsochroa renitidata Hbn.,** ♂, La Voulte, S. France (ZSM)
5b. *G. renitidata* Hbn., ♀, Sistov, Bulgaria (ZSM)
5c. *G. renitidata* Hbn., ♂, Brussa, NW. Turkey (ZSM)
5d. *G. renitidata* Hbn., ♂, Amasia, N. Turkey (ZSM)
6a. **Myinodes interpunctaria H.-S.,** ♂, El Gouina, Tunisia (ZSM)
6b. *M. interpunctaria* H.-S., ♀, El Aziza, N. Algeria (ZSM)
6c. *M. interpunctaria atlantica* Hausm., ♂, Paratype, Marrakesch, W. Morocco (ZSM)
6d. *M. interpunctaria atlantica* Hausm., ♀, Paratype, Alcolea, S. Spain (ZSM)
7a. **Myinodes shohami Hausm.,** ♂, Holotype, Quasr al Hallabad, E. Jordan (ZSM)
7b. *M. shohami* Hausm., ♀, Paratype, Romana, N. Jordan (ZSM)

PLATE 2

(all specimens natural size)

8a. **Epirranthis diversata Den. & Schiff.,** ♂, Regensburg, S. Germany (ZSM)
8b. *E. diversata* Schiff., ♀, Regensburg, S. Germany (ZSM)
8c. *E. diversata* Schiff., ♂, Koski, Finland (coll. SKOU)
8d. *E. diversata* Schiff., ♀, Älvkarleö, Norway (coll. SKOU)
9a. **Alsophila aescularia Den. & Schiff.,** ♂, Wallersberg, S. Germany (ZSM)
9b. *A. aescularia* Schiff., ♂, Toblinosee, N. Italy (ZSM)
9c. *A. aescularia* Schiff., ♂, Pyatigorsk, Caucasus, Russia (ZSM)
9d. *A. aescularia* Schiff., ♀, Königs-Wusterhausen, E. Germany (coll. GELBRECHT)
9e. *A. aescularia* Schiff., ♀, Linz, N. Austria (ZSM)
10a. **Alsophila aceraria Den. & Schiff.,** ♂, Blankenburg, C. Germany (ZSM)
10b. *A. aceraria* Schiff., ♂, Wallersberg, S. Germany (ZSM)
10c. *A. aceraria* Schiff., ♂, Dürnstein, E. Austria (ZSM)
10d. *A. aceraria* Schiff., ♀, Klosterneuburg, E. Austria (ZSM)
10e. *A. aceraria* Schiff., ♀, Wallersberg, S. Germany (ZSM), reared
11a. **Heliothea discoidaria Bsd.,** ♂, Riopar, S. Spain (ZSM)
11b. *H. discoidaria* Bsd., ♂, Castilia, C. Spain (ZSM)
11c. *H. discoidaria* Bsd., ♂, St. Ildefonso, C. Spain (MNHU)
11d. *H. discoidaria* Bsd., ♂, Albarracin, C. Spain (ZSM)
11e. *H. discoidaria* Bsd., ♀, Riopar, S. Spain (ZSM)
11f. *H. discoidaria* Bsd., ♀, Castilia, C. Spain (ZSM)
12a. **Aplasta ononaria Fuessly,** ♂, Berlin, E. Germany (ZSM)
12b. *A. ononaria* Fuessly, ♀, Naumburg, E. Germany (ZSM)
12c. *A. ononaria* Fuessly, ♀, Pforzheim, SW. Germany (ZSM)
12d. *A. ononaria* Fuessly, ♂, Schnalstal, N. Italy (ZSM)
12e. *A. ononaria* Fuessly, ♀, Sigmundskron, N. Italy (ZSM)
12f. *A. ononaria* Fuessly, ♂, Albarracin, C. Spain (ZSM)
12g. *A. ononaria* Fuessly, ♂, Riopar, S. Spain (ZSM)
12h. *A. ononaria* Fuessly, ♀, Sierra Nevada, S. Spain (ZSM)
12i. *A. ononaria* Fuessly, ♀, Riopar, S. Spain (ZSM)
12k. *A. ononaria* Fuessly, ♂, La Bessée, SE. France (ZSM)
12l. *A. ononaria* Fuessly, ♂, Oraison, SE. France (ZSM)
12m. *A. ononaria* Fuessly, ♂, Digne, SE. France (ZSM)
12n. *A. ononaria* Fuessly, ♂, Bologna, N. Italy (ZSM)
12o. *A. ononaria* Fuessly, ♀, Ulcinj, Montenegro, Yugoslavia (coll. OSWALD)
12p. *A. ononaria* Fuessly, ♀, Zachlorou, Peloponnes, Greece (ZSM)
12q. *A. ononaria* Fuessly, ♀, Kalvrita, Peloponnes, Greece (coll. GELBRECHT)
12r. *A. ononaria* Fuessly, ♀, Petrina Plan, Macedonia (ZSM)
12s. *A. ononaria* Fuessly, ♀, Nikolic, Macedonia (ZSM)
13a. **Holoterpna pruinosata Stgr.,** ♂, Trieste, N. Italy (ZSM)
13b. *H. pruinosata* Stgr., ♀, Trieste, N. Italy (NHMW)
13c. *H. pruinosata* Stgr., ♀, Ohrid, Macedonia (ZSM)
13d. *H. pruinosata* Stgr., ♀, Smyrna, W. Turkey (ZSM)
14a. **Pingasa lahayei Obth.,** ♂, Asni, Gr. Atlas, Morocco (ZSM/HERBULOT)
14b. *P. lahayei* Obth., ♂, Tizi n'Test, Gr. Atlas, Morocco (ZSM/HERBULOT)
14c. *P. lahayei* Obth., ♂, Beni-Souik, Algeria (SMNS)
14d. *P. lahayei* Obth., ♂, Wadi el Hira, Libya (coll. SKOU)

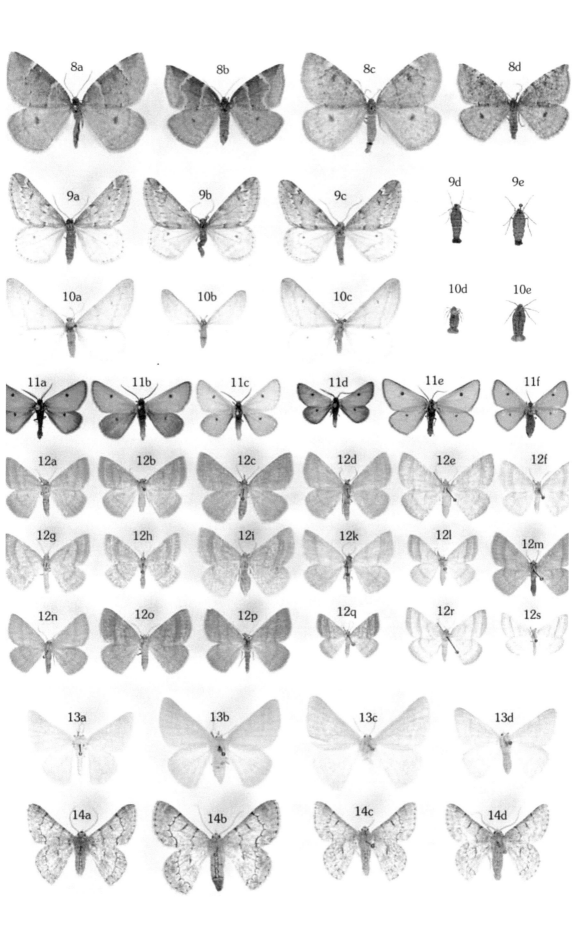

PLATE 3
(all specimens natural size)

15a. ***Pseudoterpna pruinata* Hufn.,** ♂, Malterhausen, E. Germany (coll. GELBRECHT), reared
15b. *P. pruinata* Hufn., ♀, Golmberg, E. Germany (coll. GELBRECHT), reared
15c. *P. pruinata* Hufn., ♀, Berlin, E. Germany (ZSM)
15d. *P. pruinata* Hufn., ♀, Kiel, N. Germany (ZSM), reared
15e. *P. pruinata* Hufn., ♀, Hamburg, N. Germany (ZSM), reared
15f. *P. pruinata* Hufn., ♂, Lüchow-Dannenberg, N. Germany (ZSM)
15g. *P. pruinata* Hufn., ♂, Kassel, C. Germany (ZSM), reared
15h. *P. pruinata* Hufn., ♀, Mannheim, W. Germany (ZSM), reared
15i. *P. pruinata* Hufn., ♂, Dürnstein, NE. Austria (ZSM)
15k. *P. pruinata* Hufn., ♂, Dürnstein, NE. Austria (ZSM)
15l. *P. pruinata* Hufn., ♂, Palaea Roumata, Crete (ZSM)
15m. *P. pruinata* Hufn., ♂, Kanev, C. Ukraine (coll. MÜLLER)
15n. *P. pruinata* Hufn., ♀, Kiyev, Ukraine (coll. MÜLLER)
15o. *P. pruinata* Hufn., ♂, Guberla, European Russia (ZSM)
15p. *P. pruinata* Hufn., ♂, Sarepta, European Russia (ZSM)
16a. ***Pseudoterpna coronillaria* Hbn.,** ♂, Albarracin, C. Spain (ZSM)
16b. *P. coronillaria* Hbn., ♀, Albarracin, C. Spain (ZSM)
16c. *P. coronillaria* Hbn., ♂, Gerona, NE. Spain (coll. GRÜNEWALD)
16d. *P. coronillaria* Hbn., ♂, Vannes, NW. France (ZSM)
16e. *P. coronillaria* Hbn., ♂, Vannes, NW. France (MNHU)
16f. *P. coronillaria* Hbn., ♂, Vannes, NW. France (MNHU)
16g. *P. coronillaria* Hbn., ♂, Agay, SE. France (ZSM)
16h. *P. coronillaria* Hbn., ♂, Noli, NW. Italy (ZSM)
16i. *P. coronillaria* Hbn., ♂, Noli, NW. Italy (ZSM)
16k. *P. coronillaria flamignii* Hausm., ♀, Paratype, Mtgna. Grande, C. Italy (ZSM)
16l. *P. coronillaria flamignii* Hausm., ♂, Holotype, Mtgna. Grande, C. Italy (ZSM)
16m. *P. coronillaria flamignii* Hausm., ♀, Mistretta, Sicily (ZSM)
16n. *P. coronillaria cinerascens* Z., ♂, Kalithes, Rhodos (ZSM)
16o. *P. coronillaria cinerascens* Z., ♀, Penins. Gelibolu, European Turkey (coll. GELBRECHT)
17a. ***Pseudoterpna corsicaria* Rmb.,** ♂, Corsica (ZSM)
17b. *P. corsicaria* Rmb., ♂, Esterel, SE. France (ZSM)
17c. *P. corsicaria* Rmb., ♀, Col de Vergio, Corsica (ZSM/HERBULOT)
17d. *P. corsicaria* Rmb., ♀, Corsica (ZSM)
17e. *P. corsicaria* Rmb., ♀, Saint Antoine, Corsica (coll. GRÜNEWALD)
18a. ***Geometra papilionaria* L.,** ♀, Lüchow-Dannenberg, N. Germany (ZSM)
18b. *G. papilionaria* L., ♀, Neubrandenburg, E. Germany (coll. MÜLLER)
18c. *G. papilionaria* L., ♂, Jenkofen, S. Germany (coll. GRÜNEWALD)
18d. *G. papilionaria* L., ♂, Murnau, S. Germany (coll. OSWALD)
18e. *G. papilionaria* L., ♂, Schönebeck, E. Germany (coll. MÜLLER)
18f. *G. papilionaria* L., ♂, Teberda, NW. Caucasus, Russia (coll. MÜLLER)

PLATE 4
(all specimens natural size)

19a.   ***Comibaena bajularia* Den. & Schiff.,** ♀, Lüneburg, N. Germany (ZSM)
19b.   *C. bajularia* Schiff., ♂, Günzburg, S. Germany (ZSM)
19c.   *C. bajularia* Schiff., ♀, Eisenstadt, E. Austria (ZSM)
19d.   *C. bajularia* Schiff., ♂, Eisenstadt, E. Austria (ZSM), reared
19e.   *C. bajularia* Schiff., ♂, Veldes, Slovenia (ZSM)
19f.   *C. bajularia* Schiff., ♀, Nagy Nyir, Hungary (NHMW), reared
19g.   *C. bajularia* Schiff., ♂, Digne, SE. France (ZSM)
19h.   *C. bajularia* Schiff., ♂, Palencia, N. Spain (ZSM)
19i.   *C. bajularia* Schiff., ♂, Guadarrama, C. Spain (EMEM/ZSM)
19k.   *C. bajularia* Schiff., ♀, Lozoya, C. Spain (coll. SKOU)
19l.   *C. bajularia* Schiff., ♂, Ilgaz Daglari, N. Turkey (coll. MÜLLER), reared
19m.   *C. bajularia* Schiff., ♂, Ovruch distr., N. Ukraine (coll. MÜLLER)
19n.   *C. bajularia* Schiff., ♀, Konya, C. Turkey (EMEM/ZSM)
19o.   *C. bajularia tikhonovi* Hausm., ♂, Paratype, Talyah, Aserbeidjan (ZSM)
19p.   *C. bajularia tikhonovi* Hausm., ♂, Holotype, Pyatigorsk, Caucasus, Russia (ZSM)
20a.   ***Comibaena pseudoneriaria* Whli.,** ♀, Albarracin, C. Spain (ZSM)
20b.   *C. pseudoneriaria* Whli., ♀, Sierra Nevada, S. Spain (SMNS)
21a.   ***Proteuchloris neriaria* H.-S.,** ♂, Kiten, E. Bulgaria (coll. GELBRECHT), reared
21b.   *P. neriaria* H.-S., ♀, Kiten, E. Bulgaria (coll. GELBRECHT), reared
21c.   *P. neriaria* H.-S., ♀, Latakia, Syria (SMNS)
22a.   ***Thetidia plusiaria* Bsd.,** ♂, Penalba, N. Spain (ZSM)
22b.   *T. plusiaria* Bsd., ♂, Penalba, N. Spain (ZSM)
22c.   *T. plusiaria* Bsd., ♂, Murcia, SE. Spain (NHMW)
22d.   *T. plusiaria* Bsd., ♀, Albarracin, C. Spain (ZSM)
22e.   *T. plusiaria* Bsd., ♀, Castilia, C. Spain (ZSM)
22f.   *T. plusiaria* Bsd., ♂, Albarracin, C. Spain (ZSM)
22g.   *T. plusiaria* Bsd., ♂, Albarracin, C. Spain (ZSM)
22h.   *T. plusiaria* Bsd., ♀, Orgiva, S. Spain (ZSM)
22i.   *T. plusiaria* Bsd., ♂, Cala Ratjada, Mallorca (ZSM)
22k.   *T. plusiaria* Bsd., ♂, Castilia, C. Spain (ZSM)
23a.   ***Thetidia (Aglossochloris) correspondens* Alph.,** ♂, Sarepta, European Russia (MNHU)
23b.   *T. correspondens* Alph., ♀, Sarepta, European Russia (MNHU)
23c.   *T. correspondens* Alph., ♀, Inderskischer Salzsee, Ural, Russia (ZSM/HERBULOT)
23d.   *T. correspondens* Alph., ♀, Alexander Mts., central Asia (NHMW)
23e.   *T. correspondens* Alph., ♂, Alexander Mts., central Asia (NHMW)
24a.   ***Thetidia (Antonechloris) smaragdaria* F.,** ♂, Kaiserstuhl, SW. Germany (ZSM), reared
24b.   *T. smaragdaria* F., ♀, Kaiserstuhl, SW. Germany (ZSM), reared
24c.   *T. smaragdaria* F., ♀, Rechtebachtal, W. Germany (EMEM/ZSM)
24d.   *T. smaragdaria* F., ♂, Dürnstein, E. Austria (ZSM)
24e.   *T. smaragdaria* F., ♂, Naturns, N. Italy (ZSM)
24f.   *T. smaragdaria* F., ♂, Trieste, NE. Italy (ZSM), reared
24g.   *T. smaragdaria* F., ♀, Trieste, NE. Italy (ZSM), reared
24h.   *T. smaragdaria* F., ♂, Mti. Simbruini, C. Italy (ZSM)
24i.   *T. smaragdaria* F., ♀, Mti. Simbruini, C. Italy (ZSM)
24k.   *T. smaragdaria* F., ♂, Villányi-hg, S. Hungary (coll. GELBRECHT), reared
24l.   *T. smaragdaria* F., ♀, Varna, E. Bulgaria (coll. GELBRECHT), reared
24m.   *T. smaragdaria* F., ♀, Varna, E. Bulgaria (ZSM), reared
24n.   *T. smaragdaria gigantea* Mill., ♂, Sierra de Gredos, C. Spain (ZSM)
24o.   *T. smaragdaria gigantea* Mill., ♂, Guadarrama, C. Spain (EMEM/ZSM)
24p.   *T. smaragdaria gigantea* Mill., ♀, Sierra de Gredos, C. Spain (ZSM)

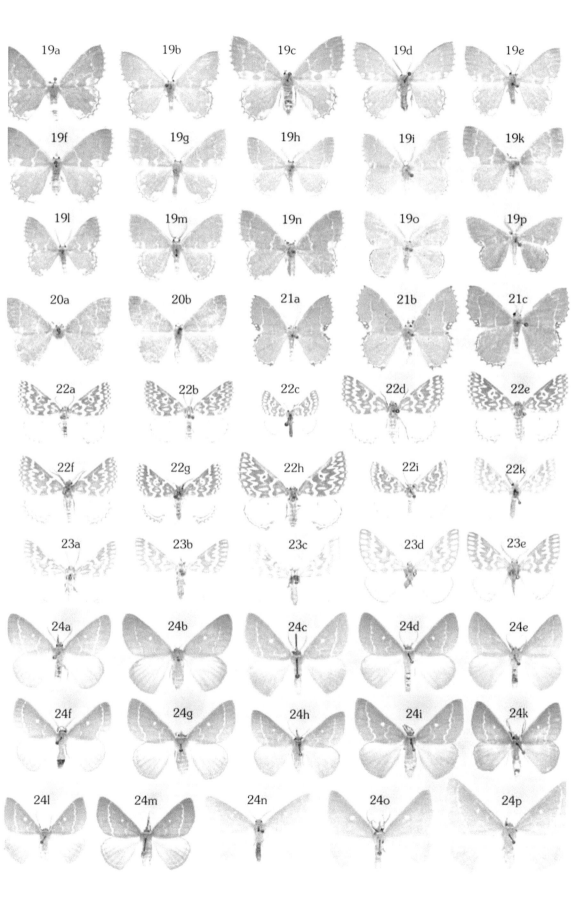

PLATE 5
(all specimens natural size)

24q. **Thetidia smaragdaria F.,** ♂, Kirowograd, Ukraine (ZSM/HERBULOT)
24r. *T. smaragdaria volgaria* Gn., ♂, Sarepta, European Russia (ZSM)
24s. *T. smaragdaria volgaria* Gn., ♂, Guberla, Urals, European Russia (ZSM)
24t. *T. smaragdaria volgaria* Gn., ♂, Sarepta, European Russia (ZSM)
24u. *T. smaragdaria volgaria* Gn., ♂, Sarepta, European Russia (ZSM)
24v. *T. smaragdaria volgaria* Gn., ♀, Sarepta, European Russia (ZSM)
24w. **Thetidia persica Hausm.,** ♂, Elburs, N. Iran (NHMW)
25a. **Thetidia sardinica Schaw.,** ♂, Gennargentu, Sardinia (ZSM)
25b. *T. sardinica* Schaw., ♂, Aritzo, Sardinia (ZSM)
25c. *T. sardinica* Schaw., ♂, Aritzo, Sardinia (ZSM)
25d. *T. sardinica* Schaw., ♀, Gennargentu, Sardinia (ZSM)
26a. **Hemistola chrysoprasaria Esp.,** ♂, Boizenburg, NE. Germany (ZSM), reared
26b. *H. chrysoprasaria* Esp., ♀, Berlin, E. Germany (coll. GELBRECHT)
26c. *H. chrysoprasaria* Esp., ♂, Berlin, E. Germany (coll. MÜLLER), reared
26d. *H. chrysoprasaria* Esp., ♀, Kassel, C. Germany (ZSM) , reared
26e. *H. chrysoprasaria* Esp., ♀, Wallersberg, S. Germany (ZSM)
26f. *H. chrysoprasaria* Esp., ♂, Aschheim, S. Germany (ZSM)
26g. *H. chrysoprasaria* Esp., ♂, Dürnstein, E. Austria (ZSM)
26h. *H. chrysoprasaria* Esp., ♀, Terlan, N. Italy (ZSM)
26i. *H. chrysoprasaria* Esp., ♂, Ainsa, Pyrenees, N. Spain (NHMW)
26k. *H. chrysoprasaria* Esp., ♂, Varna, E. Bulgaria (coll. GELBRECHT), reared
26l. *H. chrysoprasaria occidentalis* Whli., ♂, Paralectotype, Ronda, S. Spain (ZFMK)
26m. *H. chrysoprasaria occidentalis* Whli., ♀, Paralectotype, Sierra Nevada, S. Spain (ZFMK)
27a. **Hemistola siciliana Prt.,** ♂, Maiella, C. Italy (coll. ZAHM)
27b. *H. siciliana* Prt., ♂, Donnici, S. Italy (coll. SCALERCIO)
27c. *H. siciliana* Prt., ♂, Palermo, Sicily (ZSM)
27d. *H. siciliana* Prt., ♂, Mt. Pelegrino, Sicily (ZSM)
28a. **Xenochlorodes olympiaria H.-S.,** ♂, Gerona, N. Spain (coll. GRÜNEWALD)
28b. *X. olympiaria* H.-S., ♂, Trecchina, S. Italy (ZSM), reared
28c. *X. olympiaria* H.-S., ♂, Gravosa, Dalmatia, S. Croatia (ZSM)
28d. *X. olympiaria* H.-S., ♀, Gravosa, Dalmatia, S. Croatia (ZSM)
28e. *X. olympiaria* H.-S., ♀, Gravosa, Dalmatia, S. Croatia (ZSM)
28f. *X. olympiaria* H.-S., ♀, Preveza, Greece (coll. FRIEDRICH)
29a. **Eucrostes indigenata Vill.,** ♂, Homokbuckás, Hungary (ZSM)
29b. *E. indigenata* Vill., ♂, La Turbie, SE. France (ZSM)
29c. *E. indigenata* Vill., ♀, Col de Braus, SE. France (ZSM)
29d. *E. indigenata* Vill., ♂, Tivoli, C. Italy (ZSM)
29e. *E. indigenata* Vill., ♀, Tivoli, C. Italy (ZSM)
29f. *E. indigenata* Vill., ♂, Tivoli, C. Italy (ZSM)
29g. *E. indigenata* Vill., ♀, Palermo, Sicily (ZSM)
29h. *E. indigenata* Vill., ♂, Gravosa, Dalmatia, S. Croatia (ZSM)
29i. *E. indigenata* Vill., ♀, Krume, Albania (NHMW)
29k. *E. indigenata* Vill., ♂, Tripolis, Peloponnes, Greece (coll. GRÜNEWALD)
29l. *E. indigenata* Vill., ♀, Pirin, SW. Bulgaria (coll. GELBRECHT), reared
29m. *E. indigenata* Vill., ♂, Pirin, SW. Bulgaria (coll. GELBRECHT), reared
29n. *E. indigenata lanjeronica* Hausm., ♀, Holotype, Lanjaron, S. Spain (ZSM)
29o. *E. indigenata lanjeronica* Hausm., ♀, Paratype, Orgiva, S. Spain (ZSM)

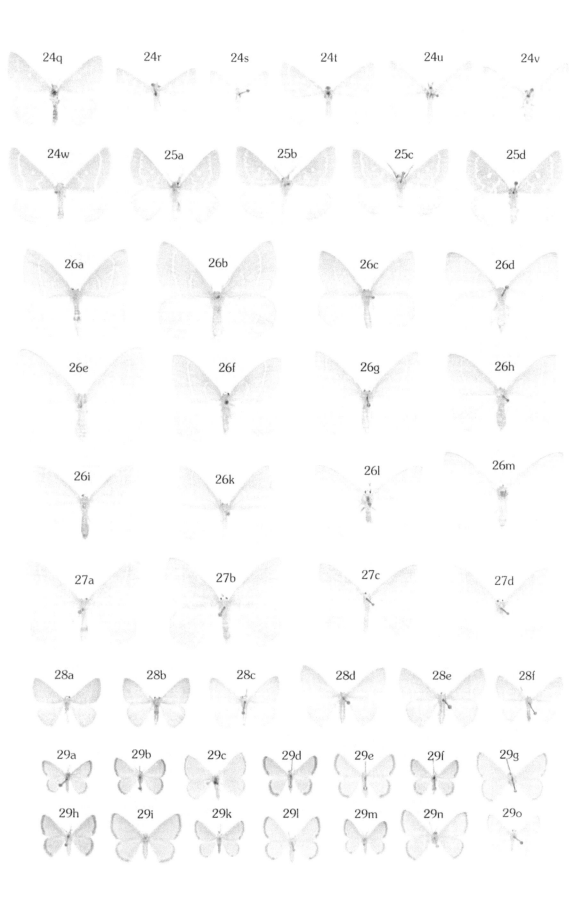

PLATE 6
(all specimens natural size)

30a. **Jodis lactearia L.,** ♂, Ratzeburg, N. Germany (ZSM)
30b. *J. lactearia* L., ♂, Rügen, NE. Germany (coll. GELBRECHT), reared
30c. *J. lactearia* L., ♀, Rügen, NE. Germany (coll. MÜLLER), reared
30d. *J. lactearia* L., ♀, Miesbach, S. Germany (ZSM)
30e. *J. lactearia* L., ♂, Deininger Moor, S. Germany (ZSM)
30f. *J. lactearia* L., ♀, Deininger Moor, S. Germany (ZSM)
30g. *J. lactearia* L., ♂, Hörsching, Austria (ZSM)
30h. *J. lactearia* L., ♂, Amfreville, N. France (ZSM/HERBULOT)
30i. *J. lactearia* L., ♂, Forêt de Benon, W. France (ZSM/HERBULOT)
30k. *J. lactearia* L., ♂, Terlan, N. Italy (ZSM)
30l. *J. lactearia* L., ♂, Kastelruth, N. Italy (ZSM)
30m. *J. lactearia* L., ♂, Cibinsgebirge, Romania (ZSM)
31a. **Jodis putata L.,** ♂, Abo, Finland (ZSM)
31b. *J. putata* L., ♀, Woblitz, E. Germany (coll. GELBRECHT), reared
31c. *J. putata* L., ♀, Lüneburg, N. Germany (ZSM), reared
31d. *J. putata* L., ♂, Oranienburg, E. Germany (coll. MÜLLER)
31e. *J. putata* L., ♀, Oranienburg, E. Germany (coll. MÜLLER)
31f. *J. putata* L., ♂, Allmannshausen, S. Germany (coll. OSWALD)
31g. *J. putata* L., ♂, Allmannshausen, S. Germany (coll. OSWALD)
31h. *J. putata* L., ♀, Brandenberg, C. Austria (ZSM)
31i. *J. putata* L., ♀, Lichtenberg, N. Austria (ZSM)
31k. *J. putata* L., ♀, Kirchschlag, N. Austria (ZSM)
31l. *J. putata* L., ♀, Neustift, N. Austria (ZSM)
31m. *J. putata* L., ♂, Bozen, N. Italy (ZSM)
32a. **Thalera fimbrialis Scop.,** ♂, Rendsburg, N. Germany (ZSM), reared
32b. *T. fimbrialis* Scop., ♂, Vogelsdorf, E. Germany (coll. GELBRECHT)
32c. *T. fimbrialis* Scop., ♀, Paitzkofen, S. Germany (ZSM), reared
32d. *T. fimbrialis* Scop., ♀, Laaber, S. Germany (ZSM), reared
32e. *T. fimbrialis* Scop., ♂, Schleißheim, S. Germany (ZSM)
32f. *T. fimbrialis* Scop., ♀, Kissingen, S. Germany (ZSM)
32g. *T. fimbrialis* Scop., ♂, Linz, N. Austria (ZSM), reared
32h. *T. fimbrialis* Scop., ♀, Linz, N. Austria (ZSM)
32i. *T. fimbrialis* Scop., ♀, Linz, N. Austria (ZSM)
32k. *T. fimbrialis* Scop., ♂, Pirin, SW. Bulgaria (coll. GELBRECHT)
33a. **Dyschloropsis impararia Gn.,** ♀, Guberli, Urals, European Russia (NHMW)
33b. *D. impararia* Gn., ♀, Guberli, Urals, European Russia (ZISP)
33c. *D. impararia* Gn., ♂, Guberli, Urals, European Russia (ZISP)
33d. *D. impararia* Gn., ♂, Sibiria (MNHU)
33e. *D. impararia* Gn., ♂, Uliassutai (MNHU)
34a. **Bustilloxia saturata iberica Hausm.,** ♂, Albarracin, C. Spain (coll. KRAUS)
34b. *B. saturata iberica* Hausm., ♂, Veleta, S. Spain (ZSM/HERBULOT)
34c. *B. saturata iberica* Hausm., ♂, Holotype, Diezma, S. Spain (ZSM)
34d. *B. saturata iberica* Hausm., ♂, Planas del Rey, N. Spain (ZSM/HERBULOT)
34e. *B. saturata iberica* Hausm., ♀, Zaragoza, N. Spain (coll. HALL)
34f. *B. saturata iberica* Hausm., ♂, Sierra Nevada, S. Spain (coll. PÖLL)
34g. *B. saturata iberica* Hausm., ♀, Granada, S. Spain (coll. HALL)
35a. **Kuchleria insignata Hausm.,** ♂, Elche, SE. Spain (ZSM)
35b. *K. insignata* Hausm., ♂, Elche, SE. Spain (ZSM)
35c. *K. insignata* Hausm., ♀, Paratype, Orgiva, S. Spain (SMNK)
35d. *K. insignata* Hausm., ♂, Paratype, Camino de Ojen, S. Spain (ZSM)
35e. *K. insignata* Hausm., ♂, Paratype, El Jadida, W. Morocco (ZSM)
35f. *K. insignata* Hausm., ♂, Essaouira, W. Morocco (ZSM/HERBULOT)
35g. **Kuchleria menadiara atlagenes Prt.,** ♀, Iminen-Tal, Gr. Atlas, Morocco (ZSM)
35h. **Kuchleria menadiara Th.-Mieg,** ♂, Tabarka area, N. Tunisia (ZSM)

224

PLATE 7
(all specimens natural size)
'

36a. **Hemithea aestivaria Hbn.,** ♂, Oranienburg, E. Germany (coll. MÜLLER), reared
36b. *H. aestivaria* Hbn., ♀, Iphofen, S. Germany (ZSM)
36c. *H. aestivaria* Hbn., ♂, Regensburg, S. Germany (ZSM)
36d. *H. aestivaria* Hbn., ♂, Lienz, Austria (ZSM)
36e. *H. aestivaria* Hbn., ♂, Lienz, Austria (ZSM)
36f. *H. aestivaria* Hbn., ♀, Marchtrenk, Austria (ZSM), reared
36g. *H. aestivaria* Hbn., ♂, Linz, Austria (ZSM)
36h. *H. aestivaria* Hbn., ♀, Nice, S. France (ZSM/HERBULOT)
36i. *H. aestivaria* Hbn., ♂, Sattnitz, SE. Austria (ZSM)
36k. *H. aestivaria* Hbn., ♂, Sattnitz, SE. Austria (ZSM)
37a. **Chlorissa viridata L.,** ♂, Cuxhaven, N. Germany (ZSM)
37b. *C. viridata* L., ♀, Lüchow-Dannenberg, N. Germany (ZSM)
37c. *C. viridata* L., ♂, Mathesdorf, Silesia, Poland (ZSM)
37d. *C. viridata* L., ♀, Mathesdorf, Silesia, Poland (ZSM)
37e. *C. viridata* L., ♀, Velden, S. Germany (ZSM)
37f. *C. viridata* L., ♂, Irlbach, S. Germany (ZSM)
37g. *C. viridata* L., ♂, Linz, N. Austria (ZSM)
37h. *C. viridata* L., ♂, Linz, N. Austria (ZSM)
37i. *C. viridata* L., ♂, Kiskörös, C. Hungary, e.l. (coll. MÜLLER), reared
37k. *C. viridata* L., ♀, Varna, Bulgaria, e.o. (coll. GELBRECHT), reared
37l. *C. viridata* L., ♂, Varna, Bulgaria, e.o. (coll. GELBRECHT), reared
37m. *C. viridata* L., ♂, Stanichno-Luganskoje, E. Ukraine (ZMUKiew)
38a. **Chlorissa cloraria Hbn.,** ♂, Bad Tölz, S. Germany (ZSM)
38b. *C. cloraria* Hbn., ♀, Regensburg, S. Germany (ZSM)
38c. *C. cloraria* Hbn., ♀, Ebelsberg, N. Austria (ZSM)
38d. *C. cloraria* Hbn., ♂, Ebelsberg, N. Austria (ZSM)
38e. *C. cloraria* Hbn., ♂, Hinterstoder, N. Austria (ZSM)
38f. *C. cloraria* Hbn., ♂, Dürnstein, E. Austria (ZSM)
38g. *C. cloraria* Hbn., ♂, Mödling, E. Austria (ZSM)
38h. *C. cloraria* Hbn., ♂, Gumpoldskirchen, E. Austria (ZSM)
38i. *C. cloraria* Hbn., ♀, Merano, N. Italy (ZSM)
38k. *C. cloraria* Hbn., ♀, Terlan, N. Italy (ZSM)
38l. *C. cloraria* Hbn., ♂, Mentone, SE. France (ZSM)
38m. *C. cloraria* Hbn., ♂, Los Ambollas, Pyr.or., S. France (ZSM/HERBULOT)
38n. *C. cloraria* Hbn., ♀, Cibinsgebirge, Romania (ZSM)
38o. *C. cloraria* Hbn., ♂, Cibinsgebirge, Romania (ZSM)
38p. *C. cloraria* Hbn., ♂, Ohrid, Macedonia (ZSM)
38q. *C. cloraria* Hbn., ♂, Pirin, SW. Bulgaria, e.o. (coll. MÜLLER), reared
38r. *C. cloraria* Hbn., ♂, Pirin, SW. Bulgaria (coll. GELBRECHT)
38s. *C. cloraria* Hbn., ♂, Crimea, S. Ukraine (ZMUKiew)
39a. **Phaiogramma etruscaria Z.,** ♀, Eisenstadt, E. Austria (ZSM)
39b. *P. etruscaria* Z., ♀, Donnerskirchen, E. Austria (ZSM)
39c. *P. etruscaria* Z., ♀, Montemaderno, N. Italy (ZSM), reared
39d. *P. etruscaria* Z., ♀, Sigmundskron, N. Italy (ZSM)
39e. *P. etruscaria* Z., ♀, Sierra Alca, E. Spain (ZSM)
39f. *P. etruscaria* Z., ♂, Tortora, S. Italy, e.o. (ZSM), reared
39g. *P. etruscaria* Z., ♂, Puyvert, S. France (ZSM/HERBULOT)
39h. *P. etruscaria* Z., ♀, La Voulte, S. France (ZSM/HERBULOT)

(continued on page 228)

PLATE 7
(all specimens natural size)
continued from page 226

39i. *P. etruscaria* Z., ♂, Gravosa, Dalmatia, S. Croatia (ZSM)
39k. *P. etruscaria* Z., ♀, Gravosa, Dalmatia, S. Croatia (ZSM)
39l. *P. etruscaria* Z., ♀, Michurin, SE. Bulgaria (coll. MÜLLER), reared
39m. *P. etruscaria* Z., ♂, Chersonissos, E. Crete (coll. FRIEDRICH)
40a. ***Phaiogramma faustinata* Mill.,** ♂, Monemvasia, Peloponnes, Greece (coll. SKOU)
40b. *P. faustinata* Mill., ♂, Cala Ratjada, Mallorca, e.o. (coll. MÜLLER)
40c. *P. faustinata* Mill., ♀, Cala Ratjada, Mallorca, e.o. (coll. MÜLLER)
40d. *P. faustinata* Mill., ♀, Valverde, S. Spain (ZSM)
40e. *P. faustinata* Mill., ♀, Gafsa, C. Tunisia (ZSM/HERBULOT)
40f. *P. faustinata* Mill., ♀, Planas del Rey, N. Spain (ZSM/HERBULOT)
41a. ***Microloxia herbaria* Hbn.,** ♂, Aranjuez, C. Spain (ZSM)
41b. *M. herbaria* Hbn., ♂, Pisa, NW. Italy (ZSM)
41c. *M. herbaria* Hbn., ♀, Mti. Sabini, C. Italy (ZSM)
41d. *M. herbaria* Hbn., ♂, Gravosa, Dalmatia, S. Croatia (ZSM), reared
41e. *M. herbaria* Hbn., ♂, Ohrid, Macedonia (ZSM)
41f. *M. herbaria* Hbn., ♀, Ohrid, Macedonia (ZSM)
41g. *M. herbaria* Hbn., ♂, Ohrid, Macedonia (ZSM)
41h. *M. herbaria* Hbn., ♀, Ohrid, Macedonia (ZSM)
41i. *M. herbaria* Hbn., ♀, Treska, Macedonia (ZSM), reared
41k. *M. herbaria* Hbn., ♀, Karadagh, Crimea, S. Ukraine (ZISP)
41l. *M. herbaria* Hbn., ♂, Karadagh, Crimea, S. Ukraine (ZISP)
41m. *M. herbaria virideciliata* Bub., ♂, Evisa, Corsica (ZSM)
41n. *M. herbaria virideciliata* Bub., ♂, Nonza, N. Corsica (ZSM), reared
41o. *M. herbaria* Hbn., trans. *advolata* Ev., ♂, Derbent, Daghestan, Caspian Sea (ZISP)
41p. *M. herbaria* Hbn., trans. *advolata* Ev., ♀, Ordubad, S. Armenia (ZISP)
41q. *M. herbaria advolata* Ev., ♀, Sarepta, European Russia (ZISP)

PLATE 8

fig.1

fig.2

fig.3

fig.4

fig.5

fig.6

fig.7

fig.8

fig.9

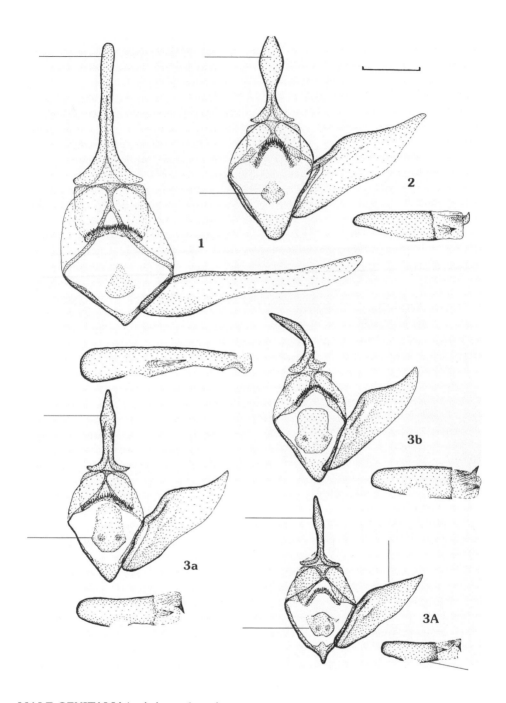

MALE GENITALIA (scale bar = 1 mm)

1. *Archiearis parthenias* L., Sweden
2. *Archiearis notha* Hbn., E. Prussia
3a. *Archiearis puella* Esp., NE. Austria
3b. *Archiearis puella mediterranea* Ganev, Bulgaria
3A. *Archiearis touranginii* Berce, France

230

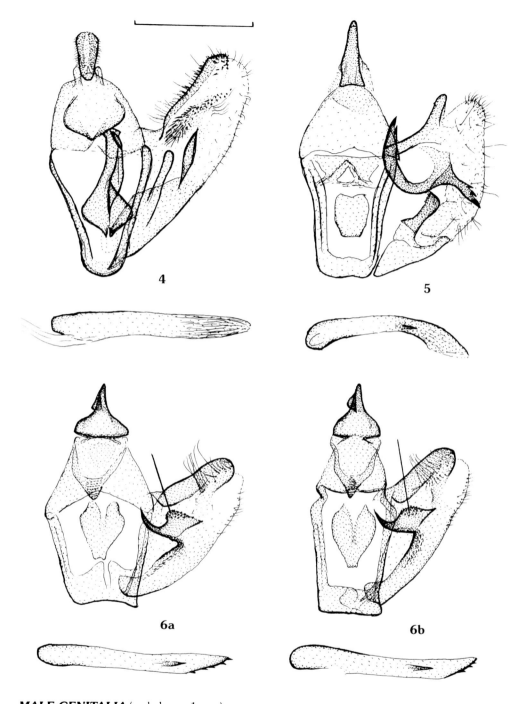

MALE GENITALIA (scale bar = 1 mm)

4. *Orthostixis cribraria* Hbn., Hungary
5. *Gypsochroa renitidata* Hbn., S. France
6a. *Myinodes interpunctaria* H.-S., Sicily
6b. *Myinodes interpunctaria atlantica* Hausm., S. Spain

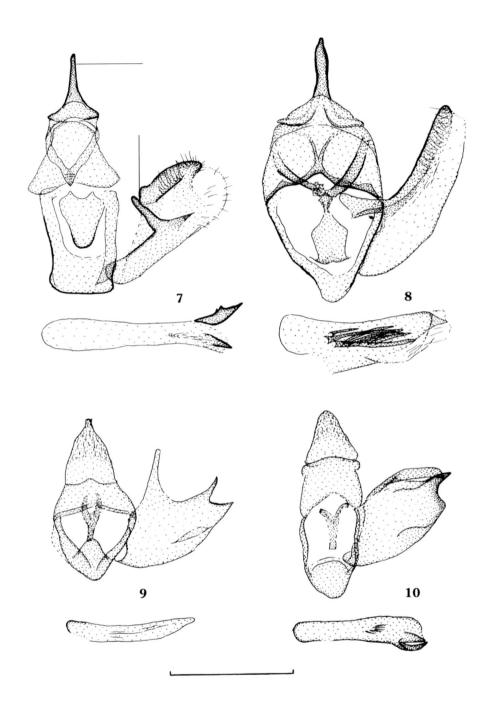

MALE GENITALIA (scale bar = 1 mm)

7. *Myinodes shohami* Hausm., N. Israel
8. *Epirranthis diversata* Schiff., S. Germany
9. *Alsophila aescularia* Schiff., NE. Poland
10. *Alsophila aceraria* Schiff., NE. Austria

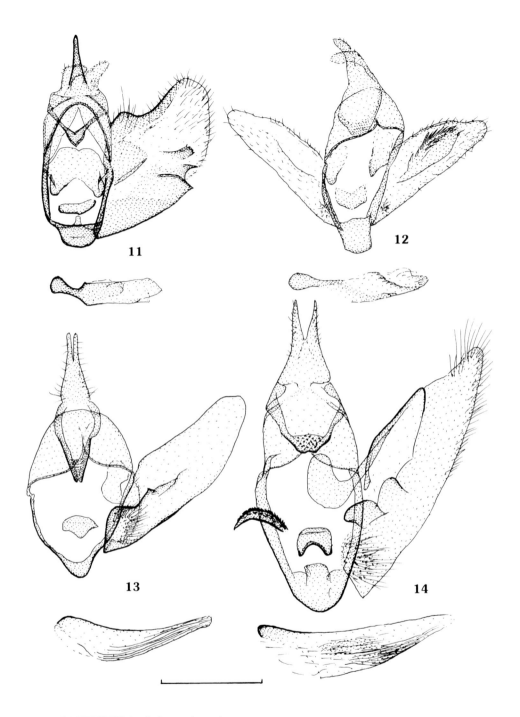

MALE GENITALIA (scale bar = 1 mm)

11. *Heliothea discoidaria* Bsd., S. Spain
12. *Aplasta ononaria* Fuessly, E. Germany
13. *Holoterpna pruinosata* Stgr., N. Israel
14. *Pingasa lahayei* Obth., Saudi Arabia

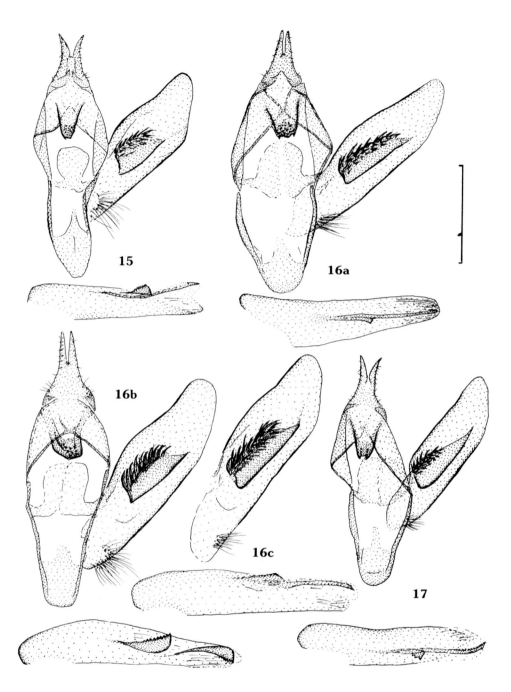

MALE GENITALIA (scale bar = 1 mm)

15. *Pseudoterpna pruinata* Hufn., E. Germany
16a. *Pseudoterpna coronillaria* Hbn., C. Spain
16b. *Pseudoterpna coronillaria flamignii* Hausm., Paratype, C. Italy
16c. *Pseudoterpna coronillaria cinerascens* Z., Rhodos (valva and aedeagus)
17. *Pseudoterpna corsicaria* Rmb., Corsica

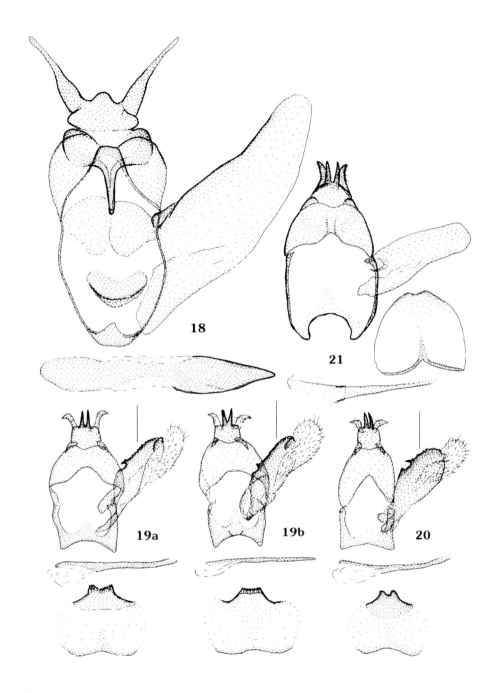

MALE GENITALIA (scale bar = 1 mm)

18. *Geometra papilionaria* L., S. Germany
19a. *Comibaena bajularia* Schiff., N. Germany
19b. *Comibaena bajularia tikhonovi* Hausm., Holotype, Caucasus
20. *Comibaena pseudoneriaria* Whli., E. Spain
21. *Proteuchloris neriaria* H.-S., Macedonia

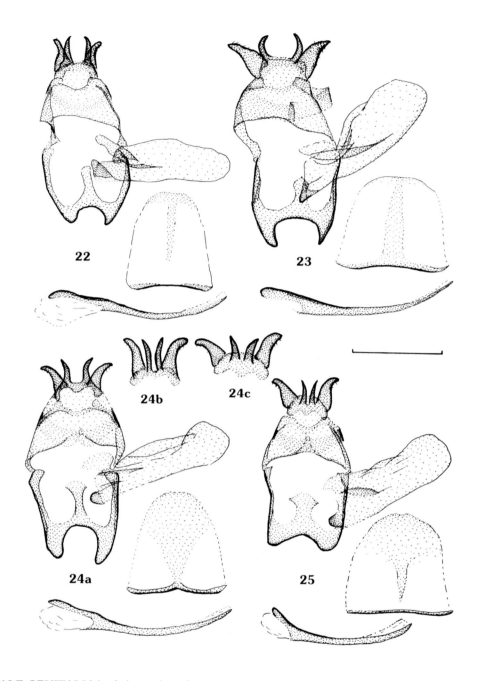

MALE GENITALIA (scale bar = 1 mm)

22. *Thetidia plusiaria* Bsd., N. Spain
23. *Thetidia correspondens* Alph., European Russia
24a. *Thetidia smaragdaria* F., SW. Germany
24b. *Thetidia smaragdaria volgaria* Gn., European Russia (uncus and socii)
24c. *Thetidia persica* Hausm., N. Iran (uncus and socii)
25. *Thetidia sardinica* Schaw., Sardinia

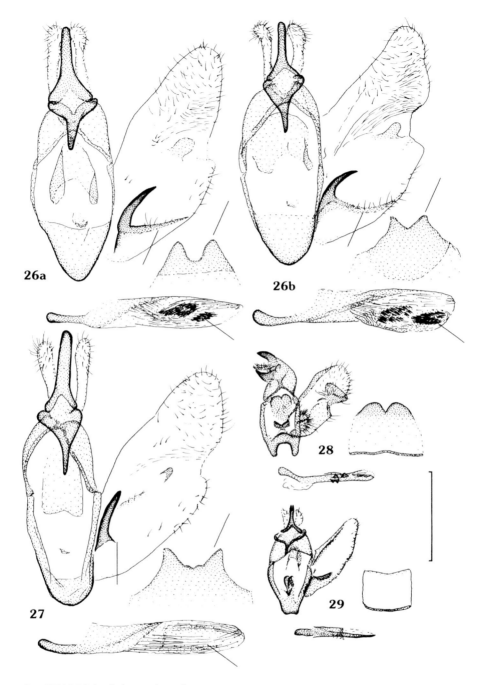

MALE GENITALIA (scale bar = 1 mm)

26a. *Hemistola chrysoprasaria* Esp., Hungary
26b. *Hemistola chrysoprasaria occidentalis* Whli., Paralectotype, S. Spain
27.   *Hemistola siciliana* Prt., Sicily
28.   *Xenochlorodes olympiaria* H.-S., N. Spain
29.   *Eucrostes indigenata* Vill., Hungary

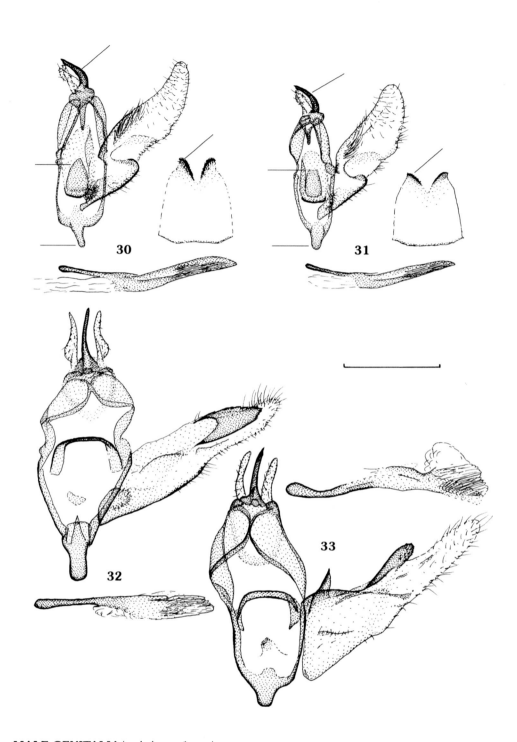

MALE GENITALIA (scale bar = 1 mm)

30. *Jodis lactearia* L., N. Germany
31. *Jodis putata* L., Finland
32. *Thalera fimbrialis* Scop., S. Germany
33. *Dyschloropsis impararia* Gn., Kirgisia

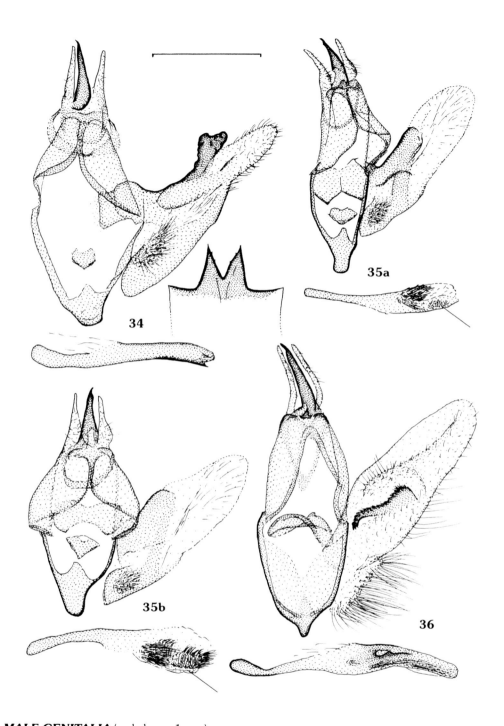

MALE GENITALIA (scale bar = 1 mm)

34. *Bustilloxia saturata iberica* Hausm., C. Spain
35a. *Kuchleria insignata* Hausm., SE. Spain
35b. *Kuchleria menadiara* Th.-Mieg, Algeria
36. *Hemithea aestivaria* Hbn., E. Germany

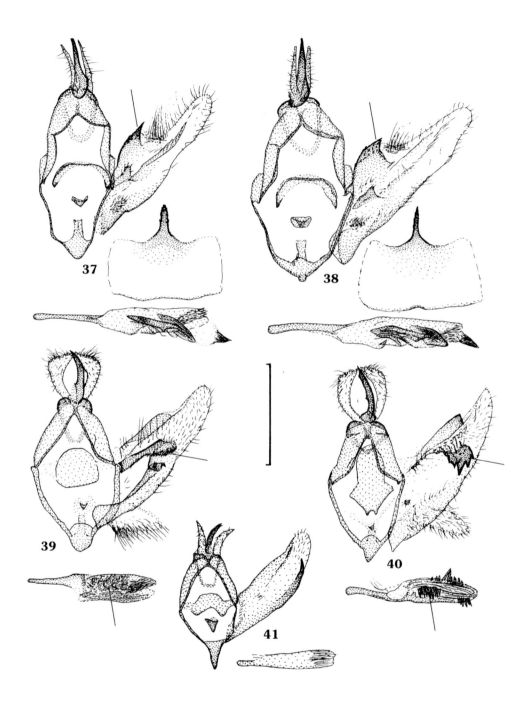

MALE GENITALIA (scale bar = 1 mm)

37. *Chlorissa viridata* L., N. Germany
38. *Chlorissa cloraria* Hbn., S. Germany
39. *Phaiogramma etruscaria* Z., N. Israel
40. *Phaiogramma faustinata* Mill., N. Israel
41. *Microloxia herbaria* Hbn., C. Turkey

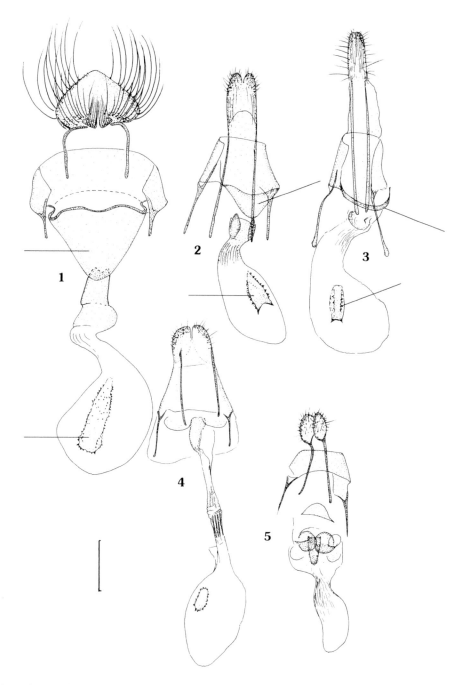

FEMALE GENITALIA (scale bar = 1 mm)

1. *Archiearis parthenias* L., S. Finland
2. *Archiearis notha* Hbn., W. Germany
3. *Archiearis puella* Esp., S. European Russia
4. *Orthostixis cribraria* Hbn., C. Turkey
5. *Gypsochroa renitidata* Hbn., S. France

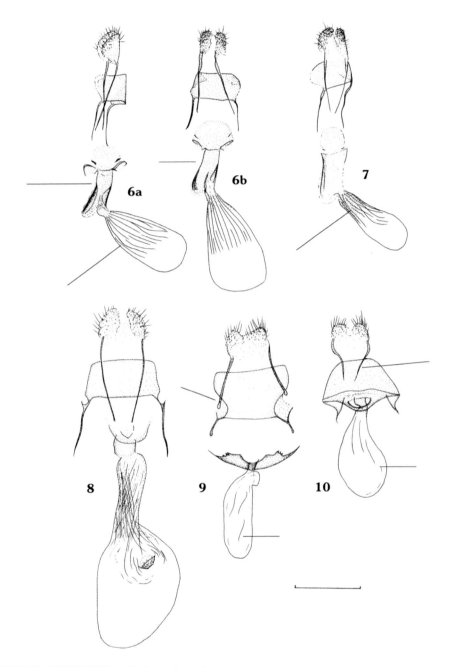

FEMALE GENITALIA (scale bar = 1 mm)

6a. *Myinodes interpunctaria* H.-S., NE. Algeria
6b. *Myinodes interpunctaria atlantica* Hausm., S. Spain
7. *Myinodes shohami* Hausm., N. Israel
8. *Epirranthis diversata* Schiff., S. Germany
9. *Alsophila aescularia* Schiff., E. Austria
10. *Alsophila aceraria* Schiff., E. Austria

242

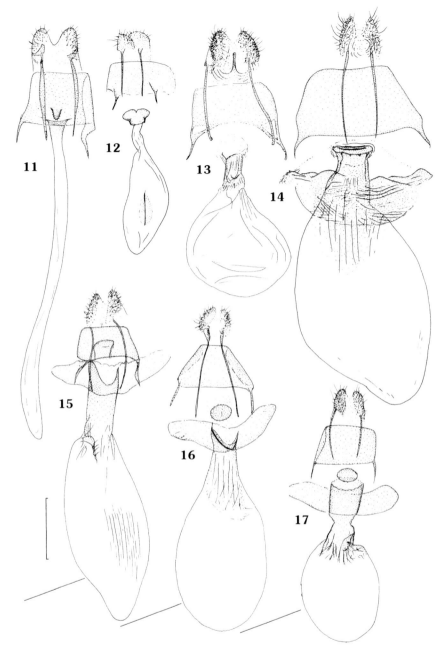

FEMALE GENITALIA (scale bar = 1 mm)

11. *Heliothea discoidaria* Bsd., C. Spain
12. *Aplasta ononaria* Fuessly, E. Germany
13. *Holoterpna pruinosata* Stgr., N. Israel
14. *Pingasa lahayei* Obth., Nigeria
15. *Pseudoterpna pruinata* Hufn., S. Germany
16. *Pseudoterpna coronillaria flamignii* Hausm., C. Italy
17. *Pseudoterpna corsicaria* Rmb., Corsica

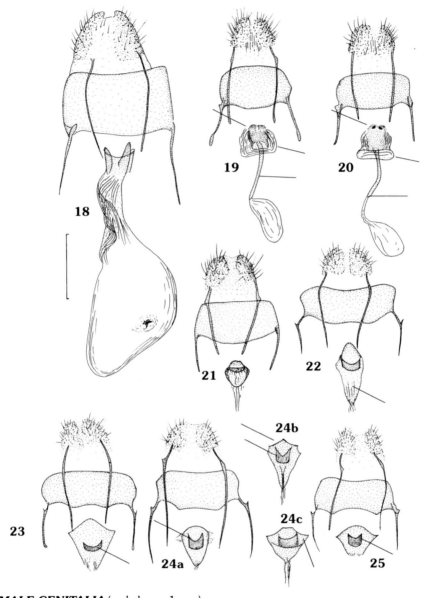

FEMALE GENITALIA (scale bar = 1 mm)

18. *Geometra papilionaria* L., S. Sweden
19. *Comibaena bajularia* Schiff., W. Bulgaria
20. *Comibaena pseudoneriaria* Whli., E. Spain
21. *Proteuchloris neriaria* H.-S., Macedonia
22. *Thetidia plusiaria* Bsd., S. Spain
23. *Thetidia correspondens* Alph., S. European Russia
24a. *Thetidia smaragdaria* F., S. France
24b. *Thetidia smaragdaria volgaria* Gn., S. European Russia (sterigma and antrum)
24c. *Thetidia persica* Hausm., N. Iran (sterigma and antrum)
25. *Thetidia sardinica* Schaw., Sardinia

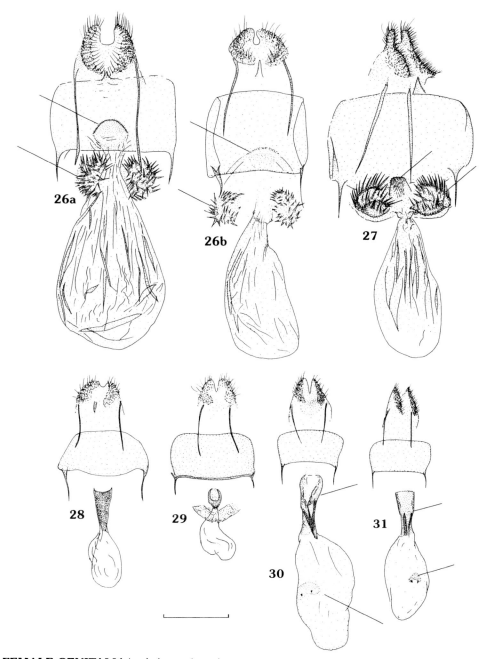

FEMALE GENITALIA (scale bar = 1 mm)

26a. *Hemistola chrysoprasaria* Esp., E. Germany
26b. *Hemistola chrysoprasaria occidentalis* Whli., S. Spain
27. *Hemistola siciliana* Prt., Syntype, Sicily
28. *Xenochlorodes olympiaria* H.-S., N. Spain
29. *Eucrostes indigenata* Vill., Sicily
30. *Jodis lactearia* L., N. Poland
31. *Jodis putata* L., S. Finland

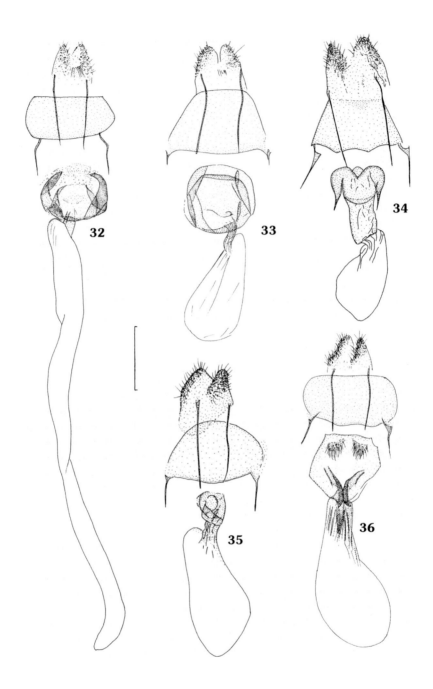

FEMALE GENITALIA (scale bar = 1 mm)

32. *Thalera fimbrialis* Scop., Sicily
33. *Dyschloropsis impararia* Gn., C. Asia
34. *Bustilloxia saturata iberica* Hausm., C. Spain
35. *Kuchleria insignata* Hausm., W. Morocco
36. *Hemithea aestivaria* Hbn., SE. Austria

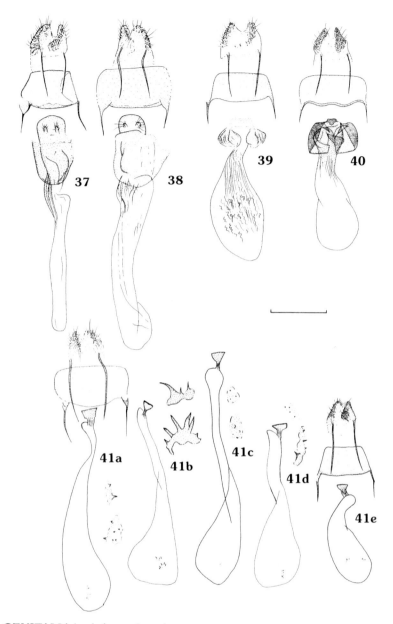

FEMALE GENITALIA (scale bar = 1 mm)

37. *Chlorissa viridata* L., N. Germany
38. *Chlorissa cloraria* Hbn., C. Greece
39. *Phaiogramma etruscaria* Z., E. Austria
40. *Phaiogramma faustinata* Mill., Crete
41a. *Microloxia herbaria* Hbn., S. Spain (signum enlarged)
41b. *Microloxia herbaria* Hbn., Macedonia (corpus bursae; signum enlarged)
41c. *Microloxia h. virideciliata* Bub., Corsica (corpus bursae; signum enlarged)
41d. *Microloxia h. advolata* Ev., European Russia (corpus bursae; signum enlarged)
41e. *Microloxia ruficornis* Warr., S. Israel

Systematic checklist of the Geometridae of Europe and adjacent areas (Vol. 1)

The checklist shows, in systematic order, the taxa (from subfamily level to subspecific level) of the European Geometridae treated in Vol. 1. Just a few synonyms are added, when in use until recently. 24 species occurring in areas immediately adjacent to Europe (as defined here) are inserted - in a smaller font - in the list with an indication of their distribution in these adjacent countries. It will be interesting to see whether, and when, some of these species will be found within Europe.

Archiearinae FLETCHER, **1953**
Archiearis HÜBNER [1823]
　parthenias (LINNAEUS, 1761)
　notha (HÜBNER, [1803])
　puella (ESPER, 1787)
　　subsp. *mediterranea* GANEV, 1984
　touranginii BERCE, 1870

Orthostixinae MEYRICK, **1892**
Orthostixis HÜBNER, [1823]
　cribraria (HÜBNER, [1799])
　　subsp. *amanensis* WEHRLI, 1932 [Turkey]
　cinerea (REBEL, 1916) [Cyprus]
　calcularia (LEDERER, 1853) [Turkey, Georgia, Armenia, Aserbeijan]

Desmobathrinae MEYRICK, **1886**
Gypsochroa HÜBNER, [1825]
　renitidata (HÜBNER, [1817])
Myinodes MEYRICK, 1892
　interpunctaria (HERRICH-SCHÄFFER, 1839)
　　subsp. *atlantica* HAUSMANN, 1994
　constantina HAUSMANN, 1994 [Morocco, Algeria, Tunisia]
　　shohami HAUSMANN, 1994
Eumegethes STAUDINGER, 1898
　tenuis STAUDINGER, 1898 [Morocco, Algeria, Tunisia, Libya]
Drepanopterula HEDICKE, 1923
　zanoni (TURATI, 1919) [Libya]
　limaria (CHRISTOPH, 1885) [Caucasus (Russia), Aserbeijan?]
Epirranthis HÜBNER, [1823]
　diversata ([DENIS & SCHIFFERMÜLLER], 1775)

Alsophilinae HERBULOT **1962**
Alsophila HÜBNER, [1825]
　aescularia ([DENIS & SCHIFFER-MÜLLER], 1775)
　aceraria ([DENIS & SCHIFFERMÜLLER], 1775)
　　(= *quadripunctaria* ESPER, 1801)

Geometrinae STEPHENS, 1829

Heliotheini EXPOSITO, **1978**
Heliothea BOISDUVAL, 1840
　discoidaria BOISDUVAL, 1840

Pseudoterpnini WARREN, **1893**
Aplasta HÜBNER, [1823]
　ononaria (FUESSLY, 1783)
Holoterpna PÜNGELER, 1900
　pruinosata (STAUDINGER, 1898)
Pingasa MOORE, [1887]
　lahayei (OBERTHÜR, 1887)
Pseudoterpna HÜBNER, [1823]
　pruinata (HUFNAGEL, 1767)
　coronillaria (HÜBNER, [1817])
　　subsp. *flamignii* HAUSMANN, 1997
　　subsp. *cinerascens* (ZELLER, 1847)
　　subsp. *algirica* WEHRLI, 1929 [Morocco, Algeria, Tunisia]
　　subsp. *axillaria* GUENÉE, [1858] [Turkey]
　　rectistrigaria WILTSHIRE, 1948 [Cyprus]
　corsicaria (RAMBUR, 1833)
　　lesuraria LUCAS, 1933 [Morocco]

Geometrini STEPHENS, **1829**
Geometra LINNAEUS, 1758
　papilionaria LINNAEUS, 1758

Comibaenini INOUE, 1961

Comibaena HÜBNER, [1823]
　bajularia ([DENIS & SCHIFFERMÜLLER], 1775)
　　(= *pustulata* HUFNAGEL, 1767)
　　subsp. *tikhonovi* HAUSMANN, 2000 [Ciscaucasus, Turkey, Georgia, Aserbeijan, Armenia]
　pseudoneriaria WEHRLI, 1926
Proteuchloris HAUSMANN, 1995
　neriaria (HERRICH-SCHÄFFER, 1852)
Thetidia BOISDUVAL, 1840
subgen. *Thetidia* BOISDUVAL, 1840
　plusiaria BOISDUVAL, 1840
subgen. *Aglossochloris* PROUT, 1912
　fulminaria (LEDERER, 1870) [Iran, Kasachstan]
　correspondens (ALPHÉRAKY, 1892)
　crucigerata (CHRISTOPH, 1887) [Armenia, Aserbeijan]
subgen. *Antonechloris* RAINERI, 1994
　smaragdaria (FABRICIUS, 1787)
　　subsp. *gigantea* (MILLIÈRE, 1874)
　　subsp. *volgaria* (GUENÉE, [1858])
　persica HAUSMANN, 1996 [Turkey, Armenia, Aserbeijan]
　sardinica (SCHAWERDA, 1934)

Hemistolini INOUE, 1961

Hemistola WARREN, 1893
　chrysoprasaria (ESPER, 1795)
　　(= *immaculata*: sensu auct. nec THUNBERG, 1784)
　　(= *biliosata*: sensu auct. nec VILLERS, 1789)
　　subsp. *occidentalis* WEHRLI, 1929
　siciliana PROUT, 1935
Xenochlorodes WARREN, 1897
　olympiaria (HERRICH-SCHÄFFER, 1852)
　　(= *beryllaria* MANN, 1853)
　　subsp. *cremonaria* (STAUDINGER, 1897) [Turkey]
　nubigena (WOLLASTON, 1858) [Madeira]
　magna WOLFF, 1977 [Madeira]
Hierochthonia PROUT, 1912
　pulverata (WARREN, 1901) [Turkey]

Comostolini INOUE, 1961

Eucrostes HÜBNER, [1823]
　indigenata (VILLERS, 1789)
　　subsp. *lanjeronica* HAUSMANN, 1996

Jodini INOUE, 1961

Jodis HÜBNER, [1823]
　lactearia (LINNAEUS, 1758)
　putata (LINNAEUS, 1758)

Thalerini HERBULOT, 1963

Thalera HÜBNER, [1823]
　fimbrialis (SCOPOLI, 1763)
Dyschloropsis WARREN, 1895
　impararia (GUENÉE, [1858])
Culpinia PROUT, 1912
　prouti (THIERRY-MIEG, 1913) [Turkey]
Bustilloxia EXPOSITO, 1979
　saturata (BANG-HAAS, 1906) [Morocco, Algeria]
　　subsp. *iberica* HAUSMANN, 1995
Kuchleria HAUSMANN, 1995
　insignata HAUSMANN, 1995
　　(= *menadiara*: (partim) sensu PROUT, 1912)
　menadiara (THIERRY-MIEG, 1893) [Algeria, Tunisia]
　　subsp. *atlagenes* (PROUT, 1935) [Morocco]
　　subsp. *ephedrae* (PROUT, 1935) [Morocco, Algeria]
　therapaena (PROUT, 1924) [Algeria, Tunisia]

Hemitheini BRUAND, 1846

Hemithea DUPONCHEL, 1829
　aestivaria (HÜBNER, 1789)
Chlorissa STEPHENS, 1831
　viridata (LINNAEUS, 1758)
　cloraria (HÜBNER, [1813])
　pretiosaria (STAUDINGER, 1877) [Turkey, Armenia, Aserbeijan]
　asphaleia WILTSHIRE, 1966 [Turkey]
Phaiogramma GUMPPENBERG, 1887
　etruscaria (ZELLER, 1849)
　　(= *pulmentaria* GUENÉE, [1858])
　faustinata (MILLIÈRE, 1868)

Microloxiini HAUSMANN, 1996

Microloxia WARREN, 1893
　herbaria (HÜBNER, [1813])
　　subsp. *virideciliata* (BUBACEK, 1926)
　　subsp. *advolata* (EVERSMANN, 1837)
　ruficornis (WARREN, 1897) [Morocco, Algeria, Tunisia, Libya]
　simonyi (REBEL, 1894) [Canary Islands, Morocco]
　schmitzi HAUSMANN, 1995 [Canary Islands, Morocco]
Acidaliastis HAMPSON, 1896
　micra HAMPSON, 1896 [Algeria, Libya]
　　subsp. *galactea* RUNGS, 1943 [Morocco]
Acidromodes HAUSMANN, 1996
　saharae (WILTSHIRE, 1985) [Algeria]
Hemidromodes PROUT, 1916
　robusta (PROUT, 1913) [Algeria?]
　　subsp. *affinis* (ROTHSCHILD, 1915) [Morocco, Algeria]

Updating and call for cooperation

The publication of a comprehensive book series like GME can only reflect the actual state of scientific research on the Geometridae. As that research will hopefully go on with greater intensity than in the past, updating from time to time seems highly desirable.

Modern electronic communication facilities can nowadays be used to create a standing forum for continuous supplementing to, and discussing of, the findings presented in the GME volumes. Three lists of data will be prepared:

- check list of species with selective nomenclatural notes
- ecological list (hostplants, parasitoids, phenology)
- faunistic list (list of countries or map)

For access to these lists an address linked with the home page of the ZSM/Lepidoptera (http://www.zsm.mwn.de) will be provided. The e-mail link (Axel.Hausmann@zsm.mwn.de) will give amateur and professional scientists the opportunity to add to the data, to deposit comments and proposals etc. All those interested in Geometridae are kindly urged to join in.

The authors hope that such a network of continuous cooperation among Geometridae specialists will improve the reference quality of the GME series during the long intervals between revised editions.

References

ABOS-CASTEL, F.P. (1995): Lepidópteros de la provincia de Huesca (España). Addenda tercera a los capítulos publicados sobre el tema en Shilap Revta. lepid. (Insecta: Lepidoptera). - Shilap Revta. lepid. 23 (89): 5-21.

AGENJO, R. (1946-1947): Catálogo ordenador de los Lepidopteros de España. - Graellsia IV (3-6); V (1-3).

AGENJO, R. (1952): Fáunula lepidopterológica almeriense, 370 pp..

ALLAN, P.B.M. (1949): Larval foodplants. - The Garden City Press, Letchworth, 126 pp..

AMSEL, H.G. (1951): Über die Variabilität der männlichen Genitalarmatur bei einigen *Crambus*-Arten (Pyralidae). - Z. Lepidopt. 1 (3): 159-163.

ANDRES, A. & A. SEITZ (1924): Die Lepidopterenfauna Ägyptens. - Senckenbergiana 5 (1/2): 1-54.

ANE (ARBEITSGEMEINSCHAFT NORDBAYERISCHER ENTOMOLOGEN Ed.) (1988): Prodromus der Lepidopterenfauna Nordbayerns. - Neue Entomologische Nachrichten 23: 1-159.

ANIKIN, V.V., SACHKOV, S.A., ZOLOTUHIN, V.V. & E.M. ANTONOVA (2000): "Fauna Lepidopterologica Volgo - Uralensis" 150 years later: Changes and additions. Part 3. Geometridae (Insecta, Leidoptera) -Atalanta 31 (1/2): 293-326.

ANTONOVA, E. M. & A. N. SMETANIN (1978): [Species composition, trophic relationships and specific zoogeographical features of loopermoths (Geometridae, Lepidoptera) in the Nature Reserve 'Trostyanets']. - Bull. Mosk. obshtsestva isp. privody, Otd. biol. 83 (1): 52-61.

ANTONOVA, E. M. (1989): [Geometrids of Il'mensky Nature Reserve: Fauna and biogeography]. - In: Nasekomye v. ekosist. Sibiri i Dalnego Vostoka. Moskow: 180-191.

BAKER, R. R. (1978): The Evolutionary Ecology of Animal Migration. - Hodder & Stoughton, London, 1,029 pp..

BANG-HAAS, A. (1907): Neue oder wenig bekannte palaearktische Makrolepidopteren. - Dt. Ent. Zeitschr. Iris 20: 69-88.

BÄNZIGER, H. & D.S. FLETCHER (1985): Three new zoophilous moths of the genus *Scopula* (Lepidoptera, Geometridae) from South-east Asia. - Journal Nat. Hist. 19: 851-860.

BAYNES, E.S.A. (1964): A revised catalogue of Irish Macrolepidoptera (Butterflies and Moths). - E.W. Classey Ltd., Hampton, Middlesex.

BAYNES, E.S.A. (1970): Supplement to a revised catalogue of Irish Macrolepidoptera. - Hampton, 28 pp..

BAYNES, E.S.A. (1973): A revised catalogue of Irish Macrolepidoptera. - Hampton, 109 pp..

BECKER, A. (1867): Noch einige Mitteilungen über Astrachaner und Sareptaer Pflanzen und Insekten. - Bull. Soc. imp. Nat. Moscou 40 (1): 105-115.

BELLIER, M. (1861): Observations sur la Faune Entomologique de la Sicile. - Annls. Soc. ent. Fr. 8 (4): 667-713.

BÉRARD, R. (2000): *Archiearis tourangini* Sand, nouvelle espèce distincte d'*Archiearis notha* Hübner. – Bull. mens. Soc. Linn. Lyon 69 (6): 142-144.

BERGMANN, A. (1951-1955): Die Großschmetterlinge Mitteldeutschlands. Verbreitung, Formen und Lebensgemeinschaften, volumes 5/1 and 5/2, Geometridae (1955). - Urania Verlag, Jena.

BLAB, J. & O. KUDRNA (1982): Hilfsprogramm für Schmetterlinge, Ökologie und Schutz von Tagfaltern und Widderchen. - Naturschutz aktuell, Kilda-Verlag, Greven.

BLAB, J. (1984): Grundlagen des Biotopschutzes für Tiere. - Kilda-Verlag Greven.

BLAB, J., NOWAK, E., TRAUTMANN, W., & H. SUKOPP (1984): Rote Liste der gefährdeten Tiere und Pflanzen in der Bundesrepublik Deutschland. - Naturschutz aktuell, Kilda-Verlag Greven.

BLASCHKE, P. (1955): Raupenkalender für das mitteleuropäische Faunengebiet. - Alfred Kernen Verlag Stuttgart, 149 pp..

BLESZYNSKI, S. (1960): Klucze do Oznaczania Owadów Polski Cz. XXVII, Zeszyt 46a, Miernikowce - Geometridae: 1-149a. - Polski Zwiazek Ent., Warszawa.

BLESZYNSKI, S. (1965): Klucze do Oznaczania Owadów; Polsi XXVII Lepidoptera 46b, Geometridae. - Polski Zwiazek Ent., Warszawa, 305 pp.

BLESZYNSKI, S. (1966): Klucze do Oznaczania Owadów; Polsi XXVII Lepidoptera 46c, Geometridae. - Polski Zwiazek Ent., Warszawa, 122 pp.

BOISDUVAL, J.A. (1940): Genera et index methodicus europaeorum lepidopterorum. - Paris, 238 pp.

BOLLAND, F. (1975): Essai de compréhension du sol et de sa flore par l'etude du peuplement d'un biotope en hétérocères. - Shilap Revta. lepid. III (10): 112-115.

BOLLAND, F. (1977): Pour mieux connaitre la faune d'Espagne, *Chlorissa faustinata*. - Shilap Revta. lepid. 17: 44-45.

BOLTE, K.B. & E. MUNROE (1979): *Hemithea aestivaria* in British Columbia: Characters of a Palearctic genus and species not previously known from North America (Lepidoptera, Geometridae: Geometrinae). - Canad. Entomologist 11 (10): 1121-1126.

BOLTE, K.B. (1990): Guide to the Geometridae of Canada (Lepidoptera), VI Subfamily Larentiinae, 1. Revision of the Genus Eupithecia. - Mem. ent. Soc. Canada 151, Ottawa, 253 pp..

BORKHAUSEN, M.B. (1794): Naturgeschichte der Europäischen Schmetterlinge nach systematischer Ordnung. 5. Theil: der Phalaenen dritte Horde: Spanner. - Frankfurt.

BOURGOGNE, J. (1963): La préparation des armures génitales des Lépidoptères. - Alexanor 3 (2): 61-70; 3 (3): 111-118; 3 (4): 153-164; 3 (5): 195-206.

BRETSCHNEIDER, R. (1939): Erfolgreiche Biston-Hybriden-Zuchten. - Ent. Zt. Frkft. 53 (1): 92-93.

BRETSCHNEIDER, R. (1951): Neue Geometriden-Formen in Sachsen. - Z. Wien. Ent. Ges. 36 (62) (1/3): 21-22.

BRETSCHNEIDER, R. (1953): Erfolgreiche Zuchten von Biston-Hybriden. - Mitt. Münchn. Ent. Ges. 43: 305-314.

BROOKS, M. (1991): A complete Guide to British Moths. - Jonathan Cape, London, 248pp..

BUBACEK, O. & H. REISSER (1926): Neue Makrolepidopteren-Formen aus den andalusischen Gebirgen. - Zeits. Österr. Ent. Ver. 11 (12): 115-120.

BUCKNER, C. H. (1969): The common shrew (*Sorex araneus*) as a predator of the winter moth (*Operophtera brumata*) near Oxford, England. - Canadian Entom. 101: 370-375.

BUDASHKIN, YU.I. & I.YU. KOSTJUK (1987): Tsheshuyekrylye, soobstshenie 2. Pyadenitsy (Lepidoptera, Geometridae). - Tsheshuyekrylye Karadaghskogo zapovednika [Butterflies and Moths of the Karadagh Reserve]. - Moscow, p. 22-31. (in Russian).

BURGERMEISTER, F. (1955): Zur Unterscheidung von *Brephos parthenias* L. und *nothum* Hb. - Z. Wien. Ent. Ges. 40:150-151.

BURROWS, C.R.N. (1940a): *Iodis lactearia*, L. - Entomol. Rec. 52: 101-103.

BURROWS, C.R.N. (1940b): *Pseudoterpna pruinata*, Hufn. - Entomol. Rec. 52: 121-125.

BUSSE, R. (1989): *Archiearis notha* am Licht (Lepidoptera: Geometridae). - Ent. Zt. Frkft. 99 (22): 334-335.

BUSZKO, J. (2000): Atlas motyli Polski III: Thyatiridae - Drepanidae - Geometridae. - Grupa Image, Warszawa, 518 pp...

BUSZKO, J., KOKOT, A., PALIK, E. & Z. SLIWINSKI (1996): Macrolepidoptera of Bialowieza Forest. - Parki nar. Rez. przyr. Bial. 15 (4): 3-46.

BYTINSKI-SALZ, H. (1934): Ein Beitrag zur Kenntnis der Lepidopterenfauna Sardiniens. - Int. Ent. Z. Guben 28: 1-41.

CALBERLA, E. (1889): Elenco dei Lepidotteri raccolti in Sicilia da sig. Enrico Calberla nel giugno e luglio 1889. - Il Naturalista Siciliano, IX (2): 42-49.

CALBERLA, H. (1890): Die Macrolepidopterenfauna der Römischen Campagna und der angrenzenden Provinzen Mittelitaliens. - Dt. Ent. Zeitschr. Iris, 3: 47-94.

CAPUSE & KOVACS, L. (1987): Catalogue de la Collection de Lépidoptères "László Diószeghy" du musée départamental Covasna, Sfintu Gheorghe: 1-397. - Inst. de Spéologie "Emile Racovitza", Bucuresti.

CARTER, D.J. & B. HARGREAVES (1987): Raupen und Schmetterlinge Europas und ihre Futterpflanzen, Verlag Paul Parey, Hamburg-Berlin.

CARTER, D.J. (1984): Pest Lepidoptera of Europe with Special Reference to the British Isles. Dr. W. Junk Publishers, Dordrecht, Boston, Lancaster, 431 pp.

CARTER, D.J. & N.P. KRISTENSEN (1999): Classification and Keys to Higher Taxa. In: KRISTENSEN, N.P. (ed.): Handbook of Zoology, vol. IV (35), Lepidoptera, Moths & Butterflies, 1: Evolution, Systematics, and Biogeography. - W. de Gruyter, Berlin & New York, (491 pp.): 27-40.

CHALMERS-HUNT, J.M. (1961): Aberrations of British Lepidoptera. - The Entomologist 94: 281-284.

CHAPELON, P. (1992): Contribution à l'inventaire des Géométrides de Saône-et-Loire (Lepidoptera Geometridae). - Alexanor 17 (8): 451-466.

CHARPENTIER, DE, T. (1821): Die Zinsler, Wickler, Schaben und Geistchen des Systematischen Verzeichnisses der Schmetterlinge der Wiener Gegend. - Braunschweig, 178 pp.

CHRÉTIEN, P. (1905): Les chenilles des santolines. - Le Naturaliste 435: 89-91.

CHRÉTIEN, P. (1910): La *Gypsochroa renitidata*. - Naturaliste Paris 32 (1910): 57-59; 67-68.

CLERCK, C.A. (1759): Icones insectorum rariorum cum nominibus erorum, locisque e C. Linnaei ... Systema Naturae allegatus. - Stockholm, 21 pp., 55 pls.

COCKAYNE, E.A. (1950): Aberrations of British Lepidoptera (Geometridae). - The Entomologist 83: 49-55.

COCKAYNE, E.A. (1952): Aberrations of British Macrolepidoptera (Geometridae). - Entomol. Rec. 64: 65-68.

COCKAYNE, E.A. (1953): *Thalera fimbrialis* - description of the larva. - Entomol. Rec. 65: 307.

COMMON, I.F.B. (1990): Moths of Australia. - E.J. Brill, and Melbourne University Press, 535 pp., 32 pls.

COOK, M.A. & M.J. SCOBLE (1992): Tympanal organs of geometrid moths: a review of their morphology, function, and systematic importance. - Systematic Entomology 17: 219-232.

COOK, M.A. & M.J. SCOBLE (1995): Revision of the neotropical genus *Oospila* Warren (Lepidoptera: Geometridae). - Bull. nat. Hist. Mus. Lond. 64 (1): 1-115.

COOK, M.A., HARWOOD, L.M., SCOBLE, M.J., G.C. MCGAVIN (1994): The Chemistry and Systematic Importance of the Green Wing Pigment in Emerald Moths (Lepidoptera: Geometridae, Geometrinae). - Biochem. Syst. and Ecology 22: 43-51.

COSTA, O.G. (1848-50): Fauna del Regno di Napoli. Geometre. - Napols, 104 pp..

CULOT, J. (1917-1919): Noctuelles et Géomètres d'Europe. Vol. 3, Géomètres I (1917; 269 pp.). Vol. 4, Géomètres II (1919; 269 pp.). - Genève.

CURTIS, J. (1823-1839): British Entomology. - London, 16 Vol. 770 pl..

DANIEL, F. (1968): Die Makrolepidopteren-Fauna des Sausalgebirges in der Südsteiermark. - Mitt. Abt. Zool. Bot. Landesmus. Joanneum 30: 87-176.

DANNEHL, F. (1927): Beiträge zur Lepidopteren-Fauna Südtirols. - Ent. Zt. Frkft. 40: 403-408; 453-468.

DANNEHL, F. (1933): Neues aus meiner Sammlung (Macrolepidoptera). - Ent. Zt. Frkft. 47 (4): 32-33.

DANTART, J. & P. ROCHE (1992): Aproximación a un catálogo de los macroheteróceros de Andorra (III) (Lepidoptera: Geometridae). - Shilap Revta. lepid. 20 (78): 125-139.

DANTART, J. (1990): Las especies ibéricas del género *Chlorissa* Stephens, 1831, y algunos datos sobre su distribución en el NE Ibérico. - Ses. Entom. ICHN-SCL, VI (1989): 151-173.

DANTART, J. (1991): Datos para el conocimiento de los Geométridos ibéricos (III): Sobre la distribución de algunas especies en el ne ibérico (Lepidoptera, Geometridae). - Treb.Soc.Cat. Lep. XI: 57-82.

DANTART, J., PEREZ DE-GREGORIO, J.J. & F. VALLHONRAT (1993): Contribució a la Fauna de Macroheteròcers de les illes Balears (Lepidoptera: Macroheterocera). - Treb. Soc. Cat. Lep. 12 (1992): 95-114.

DENIS M. & I. SCHIFFERMÜLLER (1775): Ankündung eines systematischen Werkes von den Schmetterlingen der Wienergegend. - Vienna, 324 pp., 3 pls.

DERZHAVETS, Yu.A., IVANOV, A.I., MIRONOV, V.G., MISHCHENKO, O.A., PRASOLOV, V.N. & S.Yu.SINYOV (1986): [Macrolepidoptera of the Leningrad oblast; in Russian]. - (Akad. Nauk SSSR) Trudy vsesoyuznogo entom. obsh. 67: 186-270.

DESCHKA, G. & J. WIMMER (1996): Ökologische Valenzanalyse mit Grossschmetterlingen als Indikatoren in der Gemeinde Waldhausen in Oberösterreich. - Jb. Oö. Mus.-Ver. 141/I: 341-404.

DIDMANIDSE, E.A. (1978): Butterflies of arid regions of Georgia (Lepidoptera, Heterocera). - Verlag "Metzniereba", Tbilisi (Tiflis), 319 pp.

DIERL, W. & J. REICHHOLF (1977): Die Flügelreduktion bei Schmetterlingen als Anpassungsstrategie. - Spixiana 1: 27-40.

DIETZE, K. (1911-13): Biologie der Eupithecien. - Kom. Verlag R. Friedländer u. Sohn: 173pp., 82 pls

DOGANLAR, M. & B. P. BEIRNE (1979): *Hemithea aestivaria*, a Geometrid new to North America, established in British Columbia (Lepidoptera: Geometridae: Geometrinae). - Canad. Entomologist 11 (10): 1121.

DOMINGUEZ, M. & J. BAIXERAS (1995): El ciclo biologico de *Thetidia plusiaria* (Lepidoptera: Geometridae). - Holarctic Lepidoptera 2 (1): 39-41.

DOMINGUEZ, M., YELA, J.L. & J. BAIXERAS (1994): Les Géomètres de la Région de la Alcarria (Guadalajara, Espagne). - Alexanor 18 (5): 265-274.

DONATH, H. (1987): Insektenverluste durch Strassenverkehr im Bereich eines Rotkleefeldes im Sommer 1986. Entom. Nachr. Berl. 31: 169-172.

DÖRING, E. (1955): Zur Morphologie der Schmetterlingseier. - Akademie-Verlag, Berlin, 154 pp., 61 pl..

DOUWES, P., MIKKOLA, K., PETERSEN, B. & A. VESTERGREN (1976): Melanism in *Biston betularius* from north-west Europe (Lepidoptera: Geometridae). - Ent. Scand. 7: 261-266.

DUFAY, C. (1961): Faune terrestre et d'eau douce des Pyrénées-Orientales, Fasc. 6, Lépidoptères I, Macrolépidoptères. - Vie et Milieu 12 (1) Suppl., 153 pp..

DUFAY, C. (1972): Sur la géonémie de divers Lépidoptères rares ou nouveaux pour certaines regions. - Alexanor 7 (5): 219-223.

DUKE, N.J. & A.J. DUKE (1998): An annotated list of larval host-plants utilized by Southern African Geometridae (Lepidoptera). - Metamorphosis 9 (1): 5-22.

DUPONCHEL, P.A.J. (1842): Geometridae. In: GODART & DUPONCHEL: Histoire Naturelle des Lépidoptères ou Papillons de France. Paris, Suppl. 4.

EBERT, G. (2000): Die Schmetterlinge Baden-Württembergs, Band 8-9: Nachtfalter, Geometridae. - Verlag E. Ulmer, Stuttgart (in press).

EITSCHBERGER, U., REINHARDT, R. & H. STEINIGER (1991): Wanderfalter in Europa (Lepidoptera). - Atalanta 22 (1): 1-67.

ELLISON, R.E., E.P. WILTSHIRE (1939): The Lepidoptera of the Lebanon with notes on their season and distribution. - Trans. Royal Ent. Soc. London 88 (1), 1-56.

EMBACHER, G. (1990): Prodromus der Schmetterlingsfauna Salzburgs. - Naturwissenschaftliche Arbeitsgemeinschaft Haus der Natur, Salzburg, 151 pp..

ERLACHER S.-I. & E. FRIEDRICH (1994): Verzeichnis der Spanner (Lepidoptera: Geometridae) Thüringen. - Check-Listen Thüringer Insekten 2, Jena: 55-64.

ESPER, E.J.C. (1776-[1830]): Die Schmetterlinge in Abbildungen nach der Natur mit Beschreibungen. - Erlangen, 5 vol.

EVERSMANN, E. (1844): Fauna lepidopterologica Volga-Uraliensis exhibiens lepidopterorum species. - Casani.

EXPOSITO HERMOSA, A. (1978): Catálogo provisional de la familia Geometridae. - Shilap Revta. lepid. 21: 37-44; 22: 125-130.

EXPOSITO HERMOSA, A. (1979): Sobre la familia Heliotheinae y el género *Bustilloxia* (Geometridae). - Shilap Revta. lepid. 6 (4): 286.

FABRICIUS, J.C. (1775): Systema Entomologicae, sistens Insectorum classes, ordines, genera, species, adjectis, synonymis, locis, descriptionibus, observationibus. - Flensburg, Lipsia.

FABRICIUS, J.C. (1777): Genera Insectorum eorumque characteres naturales ... adjecta mantissa specierum nuper detectarum. - Cologne, 324 pp..

FABRICIUS, J.C. (1781-[1782]): Species insectorum exhibentes eorum differentias specificas, synonyma auctorum, loca natalia, metamorphosis adjectis observationibus, descriptionibus. - Hamburg, Cologne, 2 vols.

FABRICIUS, J.C. (1787): Mantissa Insectorum sistens eorum species nuper detectas adjectis characteribus genericis, differentiis specificis emendationibus, observationibus. - Hafniae 2: 1-382.

FABRICIUS, J.C. (1794): Entomologica systematica emendata et aucta secundum classes, ordines, genera, species adjectis synonymis, locis, observationibus, descriptionibus. - Hafniae 3 (2): 1-349.

FABRICIUS, J.C. (1798): Supplementum Entomologicae systematicae. - Hafniae 3: 1-572.

FAJCÍK, J. & F. SLAMKA (1996): Die Schmetterlinge Mitteleuropas, I. Band. - Bratislava, 113 pp..

FERGUSON, D.C. (1985): The Moths of America North of Mexico, 18.1, Geometroidea, Geometridae (in part; in Dominick, R. B. et al.). - Allen Press, Inc., Lawrence, Kansas.

FIBIGER, M. & P. SVENDSEN (1981): Danske natsommerfugle. - Klampenborg, 272 pp.

FIBIGER, M. (1990): Noctuidae Europaeae, vol. 1, Noctuinae 1. - Entomological Press, Sorø, 208 pp..

FIUMI, G. & S. CAMPORESI (1988): I Macrolepidotteri. - Amministr. Prov. di Forlì, 262 pp., 10 pls.

FLETCHER, D.S. (1963): Ergebnisse der Zoologischen Nubien-Expedition 1962. - Ann. Naturhist. Mus. Wien 66: 469-470.

FLETCHER, D.S. (1979): In NYE, I. W. B: The Generic Names of Moths of the World, vol. 3. - London, 243 pp..

FORSTER, W. & T.A. WOHLFAHRT (1981): Die Schmetterlinge Mitteleuropas, 5. Band, Spanner. Franckh'sche Verlagsbuchhandlung, Stuttgart.

FOURCROY, DE, A.F. (ed.) (1785): Entomologia parisiensis, sive Catalogus Insectorum quae in Agro Parisiensi reperiuntur. Pars I. - Paris, 544 pp..

FREINA, DE, J. & T. WITT (1987): Die Bombyces und Sphinges der Westpalaearktis, Band 1. - Edition Forschung und Wissenschaft, München, 708 pp..

FRIEDRICH, E. (1984): Handbuch der Schmetterlingszucht, Europäische Arten. - Franck'sche Verlagshandlung, Kosmos-Verlag, Stuttgart, 186 pp..

FRIEDRICH, E. (1986): Breeding Butterflies and Moths, English Edition. - Harley Books, Colchester, 176 pp..

FUESSLY, I.C. (1781-1786): Archiv der Insektengeschichte. - Zürich.

GANEV, J. (1983): Systematic and Synonymic List of Bulgarian Geometridae. - Phegea 11: 31-42.

GAEDIKE, R. & W. HEINICKE (1999): Verzeichnis der Schmetterlinge Deutschlands (Entomofauna Germanica 3). - Entomologische Nachr. und Ber. (Dresden), Beiheft 5: 1-216.

GELBRECHT J. & B. MÜLLER (1987): Kommentiertes Verzeichnis der Spanner der DDR nach dem Stande von 1986 (Lep., Geometridae). - Ent. Nachr. Ber. 31 (3): 97-106.

GELBRECHT, J. (1995): Biotopansprüche ausgewählter vom Aussterben bedrohter oder verschollener Schmetterlingsarten der Mark Brandenburg (Lep.). - Ent. Nachr. Ber. 39: 183-202.

GELBRECHT, J. (1999): Die Geometriden Deutschlands - eine Übersicht über die Bundesländer (Lep.). - Entom. Nachr. Ber. 43 (1): 9-26.

GEPP, J. (1973): Kraftfahrzeugverkehr und fliegende Insekten. - Natur und Land 59: 127-129.

GERSTBERGER, M. & L. STIESY (1987): Schmetterlinge in Berlin-West, Teil II, Fördererkreis der Naturwissenschaftlichen Museen Berlins e.V. (Hrsg.), Berlin.

GERSTBERGER, M. (1979): Beitrag zur Kenntnis der mitteleuropäischen Arten der Gattung *Euphyia*. - Nachr.Bl. bayer. Ent. 28: 104-107.

GERSTBERGER, M., STIESY, L., THEIMER, F. & M. WOELKY (1991): Standardliste und Rote Liste der Schmetterlinge von Berlin (West): Großschmetterlinge und Zünsler.

GOEZE, I.A.E. (1781): Entomologische Beyträge zu des Ritter LINNÉ zwölften Ausgabe des Natursystems. - Leipzig, III. Theil, Bd. 3.

GOMEZ BUSTILLO, M. & A. EXPOSITO HERMOSA (1979): Revisión de la superfamília Geometroidea en la Península Ibérica. - Shilap Revta. lepid. 28: 287-299.

GOMEZ DE AIZPURUA, C. (1974): Catálogo de los Lepidópteros del Norte de España. - San Sebastian, 448 pp..

GOMEZ DE AIZPURUA, C. (1987): Biologia y morfologia de las orugas, Lepidoptera, tomo III, Geometridae. - Boletin de Sanidad Vegetal, Madrid, nr. 8, 230 pp..

GOMEZ DE AIZPURUA, C. (1988): Atlas Provisional de los Lepidópteros (Heterocera) de Alava, Bizkaia y Guipúzcoa, tomo VI. - Ed. Servicio Central de Publicaciones de Gobierno Vasco.

GOMEZ DE AIZPURUA, C. (1988): Catálogo de los Lepidópteros de Actividad Nocturna (Heterocera) de Alava, Bizkaia y Guipúzcoa, tomo III. - Ed. Servicio Central de Publicaciones de Gobierno Vasco.

GOMEZ DE AIZPURUA, C. (1989): Biologia y morfologia de las orugas, Lepidoptera, tomo VII, Geometridae. - Boletin de Sanidad Vegetal, Madrid, nr. 15, 224 pp..

GRAVES, P.P. (1926): Heterocera from Macedonia, Gallipoli and Central Greece. - Entomol. Rec. J. Var. 38: 152-158, 165-170.

GUENÉE, A., [1858]: Histoire naturelle des insectes (Lepidoptera), Species Général des Lépidoptères. Tom. IX. X. Uranides et Phalenites I. II. - Paris.

GUMPPENBERG, FRHR. VON, C. (1892): Systema Geometrarum zonae temperatioris septentr. - Nova Acta Ksl. Leop. Carol. Deutsche Akad. d. Naturforscher, Halle.

GUSTAVSSON, B. (ed.) (1987): Catalogus Lepidopterorum Sueciae. - Riksmuseet Stockholm: 1-140.

HABELER, H. (1996): Schmetterlinge an Bord der "Venizelos" und die Wanderfalter (Lepidoptera). - Mitt. Landesmus. Joanneum Zool. 50: 73-75.

HACKER, H. & H. KOLBECK (1996): Die Schmetterlingsfauna der Naturwaldreservate Dianensruhe, Wolfsee, Seeben, und Fasanerie. - Schriftenreihe Naturwaldreservate in Bayern 3: 77-120.

HACKER, H. (1981): Beitrag zur Lepidopterenfauna des nördlichen Fränkischen Jura, Teil 2: Geometridae. - Atalanta 12 (4): 260-284.

HACKER, H. (1989): Die Noctuidae Griechenlands (Lepidoptrea, Noctuidae). - Herbipoliana 2, Marktleuthen, 589 pp..

HACKER, H. (1996): Revision der Gattung Hadena Schrank, 1802 (Lepidoptera, Noctuidae). - Esperiana 5: 7-696.

HACKER, H. (1999): Die Typen der von E.J.Ch. Esper (1742-1810) in seinem "Die Schmetterlinge in Abbildungen nach der Natur" beschriebenen Bombycoidea, Drepanoidea, Geometroidea, Hepialoidea, Lasiocampoidea, Noctuoidea, Pyraloidea, Tineoidea (Lepidoptera) II. - Esperiana 7: 443-461.

HAGGETT, G. (1954): An account of rearing Thalera fimbrialis Scop. - Ent. Gazette 5: 95-102; 224.

HALPERIN, J. & W. SAUTER (1992): An annotated list with new records of Lepidoptera associated with forest and ornamental trees and shrubs in Israel. - Israel J. Entom. 25/26: 105-147.

HANNEMANN (1917): [Sitzung am 26. Spetember 1916]. - Int. Ent. Zeitschr. 10 (25): 146-147.

HARDONK, M. (1954): Comibaena pustulata Hufn., f. rosea Cockayne (Lep. Geom.). - Ent. Berichten 15: 20.

HARRISON, J.W.H. (1913): The Hybrid Bistoninae. - Lép. Comp. 7 (2): 343-655.

HARTIG, F. (1939) Nuovi contributi alla conoscenza della Fauna delle Isole Italiane dell'Egeo. XIII. Conoscenza attuale della fauna lepidotterologica dell'Isola di Rodi. - Boll. Lab. Ent. Agr. Portici, 3: 221-246.

HARTIG, F. & H. G. AMSEL (1951): Lepidoptera Sardinica. - Fragm. ent. 1: 1-152.

HARTWIEG (1951): Zwei neue Geometriden-Formen. - Zeitschr. F. Lepidopterol. 1 (3): 163-164.

HAUSMANN, A. (1990): Zur Dynamik von Nachtfalter-Artenspektren: Turnover und Dispersionsverhalten als Elemente von Verbreitungsstrategien. Spixiana Suppl. 16, 1-222.

HAUSMANN, A. (1992): Untersuchungen zum Massensterben von Nachtfaltern an Industriebeleuchtungen. - Atalanta 23 (3/4): 411-416.

HAUSMANN, A. (1993a): Contributo alla conoscenza della fauna siciliana: I rappresentanti della famiglia Geometridae presso la collezione dello Zoologische Staatssammlung di Monaco (Insecta: Lepidoptera). - Naturalista siciliana S. IV, XVII (1-2) 1993: 83-101.

HAUSMANN, A. (1993b): Der Aussagewert struktureller Unterschiede des 8. Sternits. Beitrag zur Systematik der italienischen Vertreter der Gattung Glossotrophia Prout, 1913 (Lepidoptera, Geometridae). - Atalanta 24: 265-297.

HAUSMANN, A. (1993c): Zur Methodik des Großschmetterling-Fangs in Malaisefallen (Lepidoptera, Macroheterocera). - Entomofauna 14 (12): 233-247.

HAUSMANN, A. (1994a): Contribution to the morphology and the taxonomy of the species belonging to the genus Myinodes Meyrick, 1892 (Lepidoptera, Geometridae). - Nota lepidopterologica 17 (2): 31-43.

HAUSMANN, A. (1994b): Beitrag zur Geometridenfauna Zyperns. - Zeits. Arb.gem. Österr. Ent. 46 (3/4): 81-98.

HAUSMANN, A. (1995a): Revision der altweltlichen Arten der Gattung Microloxia Warren, 1893 (Lepidoptera, Geometridae: Geometrinae). Atalanta 25 (3/4): 571-608.

HAUSMANN, A. (1995b): Neue Geometriden-Funde aus Zypern und Gesamtübersicht über die Fauna. - Mitt. Münch. Ent. Ges. 85: 79-111.

HAUSMANN, A. (1996a): The Geometrid Moths of the Levant and its Neighbouring Countries: Systematic List and Prodromus of Fauna (Part I: Orthostixinae-Geometrinae). - Nota lepidopterologica 19 (1/2): 91-106.

HAUSMANN, A. (1996b): The Morphology of the Geometrid Moths of the Levant and its Neighbouring Countries (Part I: Orthostixinae-Geometrinae). - Nota lepidopterologica 19 (1/2): 3-90.

HAUSMANN, A. (1997a): The Lepidoptera of Israel: Faunistic Data on Geometridae: I. Orthostixinae and Geometrinae. - Nota lepidopterologica 20 (1/2): 102-136.

HAUSMANN, A. (1997b): Zur Nomenklatur der europäischen Unterarten von *Pseudoterpna coronillaria* (HÜBNER, [1817]) (Lepidoptera: Geometridae, Geometrinae). - Nachr. entomol. Ver. Apollo, N.F. 18 (2/3): 223-225.

HAUSMANN, A. (1999a): Geometrid Moth Species from Yemen (Lepidoptera: Geometridae). - Esperiana 7: 283-305, pl. 1-5.

HAUSMANN, A. (1999b): Falsification of an entomological rule: Polymorphic genitalia in Geometrid moths (Lepidoptera, Geometridae). - Spixiana 22 (1): 83-90.

HAUSMANN, A. (2000): Die außereuropäischen Populationen von *Comibaena bajularia* ([Denis & Schiffermüller]. 1775). - Entomofauna 21 (26): 301-305.

HAUSMANN, A. & P. PARENZAN (1990): Neue und interessante Geometridenarten für die Süditalienfauna (Lepidoptera, Geometridae). - Entomofauna 11 (29), 497-503.

HAWKINS, C.N. (1953): *Thalera fimbrialis* - description of the pupa. - Entomol. Rec. 65: 307-308.

HAWKINS, C.N. (1942): Insectivorous habit of a larva of *Eupithecia oblongata*. - The Entomologist 75: 27.

HAWORTH, A.H. (1803-1828): Lepidoptera Britannica, sistens digestimen novam lepidopterorum quae in Magna Britannica repertiunter ... adjunguntur dissertationes variae ad historiam naturalam spectantes. - London, 4 parts.

HELLMANN, F. (1987): Die Macrolepidopteren der Brenta-Gruppe (Trentino - Oberitalien) (Lepidoptera). - Studi Trentini di Scienze Naturali, Acta Biologica 63: 3-166.

HELLMANN, F., BROCKMANN, E. & P.M. KRISTAL (1999): I Macrolepidotteri della Valle d'Aosta. - Monografie Mus. Reg. Sci. Nat. Saint-Pierre, Valle d'Aosta, 284 pp..

HENNIG, W. (1950): Grundzüge einer Theorie der phylogenetischen Systematik. - Deutscher Zentralverl., Berlin.

HENNIG, W. (1966): Phylogenetic Systematics. - Urbana, Univ. Illinois Press.

HERBULOT, C. (1961-1963): Mise a jour de la liste des Geometridae de France. - Alexanor (1961) 2 (4): 117-124; (1962) 2 (5): 147-154; (1963) 3 (1): 17-24; 3 (2): 85-93.

HERBULOT, C. (1987): Les types de Lépidoptères Geometrides décrits ou figurés par Pierre Rambur. - Misc. Entom. 51: 33-40.

HERBULOT, C. (1992): Combien d'espèces de Lépidoptères dans le monde? - Alexanor 17 (7): 447-448.

HERING, M. (1937) [1935-1937]: Die Blattminen Mittel- und Nord-Europas einschließlich Englands. - Verlag Gustav Feller, Neubrandenburg, 631 pp..

HERING, M. (1950): Die Oligophagie phytophager Insekten als Hinweis auf eine Verwandtschaft der Rosaceae mit den Familien der Amentiferae. - Verh. 8. Int. Ent. Kongr. Stockholm: 74-79.

HERRICH-SCHÄFFER, G.A.W. (1829-1844): Deutschlands Insekten, Fortsetzung von Panzers Fauna insectorum Germanica, 10 vol., H. 111-190.

HERRICH-SCHÄFFER, G.A.W. (1843-1856): Systematische Bearbeitung der Schmetterlinge von Europa. - Regensburg, 6 vols.

HEYDEMANN, F. (1938): Ueber einige nordwesteuropäische Lepidopteren-Formen. - Ent. Zt. Frkft. 51 (37/38): 341-343.

HEYDENREICH, G. H. (1851): Lepidopterorum Europaeorum Catalogus methodicus, Edn 3. Leipzig.

HOFFMEYER, S. (1966): De danske malere (2nd Edn). - Unversitetsforlaget i Aarhus, 361 pp..

HOFMANN, E. (1884): Die Gross-Schmetterlinge Europas. - Hofmann'sche Verlagsbuchhandlung, Stuttgart.

HOFMANN, E. (1893): Die Raupen der Gross-Schmetterlinge Europas. - Hofmann'sche Verlagsbuchhandlung, Stuttgart, 318 pp..

HOLLOWAY, J.D. (1984): The larger moths of the Gunung Mulu National Park; a preliminary asessment of their distribution, ecology, and potential as environmental indicators. - The Sarawak Mus. Journ. 30 (2): 149-190, pl. 12.

HOLLOWAY, J.D. (1993): The Moths of Borneo: Family Geometridae, Subfamily Ennominae. - London, 309 pp..

HOLLOWAY, J.D. (1996): The Moths of Borneo: Family Geometridae, Subfamilies Oenochrominae, Desmobathrinae and Geometrinae. - Malayan Nature Journal 49 (3/4): 147-326.

HOLLOWAY, J.D. (1997): The Moths of Borneo, pt. 10, Geometridae: Sterrhinae, Larentiinae. - Malayan Nature Journal 51: 1-242.

HOLLOWAY, J.D. & E.S. NIELSEN (1999): Biogeography of the Lepidoptera. In: KRISTENSEN, N.P. (ed.): Handbook of Zoology, vol. IV (35), Lepidoptera, Moths & Butterflies, 1: Evolution, Systematics, and Biogeography. - W. de Gruyter, Berlin & New York, (491 pp.): 423-462.

HORN, W., KAHLE, I., FRIESE, G. & R. GAEDIKE (1990): Collectiones entomologicae. - Akad. Landw. DDR, Berlin, 2 vols, 573 pp..

HRUBY, K. (1964): Prodromus Lepidopter Slovenska. - Bratislava, 962 pp..

HÜBNER, J. (1786-1790): Beiträge zur Geschichte der Schmetterlinge. - Augsburg, 2 vols.

HÜBNER, J. (1796-[1838]): Sammlung Europäischer Schmetterlinge. - Augsburg, 7 vols.

HÜBNER, J. (1816-[1826]): Verzeichniss bekannter Schmettlinge. - Augsburg, 431+72 pp.

HUEMER, P. (1988): Kleinschmetterlinge an Rosaceae unter besonderer Berücksichtigung ihrer Vertikalverbreitung (excl. Hepialidae, Cossidae, Zygaenidae, Psychidae und Sesiidae). - Neue Ent. Nachr. 20: 376 pp..

HUEMER, P. (1994): Schmetterlinge (Lepidoptera) im Naturschutzgebiet Rheindelta (Vorarlberg, Österreich): Artenbestand, Ökologie, Gefährdung. - Linzer biol. Beitr. 26 (1): 3-132.

HUEMER, P. & G. TARMANN (1993): Die Schmetterlinge Österreichs (Lepidoptera). - Tiroler Landesmuseum Ferdinandeum, Innsbruck.

HUFNAGEL (1767): Tabellen von den Tag-, Abend- und Nachtvögeln der Umgebung von Berlin (Geometridae: Vierte Tabelle ... der Nachtvögel hiesiger Gegend). - Berlinisches Magazin oder gesammelte Schriften für die Liebhaber der Arzneywissenschaft, Naturgeschichte und der angenehmen Wissenschaften überhaupt, Bd. 2-4.

ICZN (International Commission on Zoological Nomenclature) (1985): International Code of Zoological Nomenclature, Third Edition. - Intern. Trust Zool. Nomencl., London, 388 pp..

ICZN (International Commission on Zoological Nomenclature) (1999): International Code of Zoological Nomenclature, Fourth Edition. - Intern. Trust Zool. Nomencl., London, 306 pp..

INOUE, H. (1943): A revision of *Alsophila* (Lep., Geom.) with phenological aspect of the Japanese specimens. - Trans. Kansai Ent. Soc. 13 (2): 36-63.

INOUE, H. (1961): Geometridae, Insecta Japonica, Series 1, Part 4: 1-106. - Hokuryukan, Tokyo.

INOUE, H. (1971): Icones Heterocerorum Japonicorum in Coloribus Naturalibus (Geometridae in Part). - Osaka Hoikusha Publishing Co. Ltd.

INOUE, H. (1977): Catalogue of the Geometridae of Japan (Lepidoptera). - Bull. Fac. Dom. Sci., Otsuma Woman's University, 13: 227-346.

IPPOLITO, F. & P. PARENZAN (1998): I Macrolepidotteri del Demanio Forestale di Santo Stefano Quisquina (Agrigento). Contributi alla Conoscenza della Lepidotterofauna della Sicilia VI. - Phytophaga 8: 57-84.

ITÄMIES, J. & J. TABELL (1997): Variation in male genitalia of *Coleophora vacciniella* H.-S. (Lepidoptera, Coleophoridae). - Entom. Fennica 8: 145-150

IVINSKIS, P. (1993): Geometridae. In: Check list of Lithuanian Lepidoptera. - Ecol. Instit. Vilnius, 115-132.

JANSE, A.J.T. (1934) [1933-1935]: The Moths of South Africa, Vol. II Geometridae: 1-448 - E.P. & Commercial Printing Co. Ltd., Durban.

JANSSEN, A. (1985): Katalogus van de Antwerpse Lepidoptera, Geometridae. - Vlaamse Verein. Ent., Phegea 13-15, Suppl., 159-217.

JØRGENSEN B. & P. SKOU (1982): En klaekning af *Hemistola chrysoprasaria* (Esper, 1794). - Lepidoptera 4 (4): 117-122.

JUUL, K. (1948): Nordens Eupithecier. - Gravers Andersens Forlag, Aarhus, 147 pp..

KARSHOLT, O. & E.S. NIELSEN (1985): The Lepidoptera described by C.P. Thunberg. - Ent. scand. 16: 433-463.

KARSHOLT, O. & J. RAZOWSKI (1996): The Lepidoptera of Europe, a Distributional Checklist. - Apollo Books, Stenstrup, 380 pp.

KARSHOLT, O. & P. STADEL NIELSEN (1998): Revideret katalog over de danske Sommerfugle. - Kobenhavn, 144 pp..

KARSHOLT, O., KRISTENSEN, N.P., KAABER, S., LARSEN, K., SCHMIDT NIELSEN, E., SCHNACK, K., SKOU, P. & B. SKULE (1985): Catalogue of the Lepidoptera of Denmark. - Ent. Meddr. 52 (2-3).

KERNBACH, K. (1967): Über die bisher im Pliozän von Willershausen gefundenen Schmetterlings- und Raupenreste. - Ber. naturh. Ges. Hannover 111: 103-108, 12 figs.

KETTLEWELL, H.B.D. (1953): *Thalera fimbrialis* Scopoli in England. - Entomol. Rec. 65: 305-307.

KLOTS, A.B. (1970): Lepidoptera. In: Tuxen S. L. (ed.): Taxonomists Glossary of Genitalia in Insects. - Munksgaard, Copenhagen, 357 pp.

KNOCH, A.W. (1781-1783): Die Schmetterlinge. In Beiträge zur Insecten-Geschichte. - Leipzig, pts 1-3.

KOCAK, A.O. & S. SEVEN (1993): Über die Tagfalterfauna des Gebirges Hodulca bei Kizilcahamam (Prov. Ankara, Türkei). - Priamus 6 (3/4): 97-113.

KOCAK, A.Ö. (1982-86): On the Validity of the Species Group Names proposed by Denis & Schiffermüller 1775 in Ankündigung (sic) eines Systematischen Werkes von den Schmetterlingen der Wiener Gegend. - Priamus 2 (1): 5-42 (1982); 3 (3) 98-130; 3 (4): 133-154 (1984); 4 (1/2): 22-36 (1986).

KOCAK, A.Ö. (1990): A new species to the fauna of Turkey, *Archiearis notha* (Hübner, 1803). - Misc. Papers Centre Ent. Stud. Ankara 8: 7-8.

KOCAK, A.Ö. (1990): An annotated List of the Lepidoptera of Karadere and Bolu District (Prov. Bolu, N. Turkey). - Misc. Papers Centre Ent. Stud. Ankara 6: 1-11.

KOCH, M. (1984): Wir bestimmen Schmetterlinge, 1. einbändige Auflage. Verlag J. Neumann-Neudamm, Leipzig.

KOLOSSOW, J. (1936): Neue und wenig bekannte Formen der Spanner (Lepidoptera, Geometridae). - Ent. Nachr.Bl. 10: 149-150.

KOMAREK, O. (1950): La contribution a la conaissance de la faune Lépidopterologique de la Bohême du Nord-Est, avec les diagnoses de deux formes nouvelles. - Casopis cs. Spol. Ent. 47 (1-2): 41-45.

KOSTJUK, I.YU. (1985): Novye dannye k faune pyadenits (Lepidoptera, Geometridae) Kanevskogo zapovednika i ego okrestnostei.[New data to the fauna of the Geometrid-Moths of the Kanev nature reserve and its neighbourhoods]. - Problemy obshchei i molekulyarnoi biologii, Kiev, , vyp. 4: 78-80. (in Russian).

KOSTJUK, I.YU. (1986): Materialy k izucheniyu pyadenits (Geometridae) doliny Srednego Dnepra.[The materials to study of the Geometrid-Moths of the middle Dnieper valley]. - Problemy obshchei i molekulyarnoi biologii, Kiev, vyp. 5: 55-58. (in Russian).

KOSTJUK, I.YU. (1990): Geometridae. In: EFETOV, K.A. & YU.I. BUDASHKIN: Babotshki Kryma. - Sinferopol, (111 pp.): 86-92.

KOSTJUK, I.YU. SCHESCHURAK, P., PLJUSCH, I. & E. GALKINA (1998): [Geometridae of the Chernigov-Region, N. Ukraine]. - OOO-Nauka-Service, Neftin.

KOVACS, L. (1965): Araszolólepkék I., Geometridae I. - Fauna Hungariae 74 (8), Budapest, 55 pp..

KOZLOV, M.V. & J. JALAVA (1994): Lepidoptera of the Kola Peninsula, northwestern Russia. - Entom. Fennica 5 (2): 65-85.

KRAMPL, F. (1992): Boreal macro-moths in central Europe (Czechoslovakia) and their eco-geographical characteristics (Lepidoptera: Geometridae, Noctuidae, Notodontidae). - Acta Entom. Bohemoslov 89 (4): 237-262.

KRAUS, W. (1993): Verzeichnis der Großschmetterlinge (Insecta: Lepidoptera) der Pfalz. - Pollichia 27, Bad Dürkheim, 618 pp..

KRAUS, W. (1997): A Contribution to the knowledge of the Lepidoptera of the "Parque Natural Cabo de Gata", Nija, Almeria, Spain. - Shilap Revta. Lepid. 25: 63-64.

KRAUS, W. (1999): Beobachtungen zur Macrolepidopterenfauna der Iberischen Halbinsel, Teil 2: Artenliste Drepanidae bis Notodontidae. - Nachr. entomol. Ver. Apollo, N. F. 20 (2): 231-263.

KRISTENSEN, N.P. (1999): Historical Introduction. In: KRISTENSEN, N.P. (ed.): Handbook of Zoology, vol. IV (35), Lepidoptera, Moths & Butterflies, 1: Evolution, Systematics, and Biogeography. - W. de Gruyter, Berlin & New York, (491 pp.): 1-6.

KRISTENSEN, N.P. (ed.) (1999): Handbook of Zoology, vol. IV (35), Lepidoptera, Moths & Butterflies, 1: Evolution, Systematics, and Biogeography. - W. de Gruyter, Berlin & New York, (491 pp.): 423-462.

KRISTENSEN, N.P. & A.W. SKALSKI (1999): Phylogeny and Paleontology. In: KRISTENSEN, N.P. (ed.): Handbook of Zoology, vol. IV (35), Lepidoptera, Moths & Butterflies, 1: Evolution, Systematics, and Biogeography. - W. de Gruyter, Berlin & New York, (491 pp.): 7-26.

KRÜGER, G.C. (1913): Beiträge zur Entwicklungsgeschichte der italienischen Lepidopteren. - Ent. Mitt. Ent. Mus. 2: 106-120.

KRÜGER, G.C. (1939): Notizie sulla fauna della Sirtica occidentale: Lepidotteri. - Annali del Mus. Libico Stor. Nat. 1, 317-357.

KUDLER, J. (1978): Geometroidea. In: SCHWENKE, W. (ed.): Die Forstschädlinge Europas, Vol. 3 Schmetterlinge. - Verlag Paul Parey, Hamburg und Berlin: 218-264.

KUDRNA, O. & M. WIEMERS (1990): Lepidopterology in Europe. In: KUDRNA, O. (ed.): Butterflies of Europe, vol. 2, Introduction to Lepidopterology (557 pp.). - Aula-Verlag, Wiesbaden: 13-77.

LA GRECA, M. (1963): Le categorie corologiche degli elementi faunistici italiani. - Atti Acc. Naz. It. di Entom., Rend., 11: 231-253.

LA HARPE, DE (1853): Faune Suisse, Lépidoptères.

LAEVER, DE, E. (1966): Espèces nouvelles pour la Belgique (suite). - Lambillionea 65: 17.

LAEVER, DE, E. (1968): *Chlorissa (Nemoria) faustinata* Mill. - Mitt. Ent. Ges. Basel 18: 111-112.

LAEVER, DE, E. (1970): *Chlorissa (Nemoria) faustinata* Mill. - Mitt. Ent. Ges. Basel N.F. 20: 29.

LANDRY, J.C. (1980): Etude de l'impact d'une autoroute sur l'environnement. Arch. Sc. Geneve 33: 9-10.

LASS (1923): Biologische Beobachtungen an Geometriden. - 281-283.

LASTUVKA, Z. (1993): Katalog von Faltern der Mährisch-Schlesischen Region (Lepidoptera). - Brno, 130 pp..

LASTUVKA, Z. (1994): Lepidoptera of the Protected Landscape Area Pálava. - Univ. of Agriculture Brno, 118 pp.

LASTUVKA, Z. (ed.) (1998): Checklist of Lepidoptera of Czech and Slovak republics (Insecta, Lepidoptera). - Konvoj, 117 pp..

LATTIN, DE, G. (1951): Türkische Lepidopteren. II. - Rev. Fac. Sci. Univ. Istanbul XVI (1): 45-73.

LATTIN, DE, G. (1967): Grundriß der Zoogeographie. - VEB Gustav Fischer Verlag, Jena.

LEDERER, J. (1853): Versuch, die europäischen Lepidopteren in möglichst natürliche Reihenfolge zu stellen, nebst Bemerkungen zu einigen Familien und Arten. - Verh. Zool.-Bot. Ges. Wien, Bd. 3: 165-270.

LEMPKE, B.J. (1936) Catalogus der Nederlandse Macrolepidoptera. - Tijdschr. Entom. 79: 238ff..

LEMPKE, B.J. (1936): Au sujet d'une forme nouvelle de *Pseudoterpna pruinata* Hufn. - Ent. Berichten 212 (IX): 290-291.

LEMPKE, B.J. (1949-1950): Catalogus der Nederlandse Macrolepidoptera, Geometridae. - Tijdschrift Entom. 90: 146-196; 92: 113-218.

LERAUT, P.J.A. (1980): Liste systématique et synonymique des Lépidoptères de France, Belgique et Corse. - Suppl. à Alexanor et au Bull. Soc. ent. de France, Paris, 334 pp..

LERAUT, P.J.A. (1997): Liste systématique et synonymique des Lépidoptères de France, Belgique et Corse, 2nd edn. - Suppl. à Alexanor, Paris-Wetteren, 526 pp..

LHOMME (1923-1935): Catalogue des Lépidoptères de France et de Belgique, vol. 1, Macrolépidoptères. - Le Carriol, par Douelle (Lot).

LINNAEUS, C. (1758): Systema Naturae, I, 10th Edn. - Stockholm, 823 pp..

LINNAEUS, C. (1761): Fauna Svecica, 2nd Edn. - Stockholm, 578 pp..

LINNAEUS, C. (1767): Systema Naturae, 12th Edn. - Stockholm, 533-1327.

LUIG, J. & T. KESKÜLA (1995): Catalogus Lepidopterorum Estoniae. - Tartu, 130 pp..

LUTRAN, G. (1989): Pseudoterpna coronillaria Hübner en Corse (Lep., Geometridae). - Linneana Belgica 12 (1): 43-46.

MAC ARTHUR, R.H. & E.O. WILSON (1967): The Theory of Island Biogeography. Princeton University Press, Princeton, N.J..

MACK, W. (1985): Lepidoptera. In: Die Nordost-Alpen im Spiegel ihrer Landtierwelt, vol. 5: 1-484.

MALKIEWICZ, A. & J. SOSINSKI (1999): Systematic annotated check-list of Polish Geometridae. - Polskie Pismo Ent. 68 (2) 197-208.

MANN, J. (1864): Nachtrag zur Schmetterlingsfauna von Brussa. - Wien. Ent. Monatschr. 8: 173-190.

MANN, J. (1855): Die Lepidopteren, gesammelt auf einer entomologischen Reise in Corsika im Jahre 1855. - Verh. Zool.-Bot. Ver. Wien 5: 529-572.

MARIANI, M. (1938): Fauna Lepidopterorum Siciliae (catalogo ragionato). - Mem. Soc. ent. It., 17 (2): 129-187.

MARIANI, M. (1940-1943): Fauna Lepidopterorum Italiae. Parte I. Catalogo ragionato dei Lepidotteri d'Italia. Fasc. II e III - Giorn. Sc. Nat. Econ., 42 (3): 81-227.

MARINI, M. & I. RUSSO (1980): Interessanti reperti di Lepidotteri in Calabria. - Boll. Ist. Ent. Un. Bologna, 35: 249-265.

MARSCHNER, H. (1932): *Geometra papilionaria* L. f. *diffluata* nov. forma m. - Mitt. Münch. Ent. Ges. 22: 13-14.

MAZEL, R. & S. PESLIER (1997): Cartographie des Lépidoptères des Pyrénées-Orientales, 1, Geometridae. - Revue Assoc. Roussill. Ent., Suppl., 115 pp.

MAZZUCCO, K. (1966): *Cosymbia pupillaria* Hbn. bei 3200 m. - Z. Wien. Ent. Ges. 51: 154.

MCFARLAND, N. (1988): Portraits of South Australian Geometrid Moths. - Allen Press, Kansas, 400 pp.

MCGUFFIN, W.C. (1967): Guide to the Geometridae of Canada (Lepidoptera), I, Subfamily Sterrhinae. - Mem. ent. Soc. Canada 50, Ottawa, 67 pp..

MC GUFFIN, W.C. (1972): Guide to the Geometridae of Canada (Lepidoptera), II, Subfamily Ennominae 1. - Mem. ent. Soc. Canada 86, Ottawa, 159 pp..

MCGUFFIN, W.C. (1977): Guide to the Geometridae of Canada (Lepidoptera), II, Subfamily Ennominae 2. - Mem. ent. Soc. Canada 101, Ottawa, 191 pp..

MCGUFFIN, W.C. (1981): Guide to the Geometridae of Canada (Lepidoptera), II, Subfamily Ennominae 3. - Mem. ent. Soc. Canada 117, Ottawa, 153 pp..

MCGUFFIN, W.C. (1987): Guide to the Geometridae of Canada (Lepidoptera), II, Subfamily Ennominae 4. - Mem. ent. Soc. Canada 138, Ottawa, 181 pp..

MCQUILLAN, P.B. (1981): A review of the Australian moth genus *Thalaina* (Lepidoptera: Geometridae: Ennominae). - Transact. Royal Ent. Soc. South Austr. 105: 1-23.

MEIER, M. (1992): Nachtfalter: Methoden, Ergebnisse und Problematik des Lichtfanges im Rahmen landschafts-ökologischer Untersuchungen. - In: TRAUTNER, J. (Hrsg.): Arten- und Biotopschutz in der Planung: Methodische Standards zur Erfassung von Tierartengruppen. - Ökologie in Forschung und Anwendung 5: 203-218.

MENTZER, VON, E. (1984): Die Genera bei Denis & Schiffermüller als Nomenklaturfrage (Lepidoptera). - Nota lepidopterologica 7: 59-70.

MERZHEEVSAYA, O.I., LITVINOVA, A.N. & R.V. MOLCHANOVA (1976): Tsheshuyekrylye (Lepidoptera) Byelorussii. - Minsk, 131 pp..

MIKKOLA, K. (1979): Resting site selection by *Oligia* and *Biston* moths (Lepidoptera: Noctuidae and Geometridae). - Ann. Ent. Fenn. 45 (3): 81-87.

MIKKOLA, K. (1993): The lock-and-key mechanisms of the internal genitalia of the noctuid and geometrid moths (Lepidoptera) in relation to the speciation concepts. - Folia Baeriana (Tartu) 6: 149-157.

MIKKOLA, K. (1994): Inferences about the function of genitalia in the genus *Eupithecia*, with description of a new organ (Lepidoptera, Geometridae). - Nota lepidopterologica, Suppl. 5: 73-78.

MILLIÈRE, P. (1864-1868): Iconographie et description de chenilles de Lépidoptères inédits, vol. II, Paris, 1-506.

MILLIÈRE, P. (1873): Catalogue Raisonné des Lépidoptères des Alpes-Maritimes (2). - Mem. Soc. Sci. Nat. 1873: 141-455.

MILLIÈRE, P. (1874): Iconographie et description de chenilles de Lépidoptères inédits, vol. III, Livraison 35, Paris: 389-488.

MINET, J. (1983): Étude morphologique et phylogénétique des organes tympaniques des Pyraloidea. 1. Généralités et homologies. (Lep. Glossata). - Annls. Soc. ent. France 19: 175-207.

MINET, J. (1991): Tentative reconstruction of the ditrysian phylogeny (Lepidoptera: Glossata). - Ent. scand. 22: 69-95.

MINET, J. & M. J. SCOBLE (1999): The Drepanoid/Geometroid Assemblage. In: KRISTENSEN, N.P. (ed.): Handbook of Zoology, vol. IV (35), Lepidoptera, Moths & Butterflies, 1: Evolution, Systematics, and Biogeography. - W. de Gruyter, Berlin & New York, (491 pp.): 301-320.

MIRONOV, V.A. (1990): A systematic catalogue of geometrid moths of the tribe Eupitheciini (Lepidoptera, Geometridae) of the Fauna of the USSR I. - Entomol. Obosr. 69 (3): 656-670.

MIRONOV, V.A. (1991): A systematic catalogue of geometrid moths of the tribe Eupitheciini (Lepidoptera, Geometridae) of the Fauna of the USSR I. - Entomol. Obosr. 70 (1): 157-167.

MONTEIRO, T. & J. PASSOS DE CARVALLHO (1984): Lepidópteros do Algarve. - Anais Facul. Cien. Porto 64 (1-4): 95-219.

MONTGOMERY, S.L. (1982): Biogeography of the moth genus *Eupithecia* in Oceania and the evolution of ambush predation in Hawaiian caterpillars (Lepidoptera: Geometridae). - Entomologia gen. 8: 27-34.

MUIRHEAD-THOMSON, R.C. (1991): Trap Responses of Flying Insects. - Academic Press, London, 287 pp..

MÜLLER B. & J. GELBRECHT (1992): Veränderungen in der Spannerfauna der DDR seit 1945 (Lep., Geometridae). - Nota lepidopterologica, Suppl. 3: 70-81.

MÜLLER, B. (1996): Geometridae. In: KARSHOLT, O. & J. RAZOWSKI (ed.): The Lepidoptera of Europe, a Distributional Checklist. (pp. 218-249). - Apollo Books, Stenstrup, 380 pp.

MUNROE, E.G. (1982): Lepidoptera. In: Parker, S.B. (ed.): Synopsis and Classification of Living Organisms 2: 612-651, Mc Graw-Hill.

MURRAY, D. (1942): First stage larva. - Entomol. Rec. 54: 137-139.

NAKAJIMA, H. (1998): A Taxonomical and Ecological Study of the Winter Moths (Lepidoptera, Geometridae) from Japan. - Tinea vol. 15 (Suppl. 2), 246 pp..

NAKAJIMA, H. & R. SATO (1979): A list of food plants of the Japanese Geometridae II, Archiearinae, Oenochrominae and Geometrinae. - Japan. Heteroc. J. 100: 663-680.

NAKAMURA, M. (1987): Pupae of Japanese Geometridae I (Lepidoptera). - Tinea 12 (Suppl.): 213-219.

NESTOROVA, E. (1998): Catalogus Faunae Bulgaricae 2: Lepidoptera, Geometridae. - Pensoft Publishers, Sofia-Moscow, 193 pp..

NIELSEN, E.S., EDWARDS, E.D. & T.V. RANGSI (eds) (1996): Checklist of the Lepidoptera of Australia. Monographs on Australian Lepidoptera 4. - Csiro Publishing, Collingwood, Australia 529 pp..

NISHIDA, R., ROTHSCHILD, M. & R. MUMMERY (1994): A cyanoglucoside, Sarmentosin, from the Magpie moth, *Abraxas grossulariata*, Geometridae: Lepidoptera. - Phytochemistry 36 (1): 37-38.

NORDSTRÖM, F., WAHLGREN, E. & A. TULLGREN (1941): Svenska Fjärilar. - Stockholm, 353 pp..

NYST, R.H. (1993): *Pseudoterpna pruinata* (Hfn.) et *P. coronillaria* (Hb.) une seule et même espèce? (Lep. Geometridae). - Ent. gall. 4 (4): 159-160.

OBERTHÜR, C. (1896): Geometrae. In: Faunes entomologiques. - Études d'Ent. 20: 69-74.

OBERTHÜR, C. (1922): Les Lépidoptères du Maroc. - Ét. Lepid. Comp. 19: 1-402.

OBERTHÜR, C. (1923): Catalogue des Lépidoptères des Pyrénées-Orientales. - Ét. Lepid. Comp. 21 (2): 9-75.

OLTRA, M.-T., DOMINGUEZ, M. & J. BAIXERAS (1995): A new Iberian species of *Protapanteles* (Hymenoptera: Braconidae) associated with the endemic moth *Heliothea discoidaria* (Lepidoptera: Geometrinae). - Ent. News (Phil.) 106 (2): 87-96.

OPHEIM, M. (1972): Geometridae. In: Catalogue of the Lepidoptera of Norway, pt. 3. - Zool. Mus. Oslo: 1-36.

OSTHELDER, L. (1929): Die Schmetterlinge Südbayerns und der angrenzenden nördlichen Kalkalpen, 1. Teil Großschmetterlinge, 3. Heft, Spanner. - Mitt. Münch. Ent. Ges. 19, Supplement.

PALM, E, JENSEN, K., NIELSEN, U.S. & B.H. THOMSEN (1982): Lepidoptera VI, Nordvestsjaellands sommerfugle. - Kobenhavn, 83 pp.

PARENZAN, P. & A. HAUSMANN (1992): Nuovi interessanti reperti di Geometridi (Lepidoptera) in Italia Meridionale. - Entomofauna 13 (8), 157-172.

PARENZAN, P., HAUSMANN, A. & S. SCALERCIO (1998): Addenda e corrigenda ai Geometridae dell'Italia meridionale. - Entomologica Bari 32: 51-79.

PARENZAN, P. (1975): Contributi alla conoscenza della lepidotterofauna dell'Italia meridionale. II. Nuovi reperti di Noctuidae e Geometridae. - Entomologica, XII: 153-169.

PARENZAN, P. (1988): Nuove segnalazioni di Geometridae (Lepidoptera) per l'Italia Meridionale. - Entomologica, XXIII: 139-160.

PARENZAN, P. (1994): Contributi alla conoscenza della Lepidotterofauna dell'Italia meridionale: XVII. Geometridae. - Entomologica Bari 28: 99-246.

PASSOS DE CARVALHO, J. (1984): Contribucao para o conhecimento da Lepidopterofauna do Parque Nacional, Familia Geometridae. - Parque Nacional de Peneda-Geres, 1-40.

PATOCKA, J. & P. ZACH (1994): On the Pupae of Central European Geometridae (Lepidoptera), Subfamilies Archiearinae, Oenochrominae and Ennominae, tribe Theriini. - Biologia Bratisl. 49 (5): 739-745.

PATOCKA, J. (1978): Zur Puppenmorphologie und -taxonomie der Unterfamilie Ennominae, insbesondere der Tribus Bistonini (Lepidoptera, Geometridae). - Acta Soc. Zool. bohemoslov 42 (2): 143-151.

PATOCKA, J. (1985): Beitrag zur Kenntnis der Puppen der Unterfamilie Ennominae (Lepidoptera, Geometridae). - Biologia Bratisl. 40: 997-1012.

PATOCKA, J. (1986a): Zur Kenntnis der Puppen der mitteleuropäischen Spanner aus der Tribus Abraxini und Semiothisini (Lepidoptera, Geometridae). - Biologia Bratisl. 41: 579-595.

PATOCKA, J. (1986b): Zur Kenntnis der Puppen der Tribus Boarmiini (Lepidoptera, Geometridae) in Mitteleuropa. - Acta ent. Bohemoslov. 83: 301-315.

PATOCKA, J. (1992): Über einige Puppen der Spanner aus der Tribus Ennomini (Lepidoptera: Geometridae, Ennominae). - Ent. Ber. Amsterdam 52: 171-176.

PATOCKA, J. (1993): Über einige Puppen der Spanner aus der Tribus Boarmiini (Lepidoptera: Geometridae, Ennominae). - Ent. Ber. (Amsterdam) 53: 114-120.

PATOCKA, J. (1994): Die Puppen der Spanner Mitteleuropas (Lepidoptera, Geometridae). Charakteristik. Bestimmungstabelle der Gattungen. - Tijdschr. Ent. 137: 27-56.

PATOCKA, J. (1995): Die Spannerpuppen Mitteleuropas aus der Unterfamilie Geometrinae (Lepidoptera: Geometridae). - Nachr. entomol. Ver. Apollo, N.F. 16 (2/3): 275-296.

PEKING, H. (1953): Ei und Eiablage von Hemistola chrysoprasaria Esp. (Euchloris vernaria Hb.). - Ent. Zt. Frkft. 62 (21): 167-168.

PELLMYR, O. (1980): Morphology of the genitalia of Scandinavian brachypterous female Geometridae (Lepidoptera). - Ent. scand. 11: 413-423.

PETERSEN, W. (1909): Ein Beitrag zur Kenntnis der Gattung Eupithecia Curt. - Dt. Ent. Zeitschr. Iris 22 (4): 203-313

PIERCE, F.N. (1914): The Genitalia of the Group Geometridae of the Lepidoptera of the British Islands. - E. W. Classey Ltd. Faringdon, U.K.

PINKER, R. (1968): Die Lepidopterenfauna Mazedoniens, III Geometridae. - Prirod. Muz. Skopje, Pos. Izdanie 4, 71 pp..

PITKIN, L.M. (1993): Neotropical Emerald moths of the genera Nemoria, Lissochlora and Chavarriella, with particular reference to the species of Costa Rica (Lepidoptera: Geometridae, Geometrinae). - Bull. nat. Hist. Mus. Lond. (Ent.) 62 (2): 39-159.

PITKIN, L.M. (1996): Neotropical Emerald moths: a review of the genera (Lepidoptera: Geometridae, Geometrinae). - Zool. Journ. Linn. Soc. 118: 309-440.

PORTER, J. (1997): The Colour Identification Guide to Caterpillars of the British Isles (Macrolep.). - Viking (Penguin Books) London, 275 pp.

POWELL, J., MITTER, C. & B. FARRELL (1999): Evolution of Larval Food Preferences in Lepidoptera. In: KRISTENSEN, N.P. (ed.): Handbook of Zoology, vol. IV (35), Lepidoptera, Moths & Butterflies, 1: Evolution, Systematics, and Biogeography. - W. de Gruyter, Berlin & New York, (491 pp.): 403-422.

PRINS, DE, W. (1998): Catalogue of the Lepidoptera of Belgium. - Dokum. trav. Inst. royal Sci. nat. Belg., Brussel, 236 pp..

PROLA, C. & T. RACHELI (1979): I Geometridi dell'Italia centrale. Parte I. - Boll. Ist. Ent. Un. Bologna, 34: 191-246.

PROLA, C. & T. RACHELI (1980): I Geometridi dell'Italia centrale. Parte II. - Boll. Ist. Ent. Un. Bologna, 35: 29-108.

PROUT, L.B. (1910). Lepidoptera Heterocera. Fam. Geometridae. Subfam. Oenochrominae. In: WYTSMAN, P. (ed.): Genera insectorum 104: 1-120, 2 pls.

PROUT, L.B. (1912-1916a): Die spannerartigen Nachtfalter. In: SEITZ, A. (ed.): Die Groß-Schmetterlinge der Erde, vol. 4. - Verlag A. Kernen, Stuttgart.

PROUT, L.B. (1912b). Lepidopterorum Catalogus, Pars 8: Geometridae: Brephinae, Oenochrominae, Berlin, 94 pp.

PROUT, L.B. (1913c): Lepidopterorum Catalogus, Pars 14: Geometridae: Subfam. Hemitheinae. - W. Junk, Berlin, 192 pp.

PROUT, L.B. (1920-1941b): Die indoaustralischen Spanner. In: SEITZ, A. (ed.): Die Groß-Schmetterlinge der Erde, vol. 12. - Verlag A. Kernen, Stuttgart.

PROUT, L.B. (1930-1938c): Die spannerartigen Nachtfalter (Fauna Africana). In: SEITZ, A. (ed.): Die Groß-Schmetterlinge der Erde, vol. 16. - Verlag A. Kernen, Stuttgart.

PROUT, L.B. (1934-1935a, additions 1938a) Brephinae, Oenochrominae, Hemitheinae, Sterrhinae, Larentiinae. In: SEITZ, A. (ed.): Die Groß-Schmetterlinge der Erde, Suppl. 4. - Verlag A. Kernen, Stuttgart.

PROUT, L.B. (1934-1935d): Lepidopterorum Catalogus, Pars 61: Geometridae, Subfamilia Sterrhinae I. - Verlag W. Junk, Berlin, 486 pp..

PÜNGELER, R. (1903): Neue paläarktische Makrolepidopteren. - Dt. Ent. Zeitschr. Iris 16: 286-301.

PÜNGELER, R. (1906): Neue paläarktische Makrolepidopteren. - Dt. Ent. Zeitschr. Iris 19: 78-98.

PÜNGELER, R. (1909): Neue palaearctische Macrolepidopteren. - Dt. Ent. Zeitschr. Iris 21: 286-303.

PÜNGELER, R. (1914): Neue paläarktische Makrolepidopteren. - Dt. Ent. Zeitschr. Iris 28: 37-55.

RAINERI, V. (1985): Ricerche sui Geometridi delle Alpi liguri (Lepidoptera). - Boll. Soc. ent. It., 117 (4/7): 102-112.

RAINERI, V. (1991): Some Geometridae new to Italy (Lepidoptera). - Nota lepidopterologica 14 (3): 234-240.

RAINERI, V. (1994): Some considerations on the genus *Thetidia* and description of a new genus: *Antonechloris* gen. nov. (Lepidoptera, Geometridae). - Atalanta 25 (1/2): 365-372.

RAINERI, V. & S. ZANGHERI (1995): Checklist delle specie della fauna Italiana, pt. 90: Lepidoptera: Drepanoidea, Axioidea, Geometroidea. - Ed. Calderini, Bologna, 43 pp..

RAMBUR, M.P. (1833): Catalogue des Lépidoptères de l'île de Corse avec la désription et la figure des espèces inédites. - Annls Soc. ent. Fr. 2: 1-59, 2 pls.

RANGNOW, R. (1935): Neue Lepidopteren aus Lappland. - Ent. Rdsch. 53 (2): 21-22.

RAPPAZ, R. (1979): Les Papillons du Valais. - Martigny, 377 pp.

RAUH, W. & K. SENGHAS (1976): Schmeil-Fitschen, Flora von Deutschland, 86th ed.. - Quelle & Meyer, Heidelberg, 516 pp..

REBEL, H. (1903): Studien über die Lepidopterenfauna der Balkanländer, I. Teil Bulgarien und Ostrumelien. - Ann. K. K. Nat. hist. Hofmus.Wien 18 (2/3): 123-347.

REBEL, H. (1910): Fr. Berge's Schmetterlingsbuch nach dem gegenwärtigen Stande der Lepidopterologie, 9. Auflage. - E. Schweizerbart'sche Verlagsbuchhandlung, Nägele & Dr. Sproesser, Stuttgart.

REBEL, H. (1911): Die Lepidopterenfauna von Herkulesbad und Orsova. - Ann. K.k. Nat.hist. Hofmus. Wien 25: 253-430.

REBEL, H. (1916): Die Lepidopterenfauna Kretas. - Ann. K.K. Nat.hist. Hofmus. Wien 30: 66-172.

REBEL, H. (1924): [Geometriden aus Triest]. - Verh. Zool.-Bot. Ges Wien (1923) 73: (6)-(7).

REBEL, H. (1939): Zur Lepidopterenfauna Cyperns. - Mitt. Münchn. Ent. Ges. 29: 487-564.

REBEL, H. & H. ZERNY (1931): Die Lepidopterenfauna Albaniens (mit Berücksichtgung der Nachbargebiete). - Denkschr. Akad. Wiss. Wien 103: 161pp.

REDONDO, V.M. & F.J. GASTÓN (1999): Los Geometridae (Lepidoptera) de Aragón (España). - Monografias Soc. Entom. Arag. 3, Zaragoza, 131 pp..

REID, J. (1986): Geometridae. In: FRIEDRICH, E.: Breeding Butterflies and Moths, English Edition. - Harley Books, Colchester: 83-101.

REIPRICH, A. & I. OKÁLI (1989): Dodatky k Prodromu Lepidopter Slovenska. Vol. 3: 33-68.

REJMANEK, M. & K. SPITZER (1982): Bionomic strategies and long term fluctuations in abundance of Noctuidae (Lepidoptera). - Acta ent. bohemoslov. 79: 81-96.

RETZLAFF, H., DUDLER, H., FINKE, C., PÄHLER, R., SCHNELL, K. & W. SCHULZE (1993): Zur Schmetterlingsfauna von Westfalen, Ergänzungen, Neu- und Wiederfunde. - Mitt. Arbgem. ostwestf.-Lippischer Ent. 9 (2): 37-53.

REZBANYAI-RESER, L. (1987): *Idaea griseanova* sp.n., eine bisher verkannte Zwillingsart von *ruficostata* ZELLER, 1849, aus dem Westmediterraneum (Lepidoptera, Geometridae). - Mitt. Entom. Ges. Basel 37 (4): 141-182.

REZBANYAI-RESER, L. (1989): Ein Musterbeispiel der rezenten Arealerweiterung: *Eupithecia sinuosaria* Eversmann, 1848, bis zur Südostschweiz vorgedrungen. - Atalanta 19: 39-50.

REZBANYAI-RESER, L. (1993): Elenco critico aggiornato die macrolepidotteri del Cantone Ticino, Svizzera meridionale (Insecta, Lepidoptera). - Boll. Soc. tic. Sc. nat. (Lugano) 81 (1): 39-96.

REZBANYAI-RESER, L. (1999): Zur Morphologie, Taxonomie und Verbreitung der *Chlorissa* - Arten in der Schweiz (Lepidoptera, Geometridae). – Entom. Ber. Luzern 41: 67-94.

RIBBE, C. (1912): Beiträge zu einer Lepidopteren-Fauna von Andalusien. Geometridae. - Dt. Ent. Zeitschr. Iris 23: 299-343.

RICHARD, F. (1950): Éspèces nouvelles pour la faune belge. - Lambillionea 50 (1/2): 14-15.

RICHTER, O. (1912): Totaler Albinismus bei *Geometra papilionaria* L.. - Int. Ent. Z. Guben 5 (47): 335-336.

RIEMIS, A. (1994): Geometridae of Turkey 3. A provisional list of the Geometridae of Turkey (Lepidoptera). - Phegea 20: 131-135.

ROBINSON, G.S. (1976): The preparation of slides of Lepidoptera genitalia with special reference to the Microlepidoptera. - Entomologist's Gaz. 27: 127-132.

ROMETSCH, H. (1932): Beitrag zur Biologie von *Aplasta ononaria* Fuessl. (Lep.). - Ent. Zt. Frkft. 45 (22): 299-300.

ROSE, H.S. & DEVINDER (1985): A new species of the genus *Chlorissa* Stephens from North India (Geometrinae: Geometridae: Lepidoptera). - Journ. Bomb. Nat. Hist. Soc. 82: 614-616.

ROTHSCHILD, M. (1985): British aposematic Lepidoptera, in: HEATH, J. & A.M. EMMET (eds): The moths and butterflies of Great Britain and Ireland. - Harley Books, Colchester, vol. 2 (460pp.): 9-62.

RUHLAND (1910): Begriffsbestimmung der Aberration und Beschreibung einer Aberration der *Geometra papilionaria* L.. - Int. Ent. Z. Guben 4 (27): 145-146.

RUNGS, C.E.E. (1981): Catalogue raisonné des Lépidoptères du Maroc. Inventaire Faunistique et observations écologiques. Tome II. - Trav. Inst. Sc., Sér. Zool., n.40, Rabat, 223-588.

RUNGS, C.E.E. (1988): Liste-Inventaire systématique et synonymique des Lépidoptères de Corse, Geometridae. - Suppl. Alexanor 15: 43-49.

RUSSELL, A.G.B. (1942): *Thalera fimbrialis* (Scop.): A specimen taken at Swanage. - The Entomologist 75: 73-75.

SALKELD, E.H. (1983): A catalogue of the eggs of some Canadian Geometridae (Lepidoptera), with comments. - Mem. Ent. Soc. Canada, Nr. 126, Ottawa, 271 pp.

SATTLER, K. (1969): Das "Wiener Verzeichnis" von 1775. - Z. Wien. Ent. Ges. 54: 2-7, 3 pl..

SATTLER, K. (1991): A review of wing reduction in Lepidoptera. - Bull. Br. Mus. Nat. Hist. (Ent.) 60 (2): 243-288.

SATTLER, K. & W.G. TREMEWAN (1984): The Lepidoptera names of Denis & Schiffermüller - a case for stability. - Nota lepidopterologica 7: 282-285.

SAVENKOV, N., SULCS, I., KERPPOLA S. & L. HULDEN (1996): Checklist of Latvian Lepidoptera. - Baptria 21 (3a): 1-71.

SCHAWERDA, K. (1934): Der Formenkreis von *Euchloris smaragdaria* F. und *volgaria* Gn. (=*prasinaria* Ev.). - Int. Ent. Z. 27 (41): 461-465.

SCHMIDLIN, A. (1964): Übersicht über die europäischen Arten der Familie Geometridae (Lep.). - Mitt. ent. Ges. Basel, N.F. 14 (4, 5): 77-137.

SCHNEIDER, C. (1934): Die in Württemberg vorkommenden Arten der Geometridengattungen *Brephos* Zink. - *Timandra* Dup. - Ent. Zt. Frkft. 48 (4): 29-32.

SCHNEIDER, C. (1936): Beitrag zur Biologie von *Aplasta ononaria* Fuessli (Lep. Geom.). - Int. Ent. Z. (1935) 29 (32): 380.

SCHRANK, F. (1801-1802): Fauna Boica. - Nürnberg, 2 vols.

SCHWINGENSCHUSS, L. & F. WAGNER (1926): Beitrag zur Macro-Lepidopteren-Fauna Süddalmatiens insbesondere der Umgebung Gravosas, Geometridae. - Zeits. Österr. Ent.-Ver. 11: 35-53.

SCHWINGENSCHUSS, L. (1953): Beitrag zur Lepidopterenfauna von Niederösterreich: St. Peter i.d.Au, Seitenstetten und Umgebung, Geometridae. - Z. Wien. Ent. Ges. 38: 217-223, 251-255, 282-285.

SCOBLE, M.J. (1994): A taxonomic revision of the genera *Phrygionis* Hübner and *Pityeja* Walker (Geometridae: Ennominae, Palyadini). - Zool. Journal of the Linnean Soc. 111: 99-160

SCOBLE, M.J. (1995): The Lepidoptera: Form, Function and Diversity. - The Natural History Museum, Oxford University Press, Oxford, 404 pp.

SCOBLE, M.J. (1999): Geometrid Moths of the World, a Catalogue. - Csiro Publishing, Apollo Books, Collingwood (Australia), Stenstrup (Denmark), 1,400 pp..

SCOBLE, M.J. & E.D. EDWARDS (1990): *Parepisauris* Bethune-Baker and the composition of the Oenochrominae (Lepidoptera: Geometridae). - Ent. scand. 20: 371-399.

SCOBLE, M.J., GASTON, K.J. & A. CROOK (1995): Using taxonomic data to estimate species richness in Geometridae. - Journal Lep. Soc. 49 (2): 136-147.

SCOPOLI, G.A. (1763): Entomologia Carniolica exhibens insecta Carnioliae indigena et distributa in ordines, genera, species, varietates. Methodo Linneana. - Vindobonae, 422 pp., 43 pls.

SEPPÄNEN, E.J. (1954): Suomen Suurperhostoukkien Ravintokasvit (The food-plants of the larvae of the Macrolepidoptera of Finland). - Suomen Elaimet (Animalia Fennica) 8.

SEPPÄNEN, E.J. (1970): Suurperhostoukkien Ravintokasvit (The food-plants of the larvae of the Macrolepidoptera of Finland). - Suomen Elaimet (Animalia Fennica) 14.

SEVEN, S. (1991): Bibliographical studies on the Lepidoptera Fauna of Turkey-in-Europe. - Priamus 6 (1/2): 1-95.

SHELJUZHKO, L. (1955): Neue und wenig bekannte Noctuiden und Geometriden der Zoologischen Staatssammlung in München. - Mitt. Münchn. Ent. Ges. 44/45: 277-292.

SILVA CRUZ, DA, M.A. & T. GONCALVES (1977): Catalogo sistematico dos Macrolepidopteros de Portugal. - Publcoes Inst. Zool. Dr. Augusto Nobre 133: 1-56.

SINGH, B. (1951): Immature stages of Indian Lepidoptera, Nr. 8 Geometridae. - Indian forest Rec. 8 (7): 67-159.

SINGH, B. (1960): Description of larva of *Brephos infans* Moeschler (Lepidoptera, Geometridae, Brephinae), with a note on the relationship and affinity of larvae of subfamily Brephinae with larvae of other subfamilies of Geometridae. - Ind. Forest Rec. Entom. 9 (11): 211-214.

SKINNER, B. et al. (1981): An identification guide to the British Pugs, Lepidoptera: Geometridae. - London, 56 pp..

SKINNER, B. (1984): Colour Identification Guide to Moths of the British Isles. - Viking Books London, 269 pp..

SKOU, P. (1986): The Geometrid Moths of North Europe. - Entomonograph, Vol. 6, Copenhagen, 348 pp..

SOMMERER, M. (1983): Zum Stand der Abgrenzung von *Ectropis crepuscularia* ([Denis & Schiffermüller], 1775) und *Ectropis bistortata* (Goeze, 1781). - Entomofauna 4 (26): 446-466.

SOTAVALTA, O. (1995): Distribution atlas of Macrolepidoptera: Geometridae of eastern Fennoscandia. - Baptria 20 (1a): 1-64.

SOUTH, R. (1961): The Moths of the British Isles, Vol. 2. - Warne, London, 379 pp.

SPADA, L. (1893): Contribuzione alla Fauna marchegiana. I Lepidotteri finora trovati nel territorio di Osimo. - Il Naturalista Siciliano 12 (3/10): 53 pp.

SPITZER, K. & J. LEPS (1988): Determinants of temporal variation in moth abundance. - Oikos 53: 31-36.

SPITZER, K., REJMANEK, M. & T. SOLDAN (1984): The fecundity and long term variability in abundance of noctuid moths (Lepidoptera, Noctuidae). - Oecologia (Berlin) 62: 91-93.

SPULER, A. (1903-1910): Die Schmetterlinge Europas, 4 volumes. - Schweizerbart'sche Verlagsbuchhandlung, Stuttgart, vol. 2: 523 pp, vol. 3: 91 pl..

STAMM, K. (1981): Prodromus der Lepidopteren-Fauna der Rheinlande und Westfalens, 229 pp..

STAUDINGER, O. & REBEL, H. (1901): Katalog der Lepidopteren des palaearktischen Faunengebietes 1. - Verlag Friedländer & Sohn, Berlin.

266

STAUDINGER, O. (1861): Catalog der Lepidopteren Europas und der angrenzenden Länder. - Dresden.

STAUDINGER, O. (1871): Macrolepidoptera. In: STAUDINGER, O. & M. WOCKE: Katalog der Lepidopteren des europäischen Faunengebietes. Dresden.

STAUDINGER, O. (1879): Lepidopteren-Fauna Kleinasiens. - Horae Soc. ent. Ross. 14: 176-482.

STAUDINGER, O. (1892a) Neue Arten und Varietäten von Lepidopteren des palaearctischen Faunengebietes. - Dt. Ent. Zeitschr. Iris 4: 224-339.

STAUDINGER, O. (1892b) Neue Arten und Varietäten von paläarktischen Geometriden aus meiner Sammlung. - Dt. Ent. Zeitschr. Iris 5: 141-260.

STAUDINGER, O. (1894): Neue Lepidopterenarten und Varietäten aus dem paläarktischen Faunengebiet. - Dt. Ent. Zeitschr. Iris 7: 241-296.

STAUDINGER, O. (1895): Neue paläarktische Lepidopteren. - Dt. Ent. Zeitschr. Iris 8: 288-299.

STAUDINGER, O. (1897-1898): Neue Lepidopteren aus Palästina. - Dt. Ent. Zeitschr. Iris 10: 271-319.

STAUDINGER, O. (1899): Neue paläarktische Lepidopteren. - Dt. Ent. Zeitschr. Iris 12: 156-163.

STAUDINGER, O. (1900): Neue Lepidopteren des paläarktischen Faunengebietes. - Dt. Ent. Zeitschr. Iris 12: 352-403.

STEKOLNIKOV, A.A. & V.I. KUZNETZOV (1981): Functional morphology of the male genitalia and notes on the system of the subfamily Geometrinae (Lepidoptera, Geometridae) [in Russian]. - Ent. Obosr. 60 (3): 535-549.

STEPHENS, J.F. (1927-1835): Illustrations of British Entomology (Haustellata), vol. 1-4.

STERNECK, J. (1940): Versuch einer Darstellung der systematischen Beziehungen bei den palaearctischen Sterrhinae (Acidaliinae). - Zeitschr. Wiener Ent. Ver. 25: 6-17; 25-36; 56-59; 77-79; 98-107; 126-128; 136-142; 152-159; 161-176.

STERNECK, J. (1941): Versuch einer Darstellung der systematischen Beziehungen bei den palaearctischen Sterrhinae (Acidaliinae). Studien über Acidaliinae (Sterrhinae) IX. - Zeitschr. Wiener Ent. Ver. 26: 150-159; 176-183; 191-198; 211-216; 217-222; 222-230; 248-262.

STERZL, O. (1967): Prodromus der Lepidopterenfauna von Niederösterreich. - Verh. Zool.-Bot. Ges. Wien 107: 75-193.

SULCS, A. & I. SULCS (1978): Neue und wenig bekannte Arten der Lepidopteren-Fauna Lettlands. - Notulae Ent. 58: 141-150.

SULCS, A. & J. VIIDALEPP (1972): Verbreitung der Großschmetterlinge (Macrolepidoptera) im Baltikum, IV, Geometridae. - Dtsch. Ent. Z., N.F. 19: 151-209.

SULCS, A., VIIDALEPP, J. & P. IVINSKIS (1981): 1. Nachtrag zur Verbreitung der Großschmetterlinge im Baltikum. - Dtsch. Ent. Z., N.F. 28 (1-3): 123-146.

SURLYKKE, A., SKALS, N., RYDELL, J. & M. SVENSSON (1998): Sonic hearing in a Diurnal Geometrid Moth, *Archiearis parthenias*, Temporally isolated from Bats. - Naturwissenschaften 85 (1): 36-37.

SUTTON, S.L. & H.E. BEAUMONT (1989): Butterflies and Moths of Yorkshire. - Yorkshire Naturalist's Union, Doncaster.

SVENDSEN, P. (ed.) & M. FIBIGER (1992): The Distribution of European Macrolepidoptera. - European Faunistical Press, Copenhagen, 1992.

SVENSSON, I. (1993): Fjärilkalender, Lepidoptera-calendar. - Österslöv, 103 pp..

TABBERT. H. (2000): Schmetterlingsbeobachtungen auf See 1998 (Insecta, Lepidoptera).- Atalanta 31 (3/4): 511-514.

THOMPSON, W.R. (1944-1950): A catalogue of the parasites and predators of insect pests. Parasites of the Lepidoptera, pt. 5 - pt. 10. - The Imperial Parasite Service, Belleville, Ont., Canada.

THOMSON, E. (1967): Die Großschmetterlinge Estlands. - Helmut Rauschenbusch Verlag, Stollkamm, 203 pp.

THUNBERG, C.P. (1784-1794): Dissertatio Entomologica sistens Insecta Suecica, quorum partem orimam ... publico examini subjicit Joh. Borgstrom. - Uppsala.

THURNER, J. (1967): Lepidopteren aus Morea, ein weiterer Beitrag zur Fauna des Peloponnes (Griechenland). - Z. Wien. Ent. Ges. 52: 5-23, 50-58.

TIKHONOV, V.V. (1993): Dendrofileous Geometridae of the Northern Caucasus. - Vestnik St. Petersb. Univ. 3 (1): 29-36.

TONGE, A. E. (1932): The Ova of British Lepidoptera. Pt. III. Geometridae etc. - Trans. & Proc. Ent. & Nat. Hist. Soc. 1932-1933: 11-25.

TREITSCHKE, F. (1827/1828): Die Schmetterlinge von Europa. (Fortsetzung des OCHSENHEIMER'schen Werkes). - Leipzig, vol. 6.

TRÖGER, E. J. (1971): Über die Verwandtschaftlichen Beziehungen von Comibaena pustulata Hufn. und Euchloris (=Thetidia) smaragdaria F. (Lepidoptera, Geometridae) anhand larvaler und imaginaler Merkmale. - Proceedings XII International Congress of Entomology Moskow (1968) 1: 209-210.

TURATI, E. (1929): Eteroceri di Tripolitania. - Boll. Lab. Zool. gen. agr. Portici, XXIII: 98-128.

TURATI, E. (1913): Un record Entomologico - Materiali per una faunula dei Lepidotteri della Sardegna. - Atti Soc. Ital. Sci. Nat. 51 (3/4): 267-365.

UNCED (United Nations Conference on Environments and Development) (1992): Convention on Biodiversity, chapter 15. - Rio de Janeiro.

URBAHN, E. (1964): Habitus- und Genitalunterschiede bei Chlorissa viridata L. und C. cloraria HÜBNER (=porrinata Zeller) (Lep., Geometridae). - Entom. Zeits. 74 (24): 273-280.

URBAHN, E. (1966): Ergebnisse der Albanien-Expedition 1961 des Deutschen Entomologischen Institutes, 51 Beitrag, Lepidoptera: Geometridae. - Beitr. Entom. 16 (3/4): 407-446.

URBAHN, E. & H. URBAHN (1939): Die Schmetterlinge Pommerns mit einem vergleichenden Überblick über den Ostseeraum - Macrolepidoptera. - Stett. Ent. Zeitung 100: 185-828.

URBAHN, E. (1965): Hemistola chrysoprasaria Esp. (Geometra vernaria Hbn.) zweibrütig (Lep., Geom.). - Mitt. Dt. Ent. Ges. 24 (3): 54.

URBAHN, E. (1969): Verzeichnis der Spanner (Geometridae) der DDR. - Ent. Nachr. (Dresden) 13 (7/8): 78-86.

URBAHN, E. (1971): Zunahme von Melanismus-Beobachtungen bei Makrolepidopteren Europas in neuerer Zeit. - Mitt. Münchn. Ent. ges. 61: 1-15.

VALLE, K.J. (1946): Suurperhoset IV Mittarit (Geometrae). - Suomen Eläimet-Animalia Fennica 5, 370 pp., 22 pl..

VALLETTA, A. (1973): The Moths of the Maltese Islands. - Progress Press, Malta, 118 pp.

VALLHONRAT, F. (1980a): Pingasa lahayei Oberthür: Geomètrid Nord-Africà recollit a Andalusia. - Treb. Soc. Cat. Lep. 3: 67.

VALLHONRAT, F. (1980b): Geometridae nous o interressants per a la fauna Catalana. - Treb. Soc. Cat. Lepid. 3: 41-50.

VIGNA TAGLIANTI, A., AUDISIO, P.A., BELFIORE, C., BIONDI, M., BOLOGNA, M.A., CARPANETO, G.M., DE BIOSE, A., DE FELICI, S., PIATELLA, E., RACHELI, T., ZAPPAROLI, M. & S. ZOLA (1992): Riflessioni di gruppo sui corotipi fondamentali della fauna W-paleartica ed in particolare italiana. - Biogeographia, Lav. Soc. Ital. Biogeogr. (N.S.), 16: 159-179.

VIIDALEPP, J. (1976): A list of Geometridae (Lepidoptera) of the USSR, I. - Entom. Obosr. 55 (4): 842-852.

VIIDALEPP, J. (1977): A list of Geometridae (Lepidoptera) of the USSR, II. - Entom. Obosr. 56 (3): 564-576.

VIIDALEPP, J. (1978): A list of Geometridae (Lepidoptera) of the USSR, III. - Entom. Obosr. 57 (4): 752-761.

VIIDALEPP, J. (1979): A list of Geometridae (Lepidoptera) of the USSR, IV. - Entom. Obosr. 58 (2): 782-798.

VIIDALEPP, J. (1981): [The systematic of the genera of the subfamily Geometrinae (Lepidoptera, Geometridae), in Russian]. - Trudy vses. ent. Obshch. 63: 90-95.

VIIDALEPP, J. (1986): Subfamily Alsophilinae (Lepidoptera, Geometridae), fauna of the USSR: I. survey on the species. - Syst. Ecol. Tschesch. Dal. Wostok SSSR 1986: 57-68.

VIIDALEPP, J. (1988): Geometridae fauna of the Central Asian mountains. – Nauka Moskow 1988: 1-240.

VIIDALEPP, J. (1995): Catalogus Macrolepidopterorum Estoniae. - Abiks Loodusev. 95, Tallin - Tartu.

VIIDALEPP, J. (1996) Checklist of the Geometridae (Lepidoptera) of the former U.S.S R.. - Apollo Books, Stenstrup, 111 pp..

VILLERS DE, C. (1789): Caroli Linnaei Entomologica, Faunae Sueciae descriptionibus aucta ... generum specierumque rariorum iconibus ornata. - Lyon, 4 vols.

VIVES MORENO, A. (1994): Catálogo sistemático y sinonímico de los Lepidópteros de la Península Ibérica y Baleares (Insecta: Lepidoptera, II). - Minist. Agr., Pesca & Alim., Madrid, 775 pp..

VIVES MORENO, A. (1996): Segunda addenda et corrigenda al 'Catálogo sistemático y sinonímico de los lepidópteros de la Península Ibérica y Baleares (Segunda Parte)' (Insecta: Lepidoptera). - Shilap Revta. lepid. 24 (95): 275-315.

VOJNITS, A. (1976): Geometrinae and Sterrhinae from Mongolia (Lepidoptera, Geometrinae). - Annls Hist.-Nat. Mus. Nat. Hung. 68: 169-174.

VOJNITS, A. (1980): Araszolólepkék I., Geometridae I. - Fauna Hungariae 137, Budapest, 157 pp..

WALIA, V.K. & H.R. PAJNI (1984): A new species of the genus *Mixocera* Warren (Subfamily: Geometrinae). - Journ. Bomb. Nat. Hist. Soc. 81: 670-673.

WARDIKJAN, S.A. (1985): Atlas of the genital apparatus of geometrid moths (Geometridae, Lepidoptera) of Armenian SSR. - Akad. Nauk. Arm. Inst. Zool., Eriwan, 136pp..

WARNECKE, G. (1916): Die Geometriden-Fauna Schleswig-Holsteins, IIA. - Int. Ent. Z. 10 (17): 94-96.

WATERS, E.G.R. (1924): *Euchloris pustulata* Hufn. on beech. - Ent. Monthly Mag. 60: 64.

WEGNER, H. (1996): Neue und besonders bemerkenswerte Großschmetterlingsbeobachtungen der letzten 15 Jahre in Schleswig-Holstein. - Bombus 3 (17-20): 71-75.

WEGNER, H. (1998): Faunistische Mitteilungen aus Nordwestdeutschland. 101. (Lep. Geometridae). Eine Auflistung des aktuellen und historischen Artenbestandes in den Bundesländern Schleswig-Holstein einschließlich Hamburg (SH/HH) und Niedersachsen einschließlich Bremen (NS/HB).- Bombus 3: 137-149.

WEHRLI, E. (1926): Ein Streifzug in die andalusischen Gebirge. - Dt. Ent. Zeitschr. Iris 40: 113-129.

WEHRLI, E. (1939-1954a): Geometrinae, in SEITZ, A. [1939-1954]: Die Gross-Schmetterlinge der Erde, Suppl. 4. - Verlag A. Kernen, Stuttgart.

WEIDNER, H. (1993): Bestimmungstabellen der Vorratsschädlinge und des Hausungeziefers Mitteleuropas. - Gustav Fischer Verlag Stuttgart - Jena - New York, 328 pp.

WEIGT, H.-J. (1979): Blütenspanner-Beobachtungen 3 (Lepidoptera, Geometridae). - Dortmunder Beitr. Landeskunde 13: 3-34.

WEIGT, H.-J. (1980): Blütenspanner-Beobachtungen 4 (Lepidoptera, Geometridae). - Dortmunder Beitr. Landeskunde 14: 3-84.

WEIGT, H.-J. (1982): Lepidoptera Westfalica. - Abh. Westf. Mus. Naturkunde 44: 3-111.

WEIGT, H.-J. (1983): Lepidoptera Westfalica, Geometroidea. 55. Familie: Geometridae, Subfamilie: Boarmiinae, Tribus Boarmiini. - Abh. Westf. Mus. Naturkunde 45: 3-56.

WEIGT, H.-J. (1988): Die Blütenspanner Mitteleuropas (Lepidoptera, Geometridae: Eupitheciini), Teil 2, *Gymnoscelis rufifasciata* bis *Eupithecia insigniata*. - Dortmunder Beiträge zur Landeskunde, Naturw. Mitteilungen 22: 5-81.

WEIGT, H.-J. (1991): Die Blütenspanner Mitteleuropas (Lepidoptera, Geometridae: Eupitheciini), Teil 4, *Eupithecia satyrata* bis *indigata*. - Dortmunder Beiträge zur Landeskunde, Naturw. Mitteilungen 25: 5-106.

WEINTRAUB, J.D., LAWTON, J.H. & M.J. SCOBLE (1995): Lithiine moths on ferns: a phylogenetic study of insect-plant interactions. - Biol. J. Linn. Soc. 55: 239-250.

WHALLEY, P.E.S. (1986): A review of the current fossil evidence of Lepidoptera in the Mesozoic. - Biol. J. Linn. Soc. 28: 253-271.

WILTSHIRE, E.P. (1948): Middle East Lepidoptera, IX: Two new forms or species and thirty-five new records from Cyprus. - Ent. Rec. 60: 79-87.

WILTSHIRE, E.P. (1949): The Lepidoptera of the Kingdom of Egypt, Pt. 2. - Bull. Soc. Fouad Ier d'Ent. 33: 381-457.

WILTSHIRE, E.P. (1957): The Lepidoptera of Iraq. - Nicholas Kaye Limited, London & Bagdad.

WILTSHIRE, E.P. (1990): An Illustrated, Annotated Catalogue of the Macro-Heterocera of Saudi Arabia. - Fauna of Saudi Arabia 11: 91-250.

WIMMER, J. (1985): Beitrag zur Lepidopterenfauna von Zypern. - Jahresb. Steyr. Ent. 1985: 54-61.

WOLF, W. (1988): Systematische und synonymische Liste der Spanner Deutschlands unter beson-derer Berücksichtigung der Denis & Schiffermüller'schen Taxa (Lep., Geometridae). - Neue Ento-mologische Nachrichten 22, 1-78.

WOLFSBERGER, J. (1966): Die Macrolepidopteren-Fauna des Gardaseegebietes. - Estratto dalle Mem. Mus. Civ. Stor. Nat. Verona Vol. XIII, 1-385.

WOLFSBERGER, J. (1971): Die Macrolepidopteren-Fauna des Monte Baldo in Oberitalien. - Museo Civico di Storia Naturale di Verona, Memorie fuori serie N. 4, 1-335.

WOLFSBERGER, J., (1975): Die Macrolepidopteren-Fauna des Gardaseegebietes, 1. Nachtrag. - Boll. Mus. Civ. Stor. Nat. Verona I, 1974: 167-193.

ZECEVIC M. & S. RADOVANOVIC (1974): Leptiri Timocke Krajine (makrolepidoptera), prilog poznavanju faune leptirova Srbije. - Stamparija Bor, Zajecar, 185 pp..

ZELLER, P.C. (1847a): Verzeichniß der vom Professor Dr. Loew in der Türkei und Asien gesammelten Lepid-optera. - Isis (Oken) 1847 (1): 3-39.

ZELLER, P.C. (1847b): Bemerkungen über die auf einer Reise nach Italien und Sicilien beobachteten Schmet-terlingsarten. Isis von Oken.

ZERNY, H. (1936): Die Lepidopterenfauna des Grossen Atlas in Marokko und seiner Randgebiete. - Mem. Soc. Sc. Nat. Maroc 42 (1935): 1-163.

ZERNY, H. (1933): Lepidopteren aus dem nördlichen Libanon. - Dt. Ent. Zeitschr. Iris 47: 60-109.

ZHANG, B.-C. (1994): Index of Economically Important Lepidoptera. - CAB International, Wallingford, 599 pp..

ZHURAVLEV, S.M. (1910): Contribution sur la faune des Lépidoptères des environs d'Ouralsk et d'autres de la province de l'Oural. - Horae Soc. ent. Ross. 39: 415-463.

ZUKOWSKY, H. (1937): Reise ans Schwarze Meer und Herkulesbad. - Ent. Rdsch. Stuttg. 54: 549-553; 557-559; 565-568; 574-576.

Authorship of illustrations

Gert BROVAD (DK, København): Colour plates 1-7

Benny JØRGENSEN, (DK, Fåborg): Text-figs 34-40, 67-72.

Ruth KÜHBANDNER (D, Munich, ZSM): Text-figs 45, 89-90, 93, 102-104, 111-112, 115-123. Female gen.figs partially.

Michael LEIPNITZ, (D, Stuttgart): Text-figs 13, 16, 17, 19, 29, 113-114, 138, 160, 192, pl.8 figs 1, 6.

Marianne MÜLLER (D, Munich, ZSM): Text-figs 1-6, 35-40, 55-66, 126-134, 141-146, 149-150, 152-153, 161-167, 183-186, 193-196, 198-199, 202-203, 205-206, 210-215, 227-228, pl.8 figs 7-9.

Ole FOGH NIELSEN, (DK, Ry): text-fig. 22

Burkhard NIPPE, (D, München): Text-figs 110, 154.

Jim REID, (GB, Hertfordshire): Text-fig. 28.

Ingolf RÖDEL, (D, Bergholz-Rehbrücke): Text-figs 18, 31.

Dieter STÜNING, (D, Bonn): Text-fig. 24, pl.8 figs 2-4.

Gerhard TARMANN, (A, Innsbruck): Text-fig. 30.

Hartmut WEGNER, (D, Adendorf): Text-figs 21, 25, 27, pl.8 fig. 5.

All the other drawings, graphics and photographs are by the author.

The studies on the subfamilies Archiearinae to Geometrinae, subject of the present Volume, have been supported by the German Ministry of Science, BMBF (project BIO-LOG, "inventory of the Geometrid moth species of Europe", INGE, 01LC9904/1).

Index to subjects and taxa

(introductory chapters)

Index to scientific names

of the Systematic Account, Vol. 1

The index gives page reference to the main treatment. Numbers in semibold refer to species numbers and figures.

Printed in the United States
by Baker & Taylor Publisher Services